ANIMAL SCIENCE BIOLOGY & TECHNOLOGY

Physiology, Application, Evaluation, and Industry

ANIMAL SCIENCE BIOLOGY

MEECEE BAKER, Ph.D.

Agriculture Instructor
Greenwood High School
Millerstown, Pennsylvania

ROBERT E. MIKESELL, M.S.

Swine Geneticist
PROGENETICS Division
White Oak Mills
Elizabethtown, Pennsylvania

& TECHNOLOGY

AgriScience and Technology Series

Series Editor
Jasper S. Lee, Ph.D.

Some computer-based art by Hans & Cassady, Inc.

Order from

Interstate Publishers, Inc.
510 North Vermilion Street
P.O. Box 50
Danville, IL 61834-0050
Phone: (800) 843-4774
FAX: (217) 446-9706

Library of Congress Catalog Card No. 95-77141

ISBN 0-8134-3040-2

1 2 3 4 5 6 7 8 9 10 03 02 01 99 98 97 96 95

PREFACE

- I teach agriculture.
- I needed an animal science text.
- The text needed to build a strong foundation of scientific principles, incorporate applied management practices, emphasize selection, and review the industry based on recent data.
- In ag teacher terms, after using the text, I wanted my students to be able to explain the physiology behind cattle heat cycles, raise a market hog, deliver a sound set of reasons, and select an animal science career based on industry outlook.
- This book was written based on those needs.

Animal Science Biology and Technology was written with the agriculture student in mind. Broad in scope, the text remains detailed enough to assist the student with an agriscience project, SAE program, or judging contest. Moreover, this text approaches the discipline of animal science from a new perspective. The first unit, physiology, addresses cellular digestion, reproduction, genetics, and ethology. The second unit, application, teaches the practical uses of animal science. This segment organizes by species: swine, beef cattle, dairy cattle, sheep, and horses. The third unit instructs students in the interpretation of performance data for livestock evaluation as well as judging of meat, swine, sheep, and beef and dairy cattle. The final unit investigates livestock industries, careers, and opportunities.

With fresh new pictures and relevant line art, this comprehensive text will be attractive to any student. Each chapter in the text begins with an introductory paragraph, terms, and objectives. Chapters also include summaries, matching terms, evaluation questions, student activities, animal science facts, and laboratory investigation. The accompanying teacher's manual contains answers to all chapter-ending activities, along with author tips, puzzles, and word searches ready for copying and classwide distribution. An overhead transparency set and computer test bank also supplement the text.

About the Authors

The authors of *Animal Science Biology and Technology: Physiology, Application, Evaluation, and Industry* are the husband and wife team of MeeCee Baker and Robert E. Mikesell. Dr. Baker teaches agriculture at Greenwood High School. She earned B.S. and Ph.D. degrees in Agricultural Education from The Pennsylvania State University. Her M.S. degree in Agricultural Economics was completed at The University of Delaware. She was elected NVATA Region VI Vice President in 1992. Mr. Mikesell earned his bachelor's and master's degrees in Animal Science from The Pennsylvania State University. He currently serves as the Swine Geneticist with the PROGENETICS division of White Oak Mills. In addition, the authors have experience in raising, exhibiting, and judging livestock. They reside on the family farm where they raise beef cattle and hogs.

ACKNOWLEDGMENTS

The authors of *Animal Science Biology and Technology: Physiology, Application, Evaluation, and Industry* want to thank the many people who assisted in the production of this book. Without their help, this text would not be a reality.

The following individuals deserve special recognition for their contributions:

Our colleagues and students for their understanding, patience, and cooperation while we worked to complete this massive project.

The Department of Dairy and Animal Sciences at The Pennsylvania State University. Faculty members Stanley Curtis, Erskine Cash, William Henning, and Kenneth Kephart freely shared their insights.

Dave Swartz, our county extension agent. His contributions were invaluable in the writing of the dairy chapters.

Friends like the Brummer Family, Michelle Dubiach, Connie Fenner, Tom Heffernan, John Knapka, Kevin O'Reilly, Chris McCahren, Merle Richter, and Mike Schofield. Each provided us with technical and species-specific input.

John Sharber, agriculture teacher, Sapulpa, Oklahoma, Keith Bryan, of the Department of Dairy and Animal Sciences at The Pennsylvania State University, Jim Dyer, of the Department of Agricultural Education at the University of Illinois, and Jim Wagoner of the Department of Range Management at the University of Wyoming. They gave of their time and expertise in technically reviewing the manuscript.

Interstate Publishers for affording us the opportunity to write the text.

Finally, we want to thank our families, especially Bob "Pap" Baker. Their tolerance of our long absences and assistance in feeding the stock deserve our special gratitude.

Dedication

We dedicate this text to our bosses, Ed Burns, principal of Greenwood High School (Millerstown, PA), and John and Mark Wagner, owners of White Oak Mills. Your support and encouragement are much appreciated.

CONTENTS

UNIT III — Evaluation

UNIT IV — Industry

UNIT I

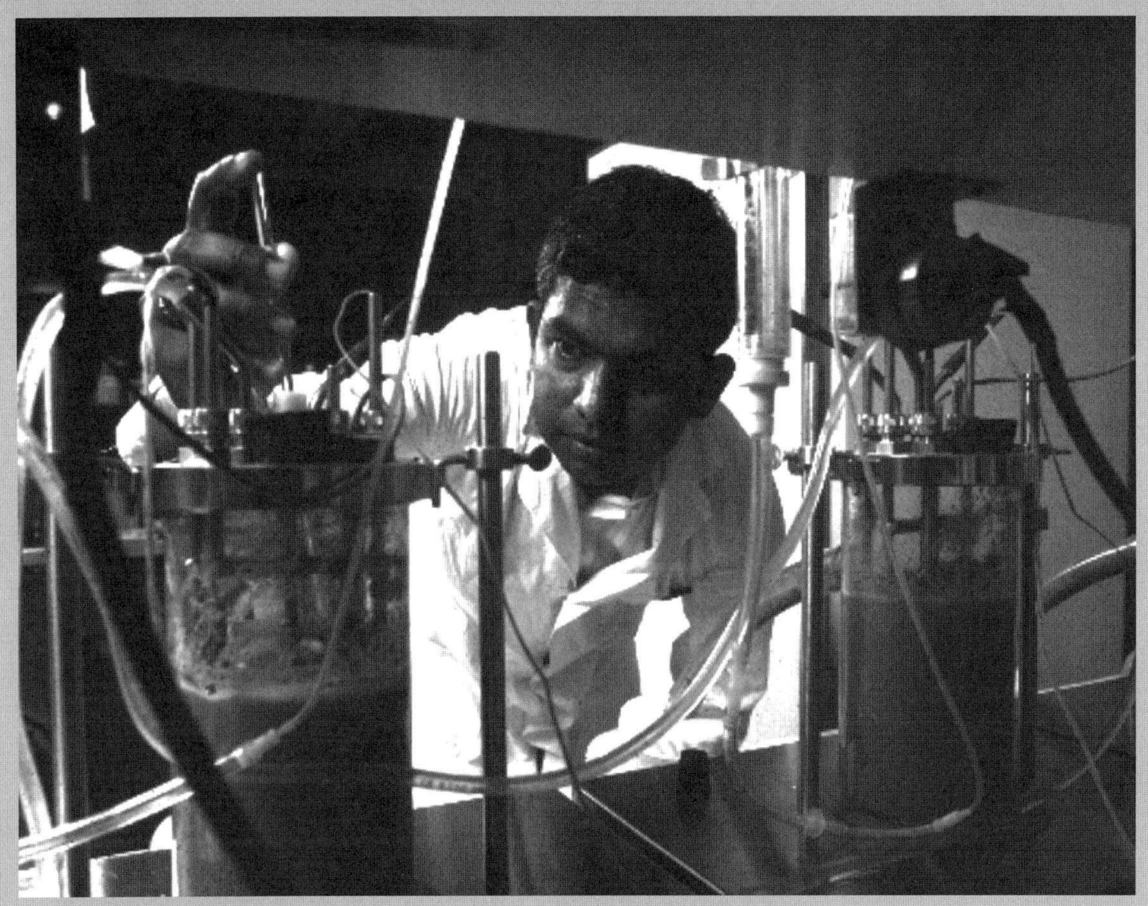

Physiology

Chapter 1

CELLULAR BIOLOGY AND ANIMAL TAXONOMY

Cells—Small but Mighty

INTRODUCTION

The basis to understanding animal science biology and technology is understanding the cell. Far from being simple, the animal cell is complex and performs a number of different functions: takes in nutrients, excretes wastes, secretes non-wastes, performs cellular work, responds to environment, and reproduces. Animal organisms vary from one-cell animals to familiar large animals, such as swine, sheep, cattle, and horses.

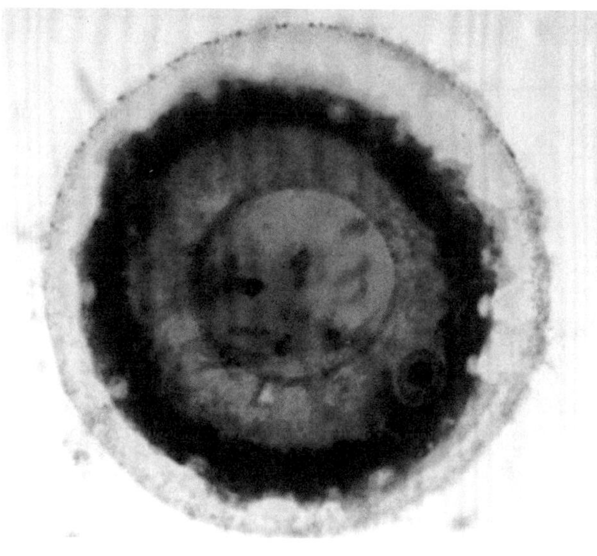

Figure 1-1.

OBJECTIVES

1. Draw, label, and describe the function of cellular structures
2. Differentiate between mitosis and meiosis
3. Discuss the relationship of tissues, organs, and systems
4. Explain how taxonomy is used to classify living organisms
5. Use binomial nomenclature to write scientific names of common livestock
6. Trace the domestication of farm animals

TERMS

binomial nomenclature
cell
cell membrane
centromere
chromosomes
cytoplasm
DNA (Deoxyribonucleic Acid)
endoplasmic reticulum
feral
gametes
genes
golgi bodies
haploid
Linnaeus
lysosomes
meiosis
mitochondria
mitosis
nucleus
ribosomes
taxonomy

ANIMAL SCIENCE FACTS

Most farm animals were first domesticated from 4,000 to 10,000 BC.

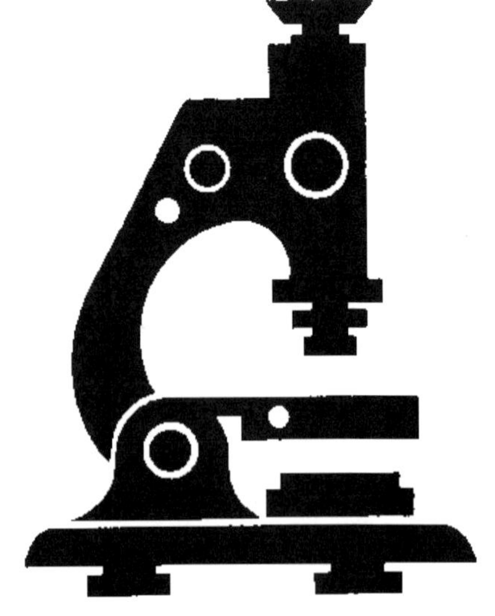

CELLULAR STRUCTURE AND FUNCTION

Cell theory states that a ***cell*** is the basic unit of life and only originates from other cells. All organisms are made of one or more cells. Cell structures and functions are similar in the following ways:

1. They must take up nutrients from their external environment.
2. They must excrete waste products into their external environment.
3. They must do some kind of "work," such as make proteins (example, liver cells), store energy (example, fat cells), carry oxygen (example, red blood cells), transport electrical impulses (example, nerve cells) store minerals (example, bone cells), or move (example, muscle cells).
4. They must reproduce themselves.

Cells can be considered as little factories with different departments responsible for different duties. The outside wall of the factory is called the ***cell membrane.*** The cell membrane is made of a thin layer of lipid (fat-like film) that separates the cell contents from the external environment. Imbedded in this lipid layer are specialized protein "doors" that allow large molecules (raw materials, such as carbohydrates or proteins) to pass into the cell. Newly made proteins (finished products) and cellular waste (waste products) pass out of the cell through these same protein "doors."

Inside the cell factory is the office, or ***nucleus,*** which is surrounded by the nuclear membrane. The nucleus controls all cell activity. ***Chromosomes*** are small strands of genetic material that reside in the nucleus. They are made of a genetic compound that controls inheritance—***DNA (Deoxyribonucleic Acid).*** See Figure 1-3. Chromosomes contain

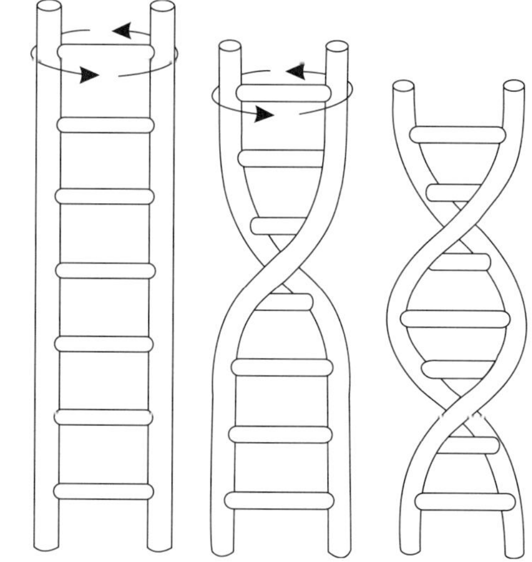

Figure 1-2. British chemist Rosalind Franklin first described the spiral shape of DNA in 1952.

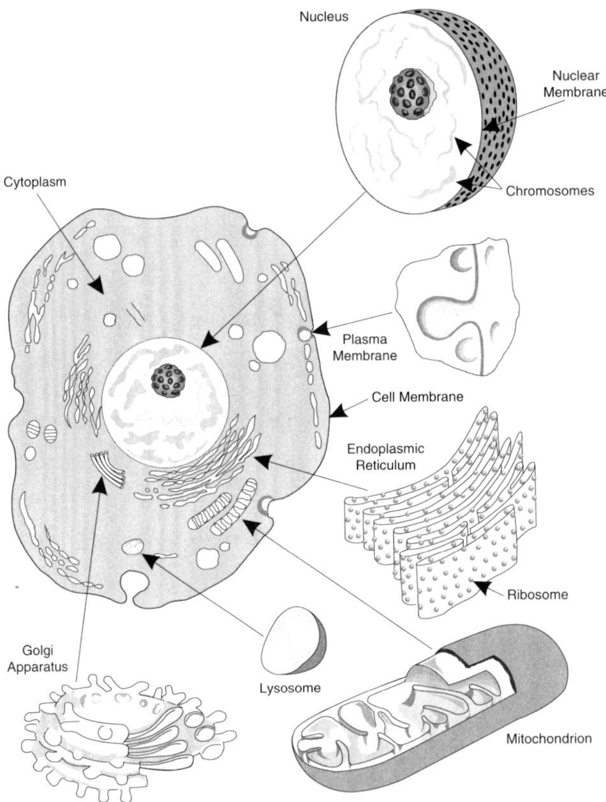

Figure 1-3. Cellular parts.

many small, coded "snippets" of DNA called ***genes.*** Genes from the chromosomes contain the blueprint for the work the cell is to do, as well as directions for the cell to replicate itself. Genes also control specific traits. Each gene contains the code for the manufacture of a specific protein. Genes contain codes for physical characteristics, such as size or hair color.

The jelly-like substance between the cell membrane and the nucleus is called ***cytoplasm.*** Within the cytoplasm are the various departments of the cell factory, which are collectively called organelles.

Raw materials entering the cell are lipids, carbohydrates, and proteins. All must be processed to be used by the cell to "feed" itself or make products for secretion. The ***endoplasmic reticulum,*** a network of membranes that connects the cell membrane to the nucleus, processes all incoming raw materials by breaking them down into simpler components. New proteins are manufactured by ***ribosomes,*** many of which are attached to the endoplasmic reticulum. The ***mitochondria,*** small, egg-shaped organelles, manufacture adenosine triphosphate (ATP), which is used as an energy source for the cell. ***Lysosomes*** are round organelles, which cause the digestion of proteins through the release of enzymes. Then, products go through final assembly and packaging in the cellular factory through a series of flat, membrane-encased ***golgi bodies.*** See Figure 1-3.

CELLULAR REPRODUCTION

Cells go through a life cycle of growth and reproduction called the cell cycle. The entire life cycle has two phases: interphase (the period between cell divisions) and cell division. Most of a cell's life is spent in interphase. Interphase is further divided into three periods: G1, S, and G2. During the first period of interphase (G1), the cell grows in size by increasing the number of organelles and the volume of cytoplasm. During the S period or second period of interphase, the genetic material replicates or copies itself so that there are two identical sets of chromatids called sister chromatids. At this point in the cycle, chromatids are still attached to each other at a central point called the ***centromere.*** During the third period of interphase, called the G2 period, the cell manufactures organelles and prepares for cell division.

MITOSIS

Mitosis, the actual division of nonsex cells, is further divided into four periods: prophase, metaphase, anaphase, and telophase. At the conclusion of these phases, what was formerly one cell becomes two. Dur-

PROPHASE
nuclear membrane
centromere
Chromosomes visible
Nuclear membrane fragments

METAPHASE
equator
Duplicated chromosomes are aligned at equator
One half of each pair of sister chromatids attach to spindle fibers

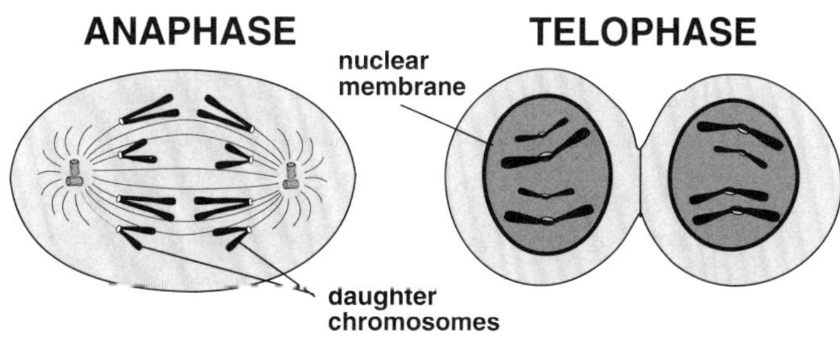

Figure 1-4. Mitosis is the division of nonsex cells.

ing prophase, the nuclear membrane disappears and the sister chromatids shorten and thicken into identifiable chromosomes. During metaphase, the sister chromatids line up along the central axis or equator of the cell. One half of each pair of sister chromatids attaches to spindle fibers, which will pull one set of chromosomes to each of the two new daughter cells. During anaphase, the chromatids separate, each moving toward opposite ends of the dividing cell. Each of the two new daughter cells forms a new nuclear membrane during telophase. Cytokinesis is when the cytoplasm and organelles evenly divide between the two new cells and a new membrane forms between them. See Figure 1-4.

MEIOSIS

Animal sex cells (ova and sperm), or ***gametes,*** divide a little differently than other cells in a process called ***meiosis*** or reduction division. In order to examine the difference, we must understand chromosome number and what happens to chromosomes at fertilization.

Each species of animal has a certain number of chromosomes present in duplicate in every cell of its body. This number is called the ***haploid*** or N number. For example, the haploid number for humans is 23. Because each chromosome is present in duplicate, the total number of chromosomes is twice the haploid number. This number is referred to as the diploid or 2N number. Therefore every human cell has 46 total chromosomes or 23 identical or homologous pairs of chromosomes. The process of meiosis reduces sex cells from 2N to N. So, when ova and sperm (gametes) meet and combine at fertilization, the resulting embryo will have 2N chromosomes, 1N from each of its parents. This explains why animals have characteristics of both parents.

The process of meiosis is divided into two steps: meiosis I and meiosis II. Both steps are similar to mitosis, except the end result is four haploid cells instead of two diploid cells.

Meiosis I begins with prophase. As with mitosis, the chromosomes thicken and become visible. Chromosomes are present in homologous pairs with a total of four sister chromatids per pair. During metaphase, the homologous pairs line up on the axis of the dividing cell opposite from each other. During anaphase, the homologous pairs of chromosomes leave each other and are pulled toward opposite poles by spindle fibers. During telophase, the cells physically divide. Each daughter cell now contains one chromosome from each pair. Meiosis I is called the reduction

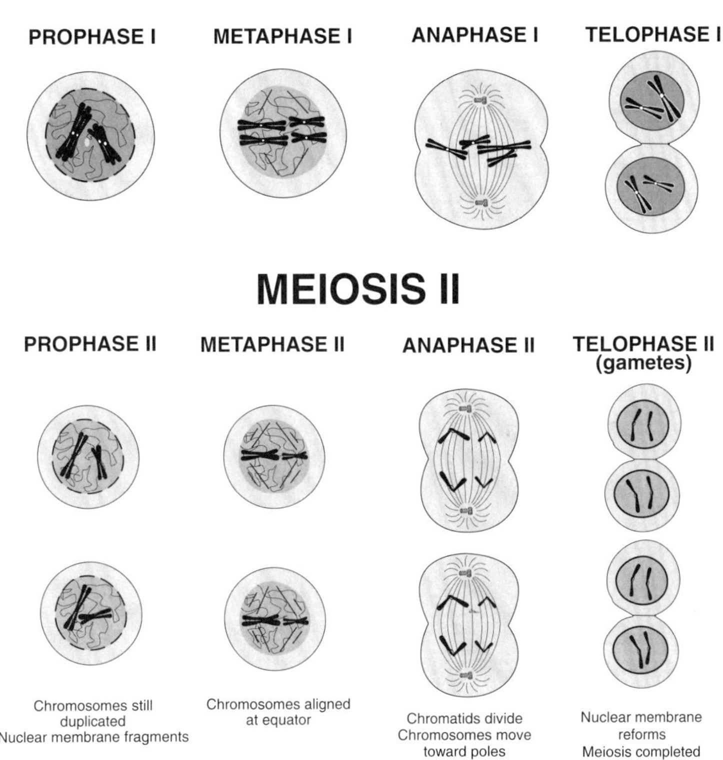

Figure 1-5. Meiosis is the division of sex cells called gametes.

division because the number of chromosomes is reduced from diploid to haploid. Remember that during mitosis, both daughter cells are diploid.

In meiosis II, each daughter cell again divides. During metaphase II of meiosis II, the sister chromatids line up on the axis of the cell. In anaphase II, the chromatids are pulled apart at the centromere by spindle fibers. The end result of meiosis II is two haploid cells from each of the daughter cells exiting meiosis I. Each of the four final haploid cells or gametes contains one strand from the original homologous pair of chromosomes. See Figure 1-5.

CELLS, TISSUES, ORGANS, AND SYSTEMS

In multicellular animals, cells are specialized. For instance, muscle cells are responsible for support and locomotion. Bone cells make up the skeleton and are responsible for structural support. Red blood cells carry oxygen. Fat cells store energy and cannot carry oxygen. Some specialized types of cells make up tissues. For example, cartilage, which is found in the human nose and ear, is a tissue that acts as a shock absorber for the body.

Cells and tissues work together to form organs, such as the liver, which manufacture substances vital for body function.

Organs and tissues combine to form systems in an animal's body. Many of these systems will be discussed further in following chapters. A partial listing of recognized systems follows:

1. The skeletal system is comprised of bone and cartilage tissue and serves as a support structure for the body.
2. The muscular system is made of muscle tissue attached to the skeletal system by tendons and allows the skeletal system to move.
3. The respiratory and circulatory systems combine to ensure that internal cells get the oxygen needed to survive. These systems are comprised of the lungs and heart, respectively, along with miles of blood vessels that carry red blood cells to internal cells.
4. The digestive system consists of many organs including the stomach, intestines, and liver, which combine to break food into particles small enough to be carried by the blood stream and used by individual cells.
5. The nervous system is made up of the brain, spinal cord, and specialized nerve cells that carry electrical impulses to the brain for processing.
6. The endocrine system encompasses many specialized cells and glands that secrete substances into the blood and keep the body chemically balanced.
7. The reproductive system is closely tied to the endocrine system. The principal gland of the female system is the ovary, which produces ova (eggs) and secretes hormones that help to sustain a pregnancy. The principal male gland is the testis, which produces sperm and hormones to ensure the sperms' proper development.

TAXONOMY

Figure 1-6. Carolus Linnaeus developed a taxonomy for classifying living organisms. (Courtesy, Stock Montage, Inc.)

In the mid 1700s, Carolus ***Linnaeus*** developed a system for scientifically classifying all organisms according to their similarities. See Figure 1-6. All organisms have since been sorted by biologists into a structured classification or ***taxonomy*** system. In descending order, the seven steps of the system are: kingdom, phylum, class, order, family, genus, and species. The top rung of the taxonomy ladder consists of very broad categories called kingdoms. There are five recognized kingdoms:

Monera (includes one-celled organisms)
Protista (includes slightly more advanced life forms, which can be one-celled or multicelled)
Fungi (includes molds, fungi, and lichens)
Plantae (includes all normally recognized plant forms)
Animalia (includes sponges, reptiles, insects, birds, mammals, and other animals)

The second rung of the ladder of classification is "phylum." Farm animals are included in the phylum *Chortdata,* which includes all animals with a backbone.

The third rung is "class." Cattle, sheep, horses, and swine are in the class *Mammalia*, which also includes humans. Mammals are characterized by hair- or fur-covered bodies, live birth, and young who are nourished by milk from their mothers.

Class *Mammalia* (mammal) is further divided into "orders" in the fourth rung of the classification ladder. Even-toed, hoofed animals, such as cattle, sheep, and swine, are in the order *Artiodactyla*. Horses are considered an odd-toed, hoofed

ANIMAL SCIENCE FACTS

Students often memorize the order of Linnaeus' hierarchy of classification by reciting a sentence in which the first letter of each word is the same first letter as the words in the taxonomy. One example is listed below. Maybe your instructor knows a different one. Can you think of your own original version?

King Philip Called Out For Giant Sandwiches.

Common Name	Hogs	Cattle	Sheep	Horses
Kingdom	Animalia	Animalia	Animalia	Animalia
Phylum	Chordata	Chordata	Chordata	Chordata
Class	Mammalia	Mammalia	Mammalia	Mammalia
Order	Artiodactyla	Artiodactyla	Artiodactyla	Perissodactyla
Family	Suidae	Bovidae	Bovidae	Equidae
Genus	*Sus*	*Bos*	*Ovis*	*Equus*
Species	*scrofa*	*taurus,* or *indicus*	*aries*	*caballus*

Figure 1-7. Taxonomy of farm animals.

animal and fall under the order *Perissodactyla*. Dogs and cats are meat-eaters and are in the order *Carnivora*.

The last three rungs of the classification ladder are "family," "genus," and "species." These three classifications further sort animals into categories with similar characteristics. See Figure 1-7.

Biologists commonly refer to organisms using the binomial system or ***binomial nomenclature.*** Binomial nomenclature, which was also proposed by Linnaeus, uses the organism's "genus" and "species" names written in Latin. For example, the binomial nomenclature name for swine is *Sus scrofa*. When using binomial nomenclature, the genus name is capitalized, but the species name is not. Both names are either underlined or italicized.

DOMESTICATION

SWINE

Domestic swine (family *Suidae*), also known as *Sus scrofa* in binomial nomenclature, were domesticated from both the European wild boar and the East Indian pig. Pigs were first imported to the Western Hemisphere by Christopher Columbus. Also, imports from established European breeds were common during colonial times. Escaped pigs easily adapted to the wild, eating whatever they could find. Wild pigs are still found in some parts of the United States, such as the Smoky Mountains of Tennessee and the hill country of Texas. Animals that were once domestic and have returned to the wild are referred to as ***feral*** species. Originally, domestic hogs were valued for their lard production, but more recently, swine breeders have concentrated on selecting the leanest, heaviest-muscled pigs.

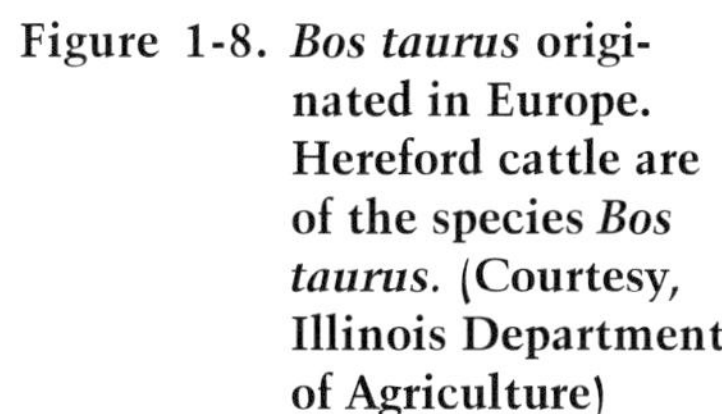

Figure 1-8. ***Bos taurus*** **originated in Europe. Hereford cattle are of the species** ***Bos taurus.*** **(Courtesy, Illinois Department of Agriculture)**

CATTLE

Two species of cattle, *Bos taurus* and *Bos indicus* make up the modern breeds of cattle. Normally, species are unable to interbreed, but these two species of cattle are an exception. Both are from the family *Bovidae*. *Bos taurus* derived cattle include short-eared cattle domesticated from the giant Aurochs, a primitive cattle of the European forests. See Figure 1-8. *Bos indicus* cattle were first domesticated in India and Africa and are characterized by their long, drooping ears and hump above the shoulders. See Figure 1-9. They are also more tolerant of heat and some diseases than their *Bos taurus* counterparts. Early agriculturalists used cattle as a source of meat, milk, and draft power. The first cattle brought to the United States were of Spanish ancestry and easily adapted to the desert Southwest. Known as Longhorns, they could fend for themselves in the wild. Other cattle imported to the colonial United States were more valued for their ability to pull a wagon or plow than for meat or milk. Later, European

Figure 1-9. ***Bos indicus*** **cattle originated in India and Africa. They are long-eared cattle, such as the Brahman. (Courtesy, Agricultural Research Service, USDA)**

breeds specializing in meat or milk production were imported to improve the common stock.

SHEEP

Sheep (family *Bovidae*) were among the first livestock to be domesticated. This is because of their useful products, productivity, relatively small size, and non-competitiveness with humans for food. Classified *Ovis aries,* there are over 200 distinct breeds of sheep scattered throughout the world. These breeds can all be traced to the wild sheep of Europe and Asia Minor. The ancestors of today's domestic sheep were imported by Spanish explorers in the 1400s. Early on, producers concentrated their efforts on those sheep that produced the finest quality wool. This is due to the fact that wool is lightweight and imperishable over time, making it ideal for production in undeveloped areas. Later, as the country's population increased and diversified, breeds were developed specifically for meat production.

HORSES

Hailing from the family *Equidae*, or *Equus caballus*, the domestic horse first served humankind in Asia several thousand years ago. Nomads used the horse for milk, meat, transportation, and as a vehicle of war. Most likely, many types of wild horses were eventually domesticated and mingled to form various breeds and types. Horses actually evolved in the United States but became extinct before the arrival of Columbus. They were reintroduced by Spanish explorers. Escaped horses turned wild (feral), were

Figure 1-10. The Chincoteague ponies of Virginia are a feral species, which were once domesticated but are now wild.

captured, and adopted by Native Americans. See Figure 1-10. Later, large draft-type horses were imported from Europe to assist with farm labor. Horses were used for all types of transportation, including pulling barges, transporting freight, carrying mail, racing, and pleasure. Except for some societies, such as the Amish, who use them daily for labor and transportation, horses in the United States are now used mainly for pleasure.

SUMMARY

The basic unit of life is the cell, which acts as a little factory. Cells take up nutrients, excrete waste, do work, and reproduce. The process of normal cell division is mitosis, while, sex cell division is called meiosis. Cells bond together to form tissues and organs, which in turn form systems in the body.

Organisms are classified into five different kingdoms, which are further divided into phyla, classes, orders, families, genera, and species. The final two classifications (genus and species) are used in binomial nomenclature to scientifically name organisms. Classification names should be italicized or underlined, and all should be capitalized except for the species name, which begins with a small case letter. A correctly written binomial name for the pig is *Sus scrofa*.

Most domesticated farm animals originated in Europe or Asia. The species have evolved to fit the needs of modern society. Formerly domesticated animals that have returned to the wild are called feral. See Figure 1-10.

CHAPTER SELF-CHECK

___ cell	1. division of nonsex cell
___ cell membrane	2. classification system of all organisms
___ nucleus	3. ova and sperm
___ chromosomes	4. basic unit of life
___ DNA	5. controls cell activity
___ genes	6. sex cell division
___ cytoplasm	7. outside of the cell
___ endoplasmic reticulum	8. scientific naming using genus and species
___ ribosomes	9. strands of genetic material

___ mitochondria	10. jelly-like substance in cells
___ lysosomes	11. small pieces of DNA contained in chromosomes
___ golgi bodies	12. membranes that processes raw materials
___ mitosis	13. manufactures protein in the cell
___ gametes	14. number of chromosomes in duplicate in every cell of a particular species
___ meiosis	15. developed a taxonomy
___ haploid	16. complete the manufacture of cellular products
___ Linnaeus	17. manufacture ATP
___ taxonomy	18. digests proteins through the release of enzymes
___ binomial nomenclature	19. wild species that were once domestic
___ feral	20. genetic compound that controls inheritance

QUESTIONS AND PROBLEMS FOR DISCUSSION

1. List the four similarities of cells.
2. What part of the cell controls all cellular activity?
3. Genes are small pieces of the larger ________________.
4. Chromosomes consist of strands of ____________________.
5. Name the two phases of the cell life cycle.
6. List the three phases of interphase.
7. What is the major difference between mitosis and meiosis?
8. The result of meiosis is four _______ cells instead of the two diploid cells, which result from mitosis.
9. Is cartilage a cell, tissue, or organ?
10. List five systems found in animals.
11. Write the seven classifications of Linnaeus' taxonomy in order, beginning with kingdom.
12. Name four farm animals that are mammals.

13. Correctly write the scientific name for swine.
14. Name one feral species found in the United States.
15. Contrast the two species of cattle.

ACTIVITIES

1. Examine a prepared slide of an animal cell. These can be obtained from a science supply company.
2. Construct a model of an animal cell using clay.
3. Use a biology text to further research animal cells.
4. Do a report on the feral ponies of Chincoteague, Virginia or the wild horses of the western United States.
5. Memorize Linnaeus' taxonomy.
6. Develop a classification system like Linnaeus' using possessions, such as video games, clothing, sports equipment, etc.

LABORATORY ACTIVITY

OBSERVATION OF AN ANIMAL CELL

Purpose

To observe an animal cell

Materials

light microscope
slide
cover slip
medicine dropper
water
toothpick
methylene blue or iodine stain

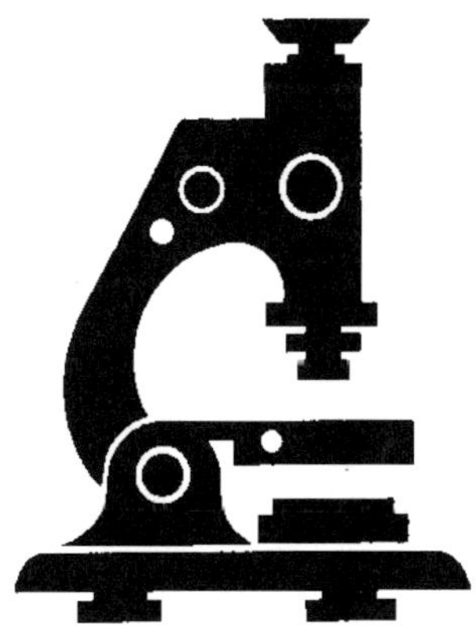

Procedure

1. Assemble the materials.
2. Place a drop of water in the middle of the slide.
3. Scrape the inside of your cheek with the toothpick.
4. Put the scraped material on the water droplet.
5. Stain the water and cheek cell mixture with a drop of methylene blue or iodine solution.
6. Place a cover slip on top of your slide.
7. Observe under low and high power.

Analysis

1. Sketch an animal cell and label the parts.
2. Sketch the cheek cell you observed under the microscope and label the parts.
3. Write a paragraph comparing the two sketches.
4. Why do you think it is important to stain the slide?

Application of Laboratory Activity

A basic understanding of cellular anatomy is crucial to further studies in animal physiology. Cells, the basic units of life, take in nutrients, excrete waste, work, and reproduce. In essence, cellular biology is the cornerstone of all phases of animal production and performance.

This laboratory activity introduces you to the structure of animal cells. Upon completion of this chapter and the accompanying laboratory activity, you should be able to sketch and identify the parts of an animal cell and describe the function of each part.

Chapter 2

BIOLOGY OF GROWTH AND DEVELOPMENT

Growing Up Big and Strong

INTRODUCTION

Many farm animals are raised solely for meat production. Therefore, producers want to raise animals that are heavily muscled without an excess of fat. Factors that influence the amount of muscle in the carcass include maturity pattern and sexual condition of the animal. The biology of muscle, fat, and bone growth, as well as two natural substances that can alter muscle to fat ratio are discussed in this chapter.

Figure 2-1. (Courtesy, White Oak Mills)

OBJECTIVES

1. Draw and label a growth curve
2. Compare and contrast bone with muscle growth
3. Describe the process of adipose tissue deposition
4. Differentiate between early and late maturing animals
5. Discuss the economic importance of the ideal slaughter point
6. Identify two substances that may someday be used in animals to increase the amount of muscle and decrease the amount of fat

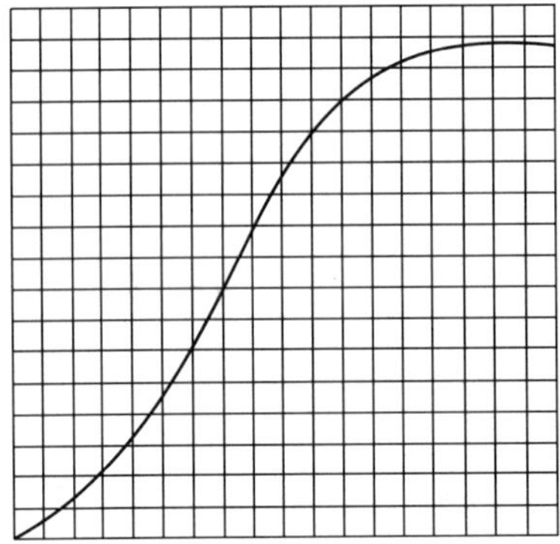

TERMS

adipocytes
adipose tissue
androgens
barrows
beta agonist
boars
bulls
calcification
castrated
collagen
conception
epiphyseal growth plate
estrogens
ewes
gilts
heifers
intact
intermuscular fat
intramuscular (IM) fat
myoblasts
myofiber
myotube
presumptive myoblasts
rams
somatotropin (ST)
steers
subcutaneous
wethers

ANIMAL SCIENCE FACTS

The Ossabaw pig is a feral species native to the island of Ossabaw, off the coast of Georgia. Since food sources are limited on the island, the pigs adapted by gorging themselves during the season when food is plentiful. As a result, they put on a thick layer (measured in inches) of fat which helps to sustain them throughout the rest of the year when feed is unavailable. The Ossabaw pig has been useful to animal scientists researching adipose tissue development. (Photo courtesy, Department of Dairy and Animal Science, College of Agricultural Sciences, The Pennsylvania State University)

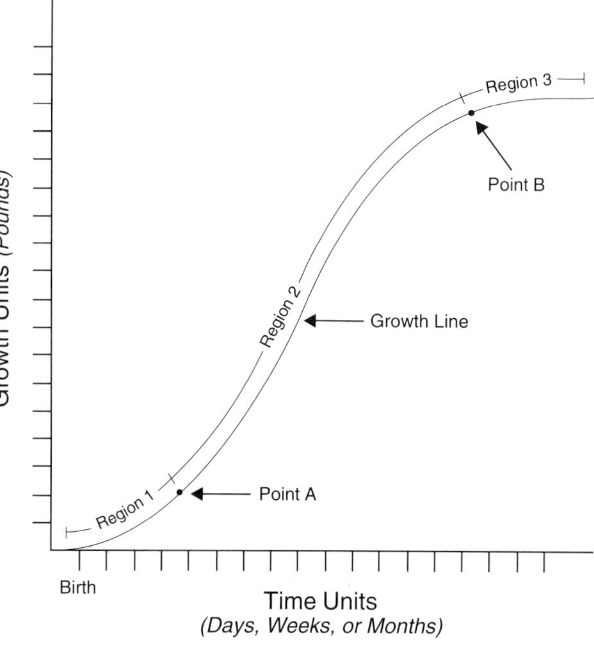

Figure 2-2. The standard S-shaped growth curve.

THE GROWTH CURVE

Animals gain weight at different rates during the growing period from ***conception*** (time of fertilization) to maturity (adulthood). For example, a 3-week-old pig may gain 1/2 pound per day, while the same pig at 20 weeks of age would gain up to 2 pounds per day. To illustrate this concept, scientists often use the standard or S-shaped growth curve. Growth units, such as pounds, are represented on the vertical axis and time units, such as days, weeks, or months, are represented on the horizontal axis. Early growth (region 1 of Figure 2-2) before and shortly after birth, is relatively slow. At a certain point after early growth, (point A of Figure 2-2) body weight gain accelerates

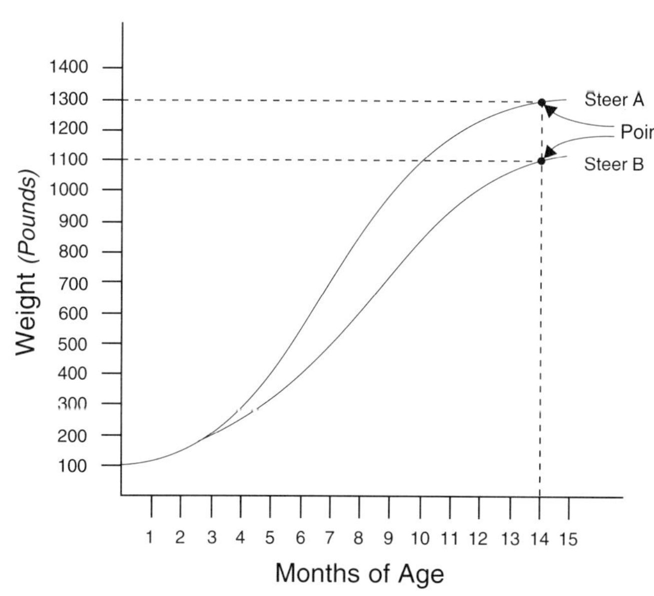

Figure 2-3. Growth curves of two steers with different mature body weights.

rapidly. This period compares to the normal teenage "growth spurt" and is shown in region 2 of Figure 2-2. As animals reach maturity (point B of Figure 2-2), body weight gain slows until a plateau (region 3 of Figure 2-2) is reached at the animal's mature body weight.

Not all animals of the same species mature at the same body weight. Compare the growth curves of the two ***steers*** (castrated male cattle) in Figure 2-3. Notice that the curve levels (point B) lower on the growth line for steer B than steer A. If you draw a line from both B points back to the growth units line, you'll see that steer B has a lower mature body weight than steer A. Even though both animals reached mature body weight at the same time, steer B matured at a much lighter body weight.

Even with animals of the same mature body weight, the rate at which they reach that weight may vary. Compare two ***barrows*** (castrated male swine) with different slopes to their growth lines. See region 2 of Figure 2-4. Again, concentrate on point B. At point B, the mature body weight is the same for both animals, but when you drop a line to the time units axis, you'll find that barrow A reached mature body weight in less time than barrow B. Therefore, barrow A must have grown faster.

Each animal has its own distinct growth curve. Some animals mature early on the time line, others late. Conversely, some animals mature at lighter weights than others. The slope of the growth line tells how fast the animal grew as it approached mature body weight. If you weighed an

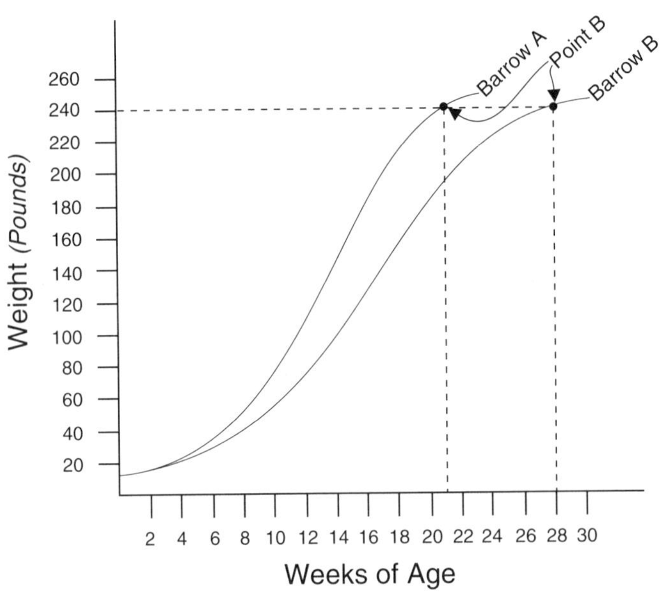

Figure 2-4. Growth curves of two barrows with the same mature body weights but different growth rates.

animal frequently from birth to maturity, you could plot a very similar curve for any normally growing animal.

ECONOMICALLY IMPORTANT TISSUES

Muscle, bone, and fat are the three tissues found in the greatest abundance in carcasses of slaughtered animals. These three tissues grow at different rates while the animal progresses from birth to maturity. See Figure 2-5. The addition of muscle tissue makes up most of the carcass weight gain in young, rapidly growing animals. As that animal gets older and larger, the amount of muscle addition slows and the amount of fat addition accelerates. At some point in the maturity process (if the animal is well fed), muscle growth virtually stops while fat development continues. This point is known as the "ideal slaughter point." At the same time, bone growth maintains a fairly constant rate from birth to maturity.

As with total body growth, the rate of muscle and fat addition varies among individual animals. Earlier maturing animals normally deposit fat at a faster rate and at a younger age than later maturing animals. Compare the muscle, fat, and bone growth curves in Figure 2-5 with Figure 2-6. Notice that the "ideal slaughter point" in Figure 2-5 is at a heavier weight and an older age than the animal in Figure 2-6. Since animals convert feed more efficiently into muscle tissue than fat, it is more economical

Figure 2-5. Percentages of muscle, bone, and fat in a large-framed, late-maturing animal.

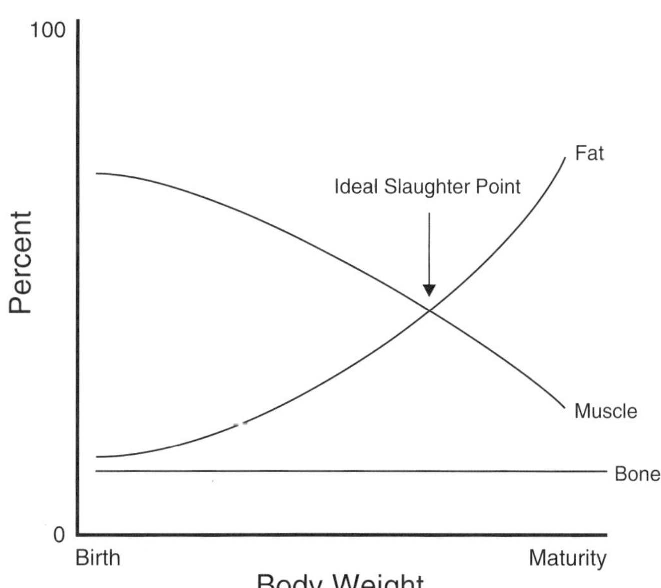

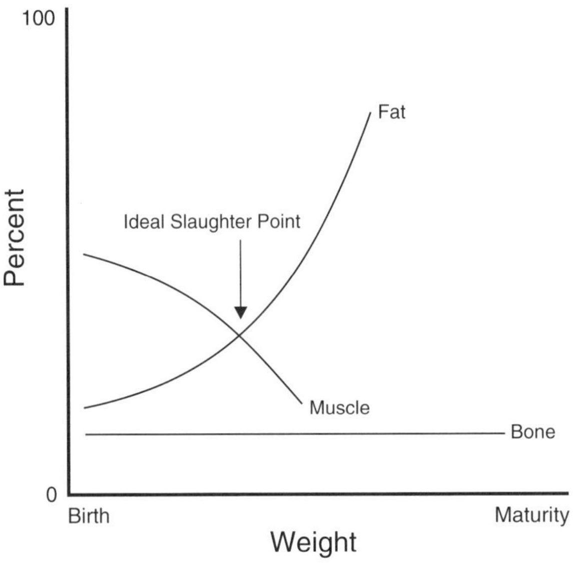

Figure 2-6. Percentages of muscle, bone, and fat in a small-framed, early-maturing animal.

for producers to raise animals that mature later and at heavier weights because feed costs per unit gain are lower.

The rate of muscle and fat addition varies depending on whether the animal is an ***intact*** (uncastrated) male, a female, or a ***castrated*** (testicles removed) male. Of these three sexual conditions, intact males—***bulls, rams,*** and ***boars*** (for cattle, sheep, and swine, respectively)—mature later and at heavier weights than either females or castrated males. The ranking of the latter two sexual conditions depends on the species of the animal. In cattle, steers (castrates) are later maturing than ***heifers*** (young females). In sheep, ***wethers*** (castrates) are later maturing than ***ewes*** (young females). In swine, ***gilts*** (young females) are later maturing than barrows (castrates). See Figure 2-7.

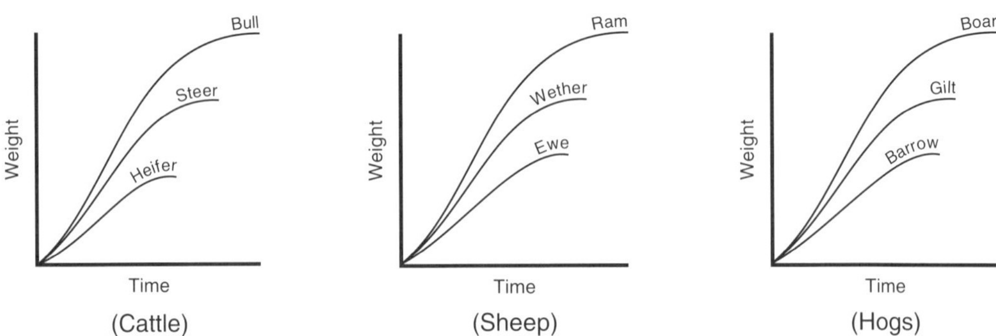

Figure 2-7. Maturity patterns among the sexes in cattle, sheep, and swine.

BIOLOGY OF MUSCLE DEVELOPMENT

There are three types of muscle. Cardiac muscle is found in the heart. Smooth muscle surrounds tubular organs, such as the intestine and esophagus. The largest and most economically important type of muscle in meat animals is skeletal muscle found attached to bones in the legs, back, shoulders, and elsewhere in the body. Skeletal muscle provides most of the meat produced for food.

The first cells that can be distinguished as premuscle cells in the embryo are called ***presumptive myoblasts.*** Presumptive myoblasts contain one nucleus and have the ability to divide. When presumptive myoblasts lose their ability to divide, they are called ***myoblasts.*** The only difference between presumptive myoblasts and myoblasts is that presumptive myoblasts can divide, while myoblasts cannot. Toward the end of fetal development, myoblasts fuse together to form myotubes. The process of fusing makes one ***myotube*** out of many myoblasts. Myotubes look like mature muscle cells because they are long, narrow, and have many nuclei. Myotubes also contain certain proteins found only in mature skeletal muscle. Like myoblasts, myotubes do not have the ability to divide. Shortly before birth, myotubes begin functioning as mature muscle cells. Mature muscle cells are called ***myofibers*** and also have many nuclei. Myofibers are discussed further in Chapter 3. Myofibers do not have the ability to divide. The number of myofibers an animal is born with is the number it will have for the rest of its life.

Muscle tissue continues to grow in width by the incorporation of satellite cells. Satellite cells are closely associated with the surface of the myofiber, but have only one nucleus. Satellite cells are very similar to presumptive myoblasts. However, they do not fuse to form myotubes. Instead, they are used by muscle tissue as a source of nuclei and organelles in exercise-induced muscle growth or muscle repair.

Muscle cells grow lengthwise by stretch-induced muscle growth. As bones grow, muscles must lengthen to keep up, so muscle cells also lengthen.

BIOLOGY OF FAT DEVELOPMENT

Fat is commonly called ***adipose tissue,*** and individual fat cells are called ***adipocytes.*** Adipocytes are very large in diameter compared with other cells in the body, but the extra size consists almost entirely of stored energy. This energy is in the form of lipid, a thick, high energy substance of about the same consistency as butter. Fat cells act as an extra gas tank for the animal. When the animal cannot consume enough energy to maintain the basic functions of life, the energy stored in adipocytes is used until more energy is eaten, or the energy demands of the animal are lessened.

Figure 2-8. Backfat is subcutaneous fat most often measured over the loin.

Mature adipocytes do not undergo mitosis, but pre-adipocyte cells do when necessary. As adipocytes store more lipid and grow, nutrients needed from the blood cannot reach the center of the cell. At that point, pre-adipocyte cells start dividing to produce a new crop of "baby adipocytes" to begin filling with lipid.

There are four major deposits of fat in the animal body. Internal fat within the body cavity surrounds the kidneys, pelvic area, and, in cattle, the heart. Internal fat is the first to be stored as the animal matures. ***Intermuscular fat*** is stored in the seams between muscles. ***Subcutaneous*** (under the skin) fat is stored between the hide and muscle and is usually called backfat. Backfat contains the largest volume of stored lipid. See Figure 2-8. Both intermuscular and backfat are stored about the same time, midway through the fat deposition process. The last fat deposit the animal develops is ***intramuscular fat.*** It is more commonly called marbling and is found inside muscle bundles. Marbling is thought to account for the flavor and juiciness of meat. The first three types of fat are undesirable in animals raised for meat production because of its low economic value relative to muscle. However, highly marbled meat is desirable to consumers because it provides flavor and juiciness. The order of fat deposition makes it difficult for producers to raise animals with highly marbled meat and small amounts of the other three types of fat. However, research suggests that animals can be genetically selected for small amounts of backfat and large amounts of marbling. See Figure 2-9.

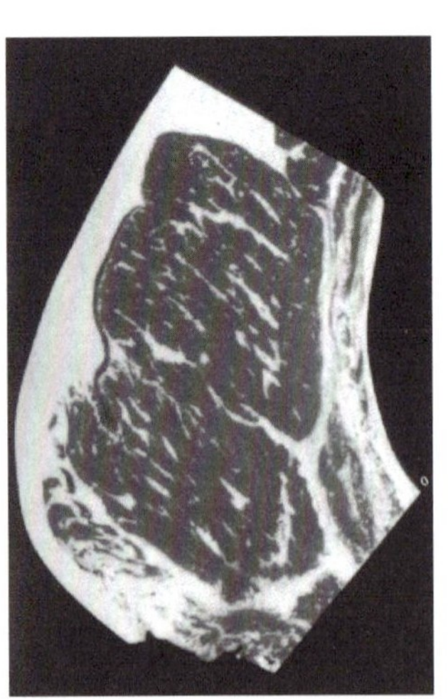

Figure 2-9. Marbling or intramuscular fat gives meat a juicy flavor.

BIOLOGY OF BONE DEVELOPMENT

Bone growth in young animals is preceded by the growth of collagen strands. ***Collagen*** is a fibrous tissue composed of three strands of tropocollagen chemically bound together in a triple helix. See Figure 2-10. Strand for strand, collagen is stronger than steel. It is formed at the end of long bones in a thin region called the ***epiphyseal growth plate.*** This growth plate contains many blood vessels that deliver raw materials to the rapidly dividing collagen cells. As new collagen is added to the end of the bone, older collagen in the interior of the bone dies and is replaced by calcium. This process is called ***calcification.*** As the animal reaches maturity, collagen stops dividing, the growth plate disappears, and the entire bone is calcified. At that point, the bone cannot grow any longer, but it may thicken as more calcium is added to the walls of the bone.

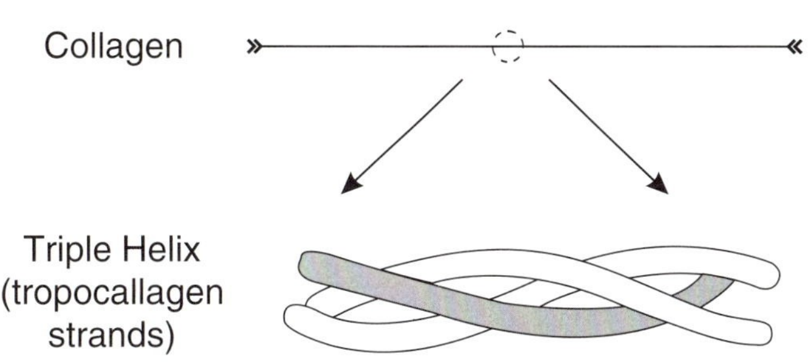

Figure 2-10. The triple helix of tropocollagen that precedes bone growth is stronger than steel.

Baby animals' skeletons are made mostly of collagen (or cartilage) that is uncalcified. This provision by nature ensures that baby animals do not break any bones during the birth process. Shortly after birth, calcification begins.

CHANGING THE MUSCLE TO FAT RATIO

Skeletal muscle and fat are the most economically important tissues in the carcasses of animals raised for meat. Scientists have long searched for methods to produce animals with a higher proportion of muscle to fat. Before learning about such methods, we must first discuss theories on the distribution of nutrients.

For purposes of this discussion, think of all of the protein in the animal's body as being present in one of two possible places: in muscle or in the blood stream. Simultaneously, protein is being made into muscle protein from blood protein, and broken down from muscle protein into blood protein. The trick is to capture a higher percentage of total body

> **ANIMAL SCIENCE FACTS**
>
> Porcine somatotropin (pST) administered by daily injection has been shown to increase growth rate by 10–20%, improve feed efficiency by 20–35%, decrease backfat by 40–45%, and increase carcass protein by about 20%. Somatotropin has not yet been approved for use in swine by the FDA.

protein in muscle rather than in the blood stream where it can be lost in the urine. In other words, if muscle manufacturing is greater than muscle break down, a net increase in muscle tissue will occur.

Likewise, glucose (energy in the blood stream) is constantly being manufactured into fat and broken down from fat. If the rate of fat production is greater than the rate of break down, an increase in adipose tissue will occur.

Both these processes happen in unison. An animal storing muscle tissue will not store much fat and an animal storing a large amount of fat will not deposit much muscle tissue. Two substances have been shown to divert protein and energy toward muscle growth rather than fat accumulation, and show some promise of receiving approval by the Food and Drug Administration (FDA) for use in farm animals.

Somatotropin (ST)

Somatotropin (ST) is a naturally occurring protein present in all animals, including humans. It is secreted by the pituitary gland, which is located at the base of the brain, and is species specific. This means that somatotropin from one species (such as pigs) will not affect any other species (such as cattle or sheep).

Somatotropin is the hormone that regulates the growth of long bones, muscle, and fat. It is secreted in relatively large amounts in young, growing animals with the amount decreasing to a lower level at maturity. The greater the amount of naturally occurring somatotropin secreted during the growth phase, the larger and later-maturing the animal will be.

Somatotropin also works to make animals heavier muscled by increasing protein formation and decreasing protein breakdown, so a net gain in muscle mass results. Similarly, somatotropin works to make animals leaner by increasing the rate of fat breakdown so a net loss of fat mass results. In lactating dairy cattle, bovine somatotropin (bST) works to increase the amount of nutrients captured by the mammary gland (udder) for milk production.

Each animal has a naturally occurring amount of somatotropin in its bloodstream. If the amount is artificially increased, an animal that is genetically programmed to be fat, light muscled, and early maturing can be changed into one that is leaner, heavier muscled, and later maturing.

Figure 2-11. Porcine Somatotrapin (pST) drastically increases lean tissue development and decreases fat depth as seen in these two pork loins.

It is impossible to collect enough natural somatotropin from dead animals to use on a large number of growing animals. Therefore, scientists have developed a way to genetically engineer bacteria to produce somatotropin in large amounts. This synthetic somatotropin is called recombinant somatotropin and has exactly the same effects as the somatotropin the animals produce themselves.

Somatotropin is a protein. If it were fed to animals, it would be digested like any other protein with no affect on the animal. Also, the effective life of somatotropin in the bloodstream is very short. Therefore, to have any measurable effect, it must be injected into the animal regularly. There is no difference in the safety or quality of meat or milk from animals injected with recombinant somatotropin when compared with meat and milk from animals that were not injected. In fact, it is impossible to tell the difference. Recombinant bovine somatotropin (rbST) has currently been approved by the FDA for use in lactating dairy cattle. It is not legal for use in animals raised for meat.

Beta Agonists

Beta agonists are synthetic substances similar to adrenaline (the hormone that courses through your body when you are suddenly frightened). As with somatotropin, they increase protein production into muscle and decrease protein breakdown from muscle. They also increase fat breakdown and decrease fat production. Beta agonists work by a different mechanism in the animal's body than somatotropin because their effects are additive when both are administered at the same time. Beta agonists are orally active and can be given to animals in the feed. They have not been approved for use in farm animals by the FDA.

Sex Hormones (Growth Promoting Implants)

Sex hormones increase growth rate and lean tissue development while decreasing fat. Blood levels of sex hormones, such as ***androgens*** (in males) and ***estrogens*** (in females), can be increased through the use of implants in cattle and sheep. Small pellets containing the hormones are implanted under the skin behind the ear. See Figure 2-12. Androgens are naturally secreted by the testes in the male and estrogens are secreted by the ovaries in the female. When males are castrated, their natural supply of androgens ceases to exist. Implants are used by producers to restore some of the more desirable effects of androgens (such as leanness, increased growth rate, and less backfat) while avoiding some of the undesirable effects (such as unpredictable temperament and strong tasting meat). Implants also have similar effects on the carcasses of females even though the estrogen factories (ovaries) are still present.

Figure 2-12. Producers use implants inserted by an implant gun to improve the feed efficiency, increase growth rate, and improve carcass composition of cattle. (Courtesy, Edward W. Osborne)

Implants are legal for use in cattle and lambs fed for slaughter, although rectal prolapses are a fairly common side effect in implanted lambs.

Genetic Selection

The best long-term technique for improving the lean content in farm animals is through genetic selection. Animals that are lean and heavily muscled tend to produce similar offspring. As agriculture becomes more competitive and producers are forced to become more efficient, the search to find new ways to make animals produce more meat and milk and less fat will undoubtedly continue.

SUMMARY

Animals mature at different rates and at different body weights. Early-maturing animals that mature at lighter weights normally are fatter and lighter muscled than later-maturing, heavier animals. Growth curves can be developed for any growing animal and can tell us much about its maturity pattern. Muscle cells, fat cells, and bone tissues develop in distinctly different ways, but all are important components of the final carcass. Consumers demand that producers raise lean, meaty animals. The understanding of how muscle tissue and fat are proportioned in the body can assist the producer in raising animals with a higher percentage of lean muscle. Somatotropin and beta agonists are two substances that have been shown to increase the muscle to fat ratio.

CHAPTER SELF-CHECK

___ conception	1. intact male swine
___ maturity	2. castrated male sheep
___ steers	3. males with testicles removed
___ barrows	4. embryonic premuscle cells that can divide
___ castrated	5. individual fat cells
___ intact	6. time of fertilization
___ bulls	7. naturally occurring growth hormone
___ rams	8. between muscles
___ boars	9. adulthood
___ heifers	10. uncastrated
___ wethers	11. intact male cattle
___ ewes	12. fat
___ gilts	13. fused myoblasts
___ presumptive myoblasts	14. synthetic substance similar to adrenaline
___ myoblasts	15. fibrous tissue stronger than steel
___ myotube	16. under skin
___ myofiber	17. mature muscle cell

___ adipose tissue
___ adipocytes
___ intermuscular fat
___ subcutaneous
___ intramuscular (IM) fat
___ collagen
___ somatotropin (ST)
___ beta agonist
___ androgens
___ estrogens

18. castrated male cattle
19. castrated male swine
20. intact male sheep
21. young female cattle
22. marbling
23. embryonic premuscle cell that cannot divide
24. young female sheep
25. young female swine
26. female sex hormones
27. male sex hormones

QUESTIONS AND PROBLEMS FOR DISCUSSION

1. Draw a standard or S-shaped growth curve.
2. True or False? Animals of the same species always grow to maturity at the same rate.
3. Name the three most common tissues found in an animal carcass.
4. When does the ideal slaughter point occur in animals?
5. Which animal tissue continues to grow at a steady rate from birth through maturity?
6. Why is it more economical for producers to raise animals that mature at later and heavier weights?
7. Which mature later, intact or castrated males?
8. In which of the three species, swine, cattle, or sheep, do females mature later than castrated males?
9. Name the three types of muscle tissue.
10. List the four areas of fat deposition.
11. Explain the calcification process of bone growth.
12. What is the only FDA approved farm animal use of somatotropin?
13. Somatotropin is secreted by the _______________ gland.

14. When beta agonists are given to animals, protein manufacture increases/decreases and fat manufacture increases/decreases. Write correct answers.

15. Implants are placed in which part of the animal's body?

ACTIVITIES

1. Plot a student growth chart of height or weight data as the growth units and years of age as the time units. Use information provided by parents, guardians, or pediatrician.

2. Keep an accurate weight record at designated time intervals for students' SAE market livestock projects. Plot growth curves using this data. Determine if animals were early or late maturing.

3. Have a classroom debate considering the pros and cons of somatotropin usage. Contact local experts to provide background information.

LABORATORY ACTIVITY

PLOTTING GROWTH CURVES

Purpose

To plot growth curves from two different animals and analyze their growth patterns

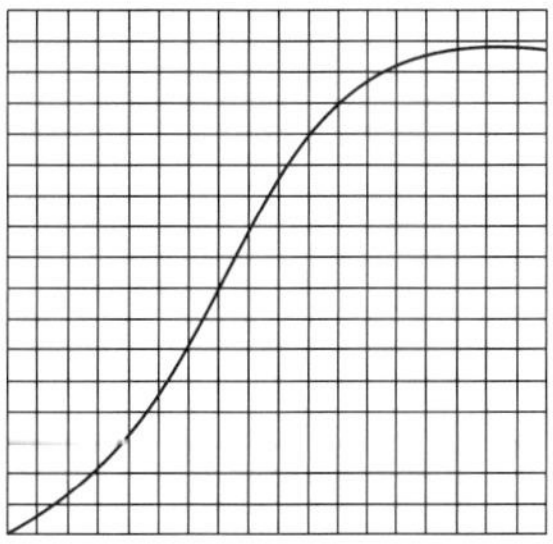

Materials

graph paper
pencil
ruler

Procedure

1. Plot the following data for two wether lambs, "Sam" and "Buck" on graph paper. Label the horizontal axis "Days of Age" and the vertical axis "Weight in Pounds."

Days of Age	Weight in Pounds: Sam	Weight in Pounds: Buck
15	8	8
30	10	9
45	12	11
60	15	13
75	24	19
90	28	23
105	31	25
120	35	27
135	43	36
150	52	45
165	61	51
180	70	59
195	75	65
210	80	71
225	92	74
240	100	80
255	102	81
270	104	82

2. Connect the dots for each lamb to form a growth curve.
3. Identify the points where growth accelerates.
4. Identify the points where growth slows as the lambs approach maturity.
5. Identify the growth line.

Analysis

1. Which lamb appeared to mature at the heavier weight?
2. Was there a difference in the time it took the lambs to reach maturity?
3. Was there any difference in the growth rate of the lambs?
4. Which lamb would you expect to be fatter upon reaching 80 pounds?

Application of Laboratory Activity

Livestock producers want to raise animals that are heavily muscled and without an excess of fat. In order to do so, they must select breeding animals based on their maturity pattern and growth rate. At the same time, they must recognize that not all animals mature in the same period of time and at the same body weight.

This laboratory activity introduces you to the concept that animals of the same species, in this case sheep, mature at different ages and at different weights. Upon completing this lab, you should have an understanding of the growth and development concepts producers must be familiar with in order to be successful.

Chapter 3

MUSCLE AND MEAT BIOLOGY

Bulking Up

INTRODUCTION

The production of meat for human consumption is the main reason for raising many livestock species. Meat is nothing but dead muscle tissue. Muscle tissue is approximately 72 percent water, 20 percent protein, 7 percent fat, and 1 percent mineral components. To better understand meat and meat products, a basic knowledge of muscle biology is critical.

Figure 3-1. (Courtesy, National Live Stock and Meat Board)

OBJECTIVES

1. Draw, label, and describe the functions of a muscle cell
2. List and distinguish among the types of muscle cells
3. Discuss two factors which can cause differences in meat quality
4. Explain post-mortem changes in muscle
5. Describe two processes that improve meat tenderness

TERMS

actin
aerobic exercise
anaerobic exercise
endomysium
epimysium
glucose
glycogen
lactic acid
myosin
perimysium
porcine stress syndrome
post mortem
rigor mortis
sarcomere
sarcoplasmic reticulum (SR)
tetany
titan
tropomyosin

ANIMAL SCIENCE FACTS

The average American's meat consumption per year is the highest in the world.

Average American meat consumption for 1992:

97 pounds of beef and veal
69 pounds of pork
1.5 pounds of lamb
81 pounds of chicken
20 pounds of turkey

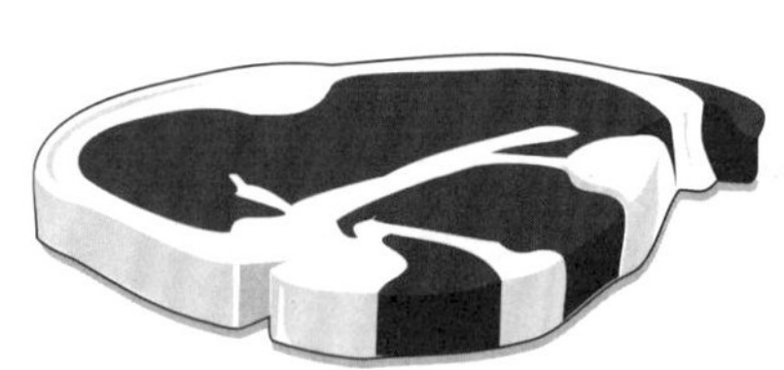

MUSCLES AND MUSCLE CELLS

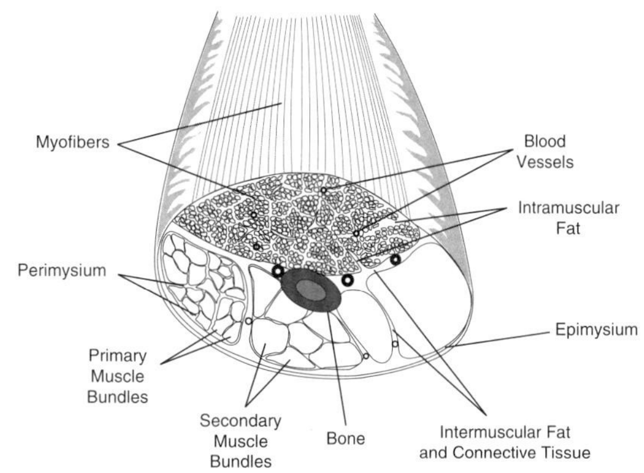

Figure 3-2. Arrangement of myofibers, muscle bundles, and connective tissue within a whole muscle.

A single muscle cell is called a myofiber. Each myofiber is individually wrapped by a thin film of connective tissue called the ***endomysium.*** Bundles of myofibers are wrapped together by a connective tissue sheath called ***perimysium.*** These bundles of myofibers make up whole muscles which are wrapped by a connective tissue sheath called ***epimysium.*** All the connective tissue surrounding myofibers, bundles of myofibers, and whole muscles consist of a protein called collagen. See Figure 3-2.

Each myofiber is divided into striped units called ***sarcomeres.*** The sarcomere is the segment of the myofiber that shortens when muscles contract and elongates when muscles relax. See Figure 3-3. Sarcomeres are located between the Z lines as shown in Figure 3-3. Sarcomeres are made of the four muscle proteins: ***myosin, actin, tropomyosin,*** and ***titan.*** The thicker filament in Figure 3-3 represents myosin and is referred to as the A band of the sarcomere. The thin filament in Figure 3-3 represents actin and is referred to as the I band of the sarcomere. While tropomyosin helps regulate muscle contraction, the purpose of titan appears to prevent overstretching of the muscle.

Of these four proteins, the thick and thin filaments (actin and myosin) are the most important to muscle contraction. As the sarcomere contracts,

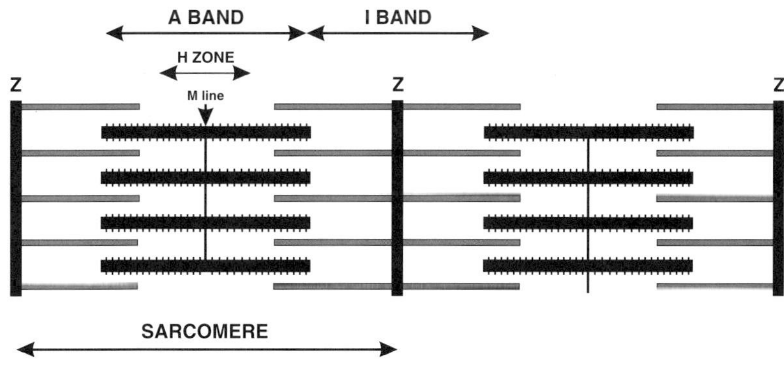

Figure 3-3. The contractile unit of a myofiber is called a sarcomere.

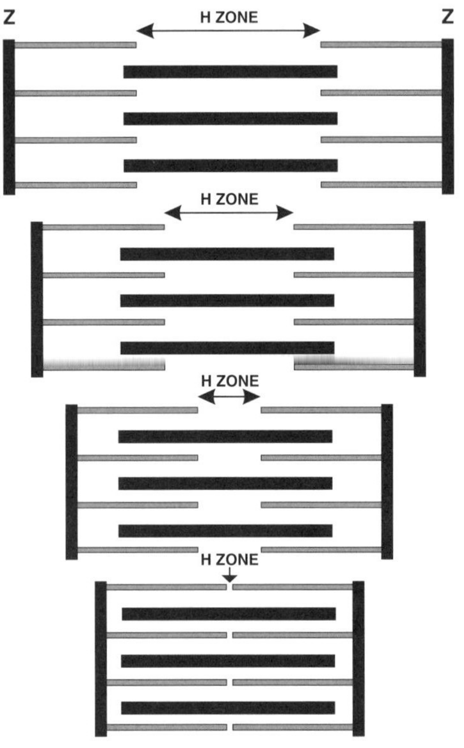

Figure 3-4. The change in shape of the sarcomere during contraction.

myosin stays stationary and attachments to myosin called "myosin feet" pull the actin filaments closer together, shortening the area called the H zone. See Figure 3-4. When the muscle relaxes, the sarcomere resumes its original position.

Calcium allows muscle contraction to occur. Each myofiber is surrounded by a reservoir of calcium called the ***sarcoplasmic reticulum (SR).*** When the muscle gets a message from a nerve that it is to contract, the SR releases calcium into the myofiber. This causes the myosin feet to pull the actin filaments closer together. When the nervous impulse ends, calcium flows back into the SR and the muscle relaxes. The relaxation process requires no energy; actin and myosin molecules simply slide back to their original positions.

The whole process of contraction is fueled by either ***glucose*** or ***glycogen.*** Glucose is blood sugar, which can be obtained directly from the blood stream. Glycogen, or stored glucose, is found within the muscle. If the muscle runs out of stored glycogen or cannot get enough glucose from the bloodstream, it goes into a state of exhaustion called ***tetany.***

TYPES OF MUSCLE CELLS

There are two main types of myofibers. The first type is called slow, white fibers. These fibers are large in diameter and have few blood vessels running through them. Slow, white fibers use stored glycogen within the muscle cell as their primary energy source. Large muscle groups consist of mostly slow, white

ANIMAL SCIENCE FACTS

In 1950, Americans ate an average of 12.6 pounds of lard (animal fat not combined with other products). As Americans became more health conscious, this figure dropped to 1.2 pounds of consumption by 1990.

fibers. The breast meat of poultry is composed almost entirely of slow, white fibers. ***Anaerobic exercise,*** which is intense and short in time (for example, weight lifting), tends to develop this type of muscle fiber.

The second type of myofiber is "fast, red fibers." These fibers get most of their energy directly from the blood. Therefore, they have many blood vessels running through them. Poultry drumsticks are composed mostly of fast, red fibers. ***Aerobic exercise,*** which is long term and less intense (for example, jogging), develop this type of muscle fiber. Think of the comparison between a drumstick (fast, red fibers) and breast meat (slow, white fibers) on a Thanksgiving turkey.

The type of myofiber that predominates in an animal is largely controlled by genetics and varies by muscle group. For example, muscle fibers present in the loin (back muscle) tend to be slow and white, while muscle fibers in the legs tend to be fast and red.

Moreover, heavily muscled animals tend to have more white, slow fibers, which are larger in diameter and take up more space.

MUSCLE TO MEAT, THE DEATH PROCESS

When an animal is slaughtered, blood stops flowing to muscle tissue. Thus, the immediate energy supply of glucose is cut off. Muscle cells continue to attempt contraction using glycogen as an energy source. This process produces ***lactic acid*** as a waste product. Lactic acid is normally

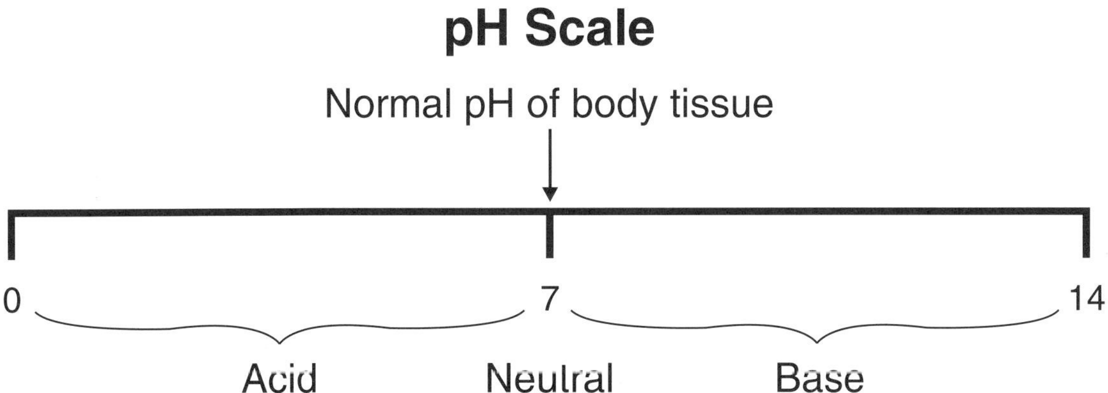

Figure 3-5. The pH scale. Numbers greater than seven indicate basic substances. Numbers less than seven indicate acidic substances.

carried away by the blood stream, but since that avenue is no longer available, the acid builds up in the muscle tissue causing the pH of the muscle to decline. Normal pH in a myofiber is 7.0. When the pH drops to 5.5 due to lactic acid build-up, muscles cease to use glycogen. See Figure 3-5. Furthermore, the lactic acid build-up prohibits sarcomeres from moving and the muscle becomes stiff. This state of ***post mortem*** or after death muscle stiffness is called ***rigor mortis.***

DIFFERENCES IN MEAT QUALITY

The treatment of animals before and after slaughter can affect the final quality of meat. If muscles undergo strenuous exercise immediately before slaughter, muscle glycogen reserves are depleted. Therefore, there is not enough glycogen present in the muscle after slaughter to reduce the pH to normal post mortem levels. Meat from such animals exhibits a dark, firm, and dry appearance and is known as DFD. See Figure 3-6.

On the other hand, if muscle pH drops too rapidly after slaughter, meat develops a pale, soft, and watery appearance. Pale, soft, and exudative or watery meat (PSE) is often associated with pigs that have a condition known as ***porcine stress syndrome*** or PSS. See Figure 3-6. Pigs with this condition have a genetic abnormality in which they have leaky calcium channels between the SR and myofiber. In other words, calcium can flow from the SR into muscle causing a contraction without a nervous impulse. After slaughter, glycogen is converted to lactic acid more rapidly than normal, causing a quick drop in pH and ultimately, PSE meat.

Both DFD meat and PSE meat are very unappealing to consumers and undesirable to packers.

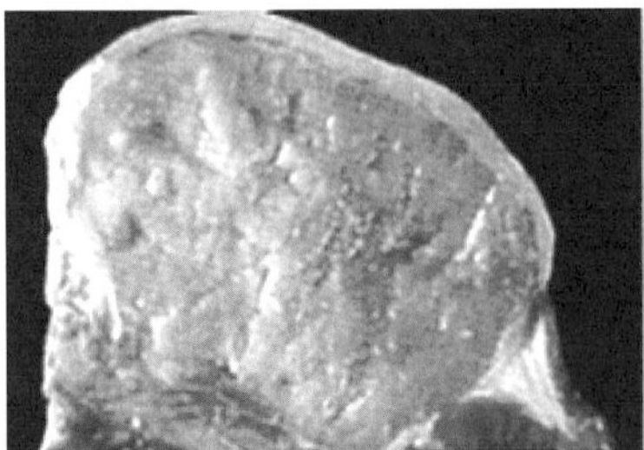
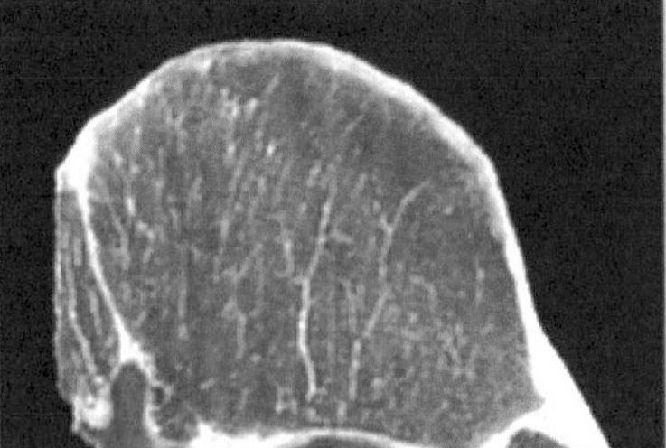
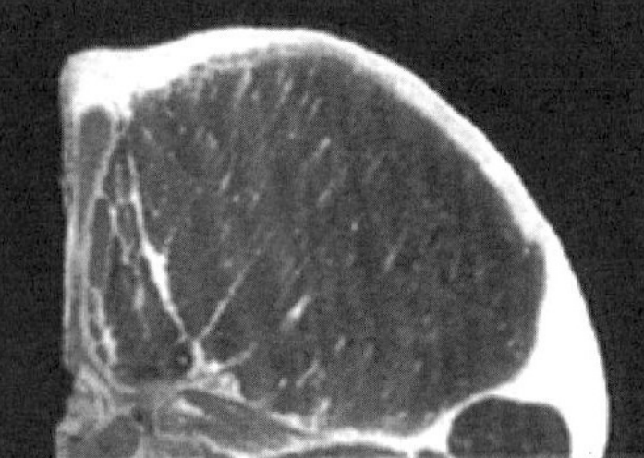

Figure 3-6. Pork chop on the left is pale, soft, and exudative (watery). Pork chop on the right is dark, firm, and dry. Pork chop in the middle is ideal in color and texture.

METHODS OF MEAT QUALITY IMPROVEMENT

Muscles are like ropes. If all the fibers of the rope are continuous, the rope is very strong. Likewise, if all the myofibers and sarcomeres of a muscle are intact, the muscle is very strong — or tough to chew. However, there are several ways to break the myofiber and sarcomere structure and, in the process, make meat more tender.

Meat becomes more tender with age. In fine restaurants, you will often see "aged beef" on the menu. Culinary experts identify this to mean "tender beef." The reason aging makes meat more tender is because of enzymes, which work by breaking some of the Z lines between sarcomeres. The longer meat is aged (held in a cooler after slaughter for up to 14 days), the more Z lines are broken by enzymes and the more tender the meat.

Freezing is a second way to tenderize meat. Remember that muscle is 72 percent water. When meat is frozen, the water present in the meat also freezes. When water freezes, it expands, breaking apart sarcomeres, and making the meat more tender when it is thawed and cooked.

In some large beef packing plants, an electrical current is passed through the carcass immediately after slaughter. The current causes all the muscles in the carcass to contract sharply and increases the rate of glycogen depletion. The pH of stimulated carcasses drops faster than normal, but in beef, this generally improves the color of the lean meat and, for some unknown reason, improves tenderness.

SUMMARY

Individual muscle cells are called myofibers. Myofibers are divided into striped units called sarcomeres, which shorten when muscles contract. Actin and myosin are the two most important proteins found in the sarcomere. They cause muscle contractions. Contractions of muscles use either stored energy in the form of glycogen, or energy from the bloodstream in the form of glucose. Aerobic muscle fibers use mostly blood glucose and are called fast, red fibers. Anaerobic muscle fibers use mostly stored glycogen and are called slow, white fibers.

After slaughter, muscle continues to mobilize stored and immediately available energy reserves. This produces lactic acid and causes the pH of the meat to fall. Preslaughter animal handling, genetics, and the amount of energy available to muscle cells after death can alter the quality of the final meat product. Dark, firm, and dry meat and pale, soft, and watery pork are examples of muscle quality problems. Aging, freezing, and electrical stimulation are examples of ways to improve meat quality.

CHAPTER SELF-CHECK

___ myosin	1. swine genetic abnormality
___ endomysium	2. stored glucose
___ perimysium	3. waste product of contraction
___ epimysium	4. thin filament muscle protein
___ sarcomere	5. surrounds whole muscles
___ actin	6. protein that prevents overstretching
___ tropomyosin	7. intense, short term exercise
___ titan	8. reservoir of calcium
___ sarcoplasmic reticulum	9. after death
___ glucose	10. surrounds individual muscle cells
___ glycogen	11. muscle stiffness after death
___ tetany	12. surrounds bundles of myofibers
___ anaerobic exercise	13. less intense, long term exercise
___ aerobic exercise	14. muscle exhaustion
___ lactic acid	15. blood sugar
___ post mortem	16. protein that regulates muscle contraction
___ rigor mortis	17. thick filament muscle protein
___ porcine stress syndrome	18. striped divider of myofiber

QUESTIONS AND PROBLEMS FOR DISCUSSION

1. A single muscle cell is a ______________.
2. List, in order, the connective tissues that surround myofibers, bundles of myofibers, and whole muscle.
3. The ______________ shorten when muscles contract and elongate when muscles relax.

4. Which muscle protein helps regulate contraction?
5. Calcium, which causes muscle contraction, is stored in the ____________________.
6. _________ is blood sugar, while _________ is reserve sugar that is stored in the muscle.
7. Tetany occurs when muscles run out of _________ and can't get enough glucose from the blood stream.
8. Name the type of myofiber that a weight lifter would develop.
9. Give an example of poultry meat that contains mostly fast, red myofibers.
10. Normal muscle pH is _____.
11. Explain why rigor mortis occurs.
12. Describe the appearance of meat from a hog that has porcine stress syndrome.
13. What intramuscular occurrence makes meat more tender?
14. The text describes several methods used to tenderize meat. Name two.
15. What does electric current do to a beef carcass?

ACTIVITIES

1. Inspect the muscle tissue in a chicken breast and drumstick. Note the differences in color.
2. Interview a local butcher or federal meat inspector. Ask about the aging process and its effect on meat tenderness. If possible, arrange a tour of a butcher shop.
3. Do a report on porcine stress syndrome. Investigate how the swine industry is dealing with this problem.
4. Contact your local extension office for information on safe preparation and cooking of meat. Using this information, make a poster to display in school.

LABORATORY ACTIVITY

COMPARING THE TENDERNESS OF STEAK

Purpose

To compare the tenderness of a steak that has been frozen and thawed with the tenderness of a steak that has never been frozen

Hypothesis

Select a hypothesis from below for the outcome of this activity. Record this in the analysis section.

Examples

1. The steak that has been frozen and thawed will be more tender.
2. The fresh steak will be more tender.
3. There will be no difference in the tenderness of the steaks.

Materials

Two steaks from the same animal (cut from the same area of the carcass)
Microwave to thaw steak
Grill to cook steaks
Plate for steaks prior to cooking
Plate for steaks after cooking
Knife and fork to cut steaks into bite-sized pieces
Toothpicks for serving steaks

Safety Precaution

Care must be taken when preparing and cooking meat to avoid contamination and possible food poisoning from agents such as *E. coli*. Meat can be safely thawed in a refrigerator or a microwave, but not at room temperature. Furthermore, different plates should be used for raw and cooked meat. Steaks should be browned on both sides, while hamburger or ground meat should be cooked thoroughly leaving no pink color in the middle. Meat products that have been exposed to air (all ground meat and the outside of steaks) can harbor *E. coli* and other bacteria that can cause food poisoning. Therefore, ground meat should be thoroughly cooked, while steak should be cooked on the outside. Exercise caution when using the grill. Be sure to follow the directions for use closely and avoid burns by staying clear of the heating mechanism.

Procedure

1. Select two fresh steaks from the same animal and carcass area.
2. Keep one steak refrigerated.
3. Freeze and thaw the other steak. Use a refrigerator or microwave for thawing.
4. After completing procedure #3, grill both steaks using the same method and amounts of time.
5. After placing the steaks on fresh plates, cut the steaks into bite-sized pieces.
6. Place toothpicks in the samples. Arrange samples on two different platters.
7. Ask non-biased taste testers to rate the tenderness of each sample using a scale of one (toughest) to five (most tender).
8. Record your results.

Analysis

1. What was your original hypothesis?
2. What was the average tenderness rating for each sample?
3. What are your conclusions concerning the tenderness of the sample steaks?
4. How do the conclusions compare with your original hypothesis?

Application of Laboratory Activity

In today's market, consumers are concerned more than ever with meat quality (tenderness, juiciness, and flavor). In response to this concern, livestock producers use various methods to sustain and improve the quality of meat. Freezing is one method used to increase the tenderness of meat.

This laboratory activity demonstrates how freezing and thawing affects the tenderness of meat by rupturing sacromeres. A significant amount of the meat purchased by consumers has been frozen and thawed during storage, shipment, and handling.

Chapter 4

BIOLOGY OF DIGESTION

I Can't Believe I Ate the Whole Thing!

INTRODUCTION

Animals use feed to produce products (meat, wool, milk, or eggs) and energy for work. The process of digestion reduces feed particle size and simplifies chemical composition of the feed for absorption by the animal. After absorption, the bloodstream transports the digested nutrients to body cells. There they are used to sustain life or are manufactured into animal products. Cattle and sheep have a special stomach chamber called a ***rumen.*** It contains multitudes of bacteria that digest some feedstuffs that other animals, such as swine and poultry, cannot digest. Horses also have a population of microbes to assist in digestion.

Figure 4-1.

OBJECTIVES

1. List the six essential nutrients and their functions within the body
2. Describe the process of digestion
3. Compare and contrast digestion in pigs, poultry, cattle, sheep, and horses.

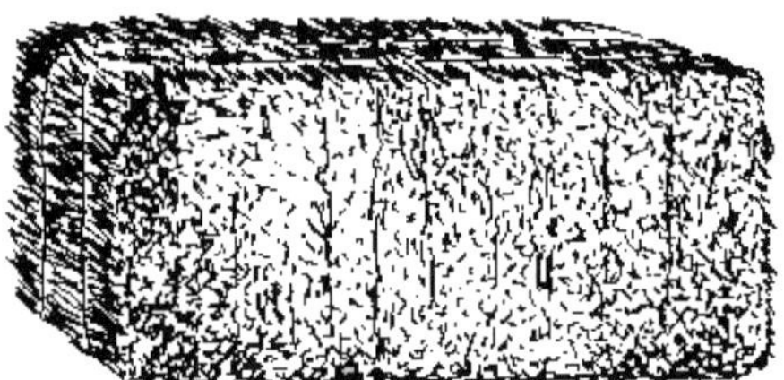

TERMS

abomasum
amino acid
cecum
chyme
complex carbohydrates
crop
gizzard
hydrochloric acid
macrominerals
microminerals
monogastric
omasum
peptidases
peristalsis
proventriculus
reticulorumen
rumen
ruminant
simple carbohydrates
symbiotic
villi

ANIMAL SCIENCE FACTS

The rumen (large stomach chamber) in cattle has been known to hold as much as 50 gallons. This volume equals the same amount as 800 one-half pint milk cartons.

ESSENTIAL NUTRIENTS

There are six essential nutrients that must be present in animal diets to sustain life: water, protein, carbohydrates, fats, vitamins, and minerals. Some must be ingested in large amounts, some in small amounts, but all are critical.

WATER

Water is the most abundant nutrient in the body. Between 40 and 80 percent of total body weight is comprised of water. It is the medium in which all chemical reactions occur within the cells of the body. Water is also a major component in many bodily fluids, including blood and the lubrication fluid that surrounds bone joints. Water is lost through urine, water vapor from breathing, feces, and in some cases, perspiration. Animals will freely consume enough water to support bodily functions if a clean, fresh supply is available at all times. See Figure 4-2.

Figure 4-2. Animals need access to plenty of clean water at all times.

PROTEIN

Protein promotes the growth and repair of body cells. Protein consists of strings of building blocks called ***amino acids.*** Nitrogen is the key element in amino acids. With the exception of animals that have a special stomach chamber called a rumen, amino acids cannot be manufactured within the body but must be consumed in the feed. Some amino acids are critical to bodily functions. They are called essential amino acids. A deficiency in one or more essential amino acids can cause reduced growth rate, or, in extreme cases, death.

Figure 4-3. Young alfalfa hay is a good source of forage protein.

Common plant feeds that contain high levels of protein include soybean meal, young alfalfa hay, and distillers dried grains (by-products from distilleries). See Figure 4-3. Most feeds derived from animal products are also high in protein. Animal protein feeds include fish meal, meat meal, and milk products.

Bacteria living within a unique stomach chamber called the rumen in cattle and sheep ***(ruminants)*** have the ability to manufacture new proteins if fed nitrogen and an energy source. This process will be discussed in further detail later in the chapter.

During digestion, feed proteins (from either plant or animal sources) are disassembled into individual amino acids. Amino acids are transported by the blood stream to body cells where they are reassembled into animal proteins, such as muscle, hair, or collagen. Animal manufactured protein is also present in milk and eggs. In the event of reduced energy producing intake, animals have the ability to convert muscle protein to an energy source for body cells.

Figure 4-4. Small grains are an example of a feed high in starchy carbohydrates.

CARBOHYDRATES

The major element in carbohydrates is carbon. ***Simple carbohydrates***, such as starches and sugars, are used as a quick energy source by the animal. See Figure 4-4. Simple carbohydrates are made up of short chains of several glucose molecules chemically bound together. The digestion process disassembles these glucose chains to single molecules of glucose. Glucose is then absorbed by the blood stream, which carries it to body cells. Glucose is used as fuel for chemical reactions that produce animal products, such as milk, muscle, or eggs, or energy for work.

Two ***complex carbohydrates,*** cellulose and hemicellulose, cannot be digested by simple-stomached animals, such as swine and chickens, but can be digested by ruminants and horses. Bacteria within the digestive tract disassemble complex carbohydrates into simpler compounds that can be used by ruminants and horses.

FATS

Fats are a concentrated energy source that can be fed to animals in limited amounts. In fact, fat contains 2 1/4 times as much energy as carbohydrates or proteins. Fats are used in animal diets to boost energy levels derived from feed without increasing the volume. They are also important carriers of some vitamins.

VITAMINS

The two classes of vitamins are fat soluble and water soluble. Fat soluble vitamins include vitamins A, D, E, and K. Vitamin A is especially important to eyesight and maintenance of skin cells. Vitamin D is essential to bone and tooth development. Vitamin E is important to red blood cell structure and keys the energy metabolism of other cells. Vitamin K is an essential blood clotting factor. Fat soluble vitamins are stored in fat tissue within the bodies of animals. These vitamins are found in plant-based feeds in a preliminary or precursor form. Animals have the ability to convert this precursor form to a final, useable vitamin.

Water soluble vitamins include biotin, choline, folacin, inositol, niacin, the B-vitamin family, and vitamin C. The water soluble vitamins assist in many bodily functions. Some can be made available within the animal's body from precursors. Others are manufactured by bacteria within the rumen. Water soluble vitamins cannot be stored within the body and must be consumed or manufactured on a regular basis.

ANIMAL SCIENCE FACTS

Calves are not born with fully functioning rumens. Their rumens gradually develop over the first few months of life.

Figure 4-5. Many producers allow cattle and sheep to consume vitamins and minerals free-choice (ad libitum).

Vitamin needs are normally met in farm livestock by feeding vitamin-mineral premixes either free-choice or mixed with other feeds. See Figure 4-5.

MINERALS

There are two groups of minerals. Major minerals ***(macrominerals)*** are needed in the diet in relatively large amounts (grams per day). Trace minerals ***(microminerals)*** are required in smaller amounts (thousandths or millionths of a gram per day).

The macrominerals include salt (a mixture of sodium and chlorine), calcium, phosphorus, magnesium, potassium, and sulfur. Salt and potassium are involved in maintaining fluid balance inside and outside of cells. Calcium, phosphorus, and magnesium are important to bone structure and development. Sulfur is an important element in some amino acids.

Microminerals include chromium, cobalt, copper, fluorine, iodine, iron, manganese, molybdenum, selenium, silicon, and zinc. The trace amounts of these minerals are often present in normally fed grains and forages, but producers routinely supplement them in the diet to be sure enough are consumed. See Figure 4-5.

THE DIGESTION PROCESS

Different parts of the digestive tract and accessory organs serve different purposes during the digestion process. The mouth and teeth function as the first step in particle reduction. Teeth physically break apart food particles while enzymes in saliva begin to break down simple carbohydrates. Food then moves down the esophagus to the stomach, in a process started by swallowing. The esophagus is simply a tube made of smooth muscle linking the mouth with the stomach. Food moves through the esophagus by involuntary smooth muscle contractions called ***peristalsis.***

When food particles get to the stomach, they are mixed with ***hydrochloric acid***—an extremely acidic substance that begins to break down proteins into shorter chains of amino acids. The low pH of the stomach is caused by hydrochloric acid. The acid kills most bacteria that were ingested with the feed and prevents them from infecting the lower portions of the digestive tract.

Partially digested material, called ***chyme,*** then moves into the small intestine. The small intestine is divided into three parts: duodenum, jejunum, and ileum. Much of the remaining digestive activity occurs in the duodenum, while food is absorbed into the bloodstream from the jejunum and ileum. In the duodenum, enzymes produced by the pancreas called ***peptidases*** are mixed with the chyme. Peptidases further reduce the amino acid chains that escape from the stomach. The reduced amino acids can then be absorbed by the many small blood vessels found in the walls of the intestine. Amylases are also secreted by the pancreas to further break down carbohydrates into simple glucose. Lipases secreted by the pancreas, and bile secreted by the liver and stored in the gall bladder, break up fat molecules into a form that can be absorbed. Pancreatic enzymes and bile are both added to the small intestine in the duodenum.

The absorption process is aided by the structure of the small intestine. The interior of the small intestine contains many folds, which increase the surface area, thereby exposing more blood vessels to absorb digested food particles. From these folds, microscopic finger-like projections called ***villi*** further increase the surface area for absorption.

Indigestible components of the diet, as well as water, enter the large intestine or colon after leaving the small intestine. At the beginning of the colon is the organ called the ***cecum.*** The human version of the cecum is called the appendix. The cecum has no known function in cattle, sheep, or hogs but supports a population of digestive bacteria in horses. In the colon, water is absorbed by the blood stream, and undigested material is packaged for excretion.

TYPES OF DIGESTIVE SYSTEMS

The basic digestive tract of all species is the one described in the last section. However, most species of farm animals have slight modifications to this theme.

Swine

Swine are ***monogastric*** (one- or simple-stomached) animals that have digestive systems most similar to the one just described. See Figure 4-6.

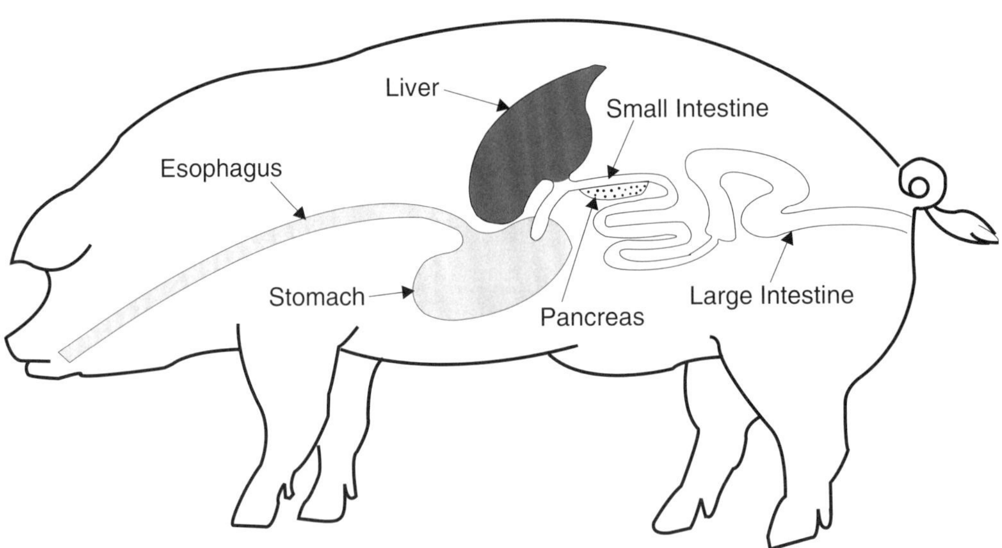

Figure 4-6. The swine digestive tract is an example of a simple-stomached or monogastric animal.

The digestive tracts of pigs are designed to use high energy feeds. Therefore, diets for pigs are high in starches, such as corn and other grains. Monogastric animals, like pigs, are not especially suited to diets containing large amounts of forages, such as hay. The human and swine digestive systems are remarkably similar.

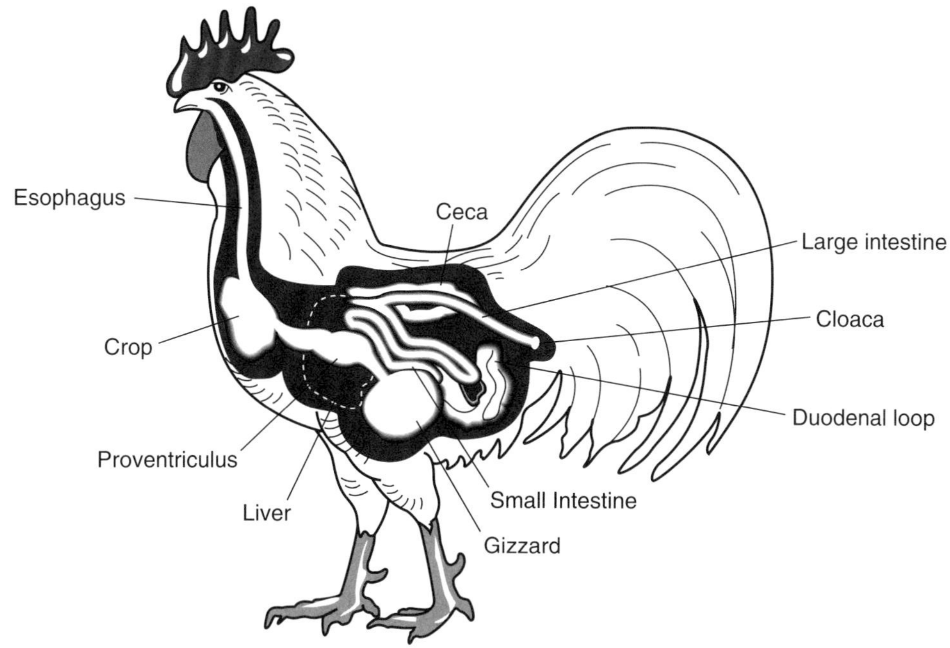

Figure 4-7. The poultry digestive tract is another example of a monogastric digestive system.

Poultry

Poultry are also monogastric. Their digestive tract is different from that of a pig or human. See Figure 4-7. The upper portion of the digestive tract of poultry includes two additional structures: the ***crop*** and ***gizzard.*** The crop, located at the base of the neck, serves as a storage area for recently ingested feed. From the crop, food passes to the ***proventriculus,*** or true stomach. The proventriculus serves the same functions of acid-mixing as the true stomach in other species. Food particles pass from the proventriculus to the gizzard. Since poultry have no teeth, the gizzard is the structure used to grind coarse feed particles.

Cattle and Sheep (Ruminants)

The digestive systems of cattle and sheep (ruminants) differ from monogastrics. Ruminants have adapted to diets containing large amounts of forages. See Figure 4-8. Forages are plant-based feeds, which include stems and leaves. Some examples of forages are hay, pasture grasses, legumes, and silage. The stems and leaves of these feeds contain sources of complex carbohydrates, cellulose, and hemicellulose, which monogastrics cannot digest. However, ruminants can digest these complex carbohydrates with the aid of a remarkable stomach chamber called the rumen and "houseguest bacteria."

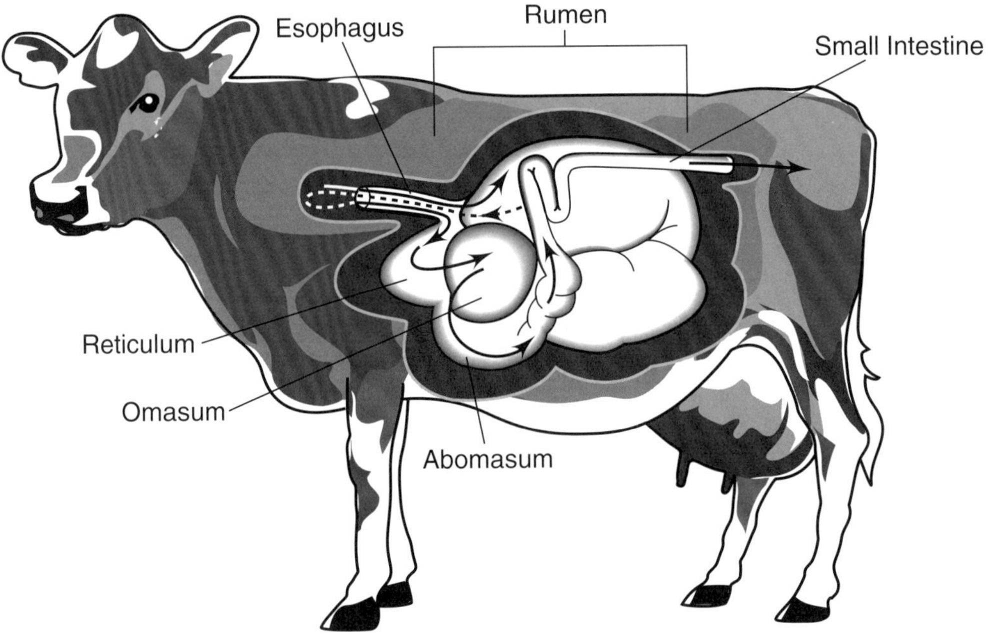

Figure 4-8. The ruminant digestive system is designed to digest large amounts of forages.

The rumen is the first part of a four-compartment stomach that is located in the digestive tract. It precedes the small intestine. In cattle, the rumen is by far the largest of the four compartments, holding up to 50 gallons. The first two stomach parts, the rumen and reticulum are often referred together as the ***reticulorumen.*** The reticulum is a small pouch located on the side of the rumen. It functions as a trap for foreign materials (nails, glass, or wire) which are mistakenly eaten by the animal. The third compartment of the ruminant stomach is referred to as the ***omasum*** or "manyplies." The omasum absorbs water from the chyme. The fourth compartment, the true stomach or ***abomasum,*** contains high levels of acid similar to those of the aforementioned species.

Ruminants are equipped with a dental pad in lieu of top incisor teeth. Ruminants use this adaptation to tear off large bites of forage in a single bite. Since the meal is not immediately chewed, ruminants can eat a large amount of forage in a short period. The forage is then stored in the reticulorumen until the animal has some spare time. Later, while the animal is resting, the forage is regurgitated, chewed, and reswallowed. This process called rumination (chewing a cud) serves to reduce feed particle size. Rumination also mixes the regurgitated forage with saliva. Saliva has a buffering effect on the rumen, which helps keep the pH near 7. This level of acidity is where the rumen "houseguest bacteria" are most comfortable. Starchy feeds, such as corn, can be digested by rumen bacteria, but some forage must be present in the diet to ensure proper rumen function.

Other than serving as a storage area for ingested forages, the reticulorumen serves several other purposes. The reticulorumen is a large fermentation vat containing billions of bacteria These live-in guests digest complex carbohydrates that the animal's own digestive system cannot. The products of the digestion or fermentation process are known as volatile fatty acids or VFAs. VFAs are absorbed by the lining of the rumen and serve as an energy source for the animal. All the time bacteria are digesting cellulose, they are multiplying, growing, and producing bacterial proteins from nitrogen contained within the forage. When food moves to the omasum and then on into the abomasum, many bacteria are also moved along. When these bacteria are deposited in the abomasum, acid kills them. These dead bacteria provide the ruminant with large amounts of high quality protein. Except in specially prepared diets, none of the protein entering the abomasum is the same as when it entered the animal. Bacteria have disassembled and reassembled all of it.

The relationship between ruminants and bacteria is called a ***symbiotic*** relationship. Symbiotic relationships between two species are mutually beneficial. In this case, the cattle provide bacteria with food and an ideal environment in which to grow and reproduce. The bacteria provide ruminants with VFAs and microbial protein.

Horses

Horses are also adapted to digesting high forage diets, but in a slightly different way. The digestive tract of a horse is similar to that of a monogastric until the end of the small intestine. There, the cecum is enlarged and contains a population of bacteria, which ferment forages. See Figure 4-9. VFAs are absorbed by the lining of the cecum, but microbial protein is lost in the feces.

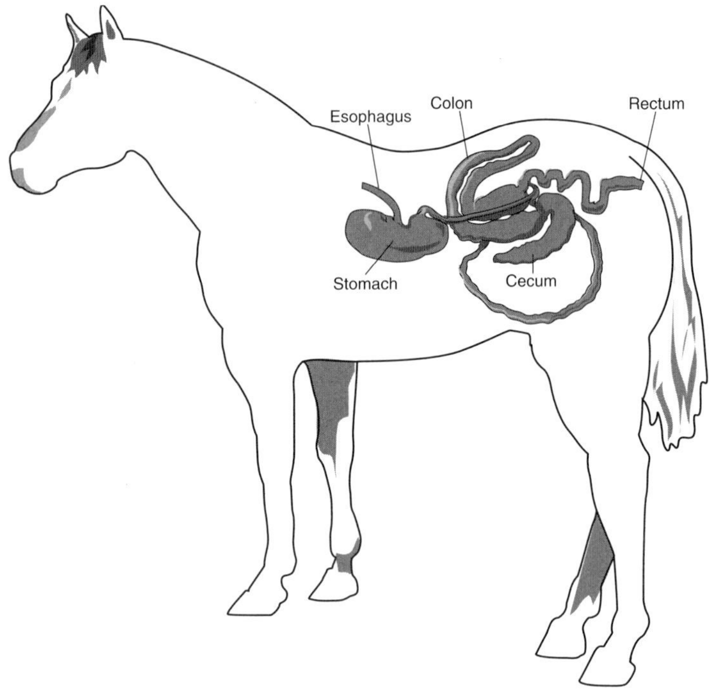

Figure 4-9. The equine digestive system can also digest forages, but in a way different from ruminants.

SUMMARY

Animals must consume and digest food to produce animal products, such as meat, wool, milk, and eggs, and to do work. Six essential nutrients must be present in the diet: water, protein, carbohydrates, fats, vitamins, and minerals. All are needed by the animal to provide building materials and fuel for cellular growth and production. Feedstuffs are ingested by animals and are broken down into simpler compounds as they pass through the digestive system. These simpler compounds are absorbed into the bloodstream and travel to cells where they are reassembled into cellular products. Swine have a very basic digestive system that consists of the mouth, teeth, esophagus, stomach, small intestine, cecum, and large intestine. All other species have this same system with minor modifications. Poultry store undigested feed in the crop and grind it in the gizzard before it enters the true stomach. Cattle and sheep have a special stomach called the rumen, which contains a population of microbes that digest complex carbohydrates. Horses also have a population of microbes housed in the cecum, an appendage of the large intestine.

CHAPTER SELF-CHECK

___	amino acid	1.	mutually beneficial relationship
___	ruminant	2.	strong digesting agent in stomach
___	simple carbohydrates	3.	protein building blocks
___	complex carbohydrates	4.	true stomach in ruminants
___	macrominerals	5.	one-stomached animal
___	microminerals	6.	true poultry stomach
___	peristalsis	7.	stores poultry food
___	hydrochloric acid	8.	manyplies
___	chyme	9.	first two stomach parts
___	peptidases	10.	grinds poultry food
___	cecum	11.	animal with four-part stomach
___	monogastric	12.	sugars and starches used as quick energy
___	crop	13.	cellulose and hemicellulose
___	gizzard	14.	required in diet in smaller amounts (thousandths or millionths of a gram per day)
___	proventriculus	15.	located between small and large intestine
___	reticulorumen	16.	required in the diet in grams per day
___	omasum	17.	partially digested food
___	abomasum	18.	involuntary muscle contraction
___	symbiotic	19.	enzymes produced by pancreas

QUESTIONS AND PROBLEMS FOR DISCUSSION

1. What are the six essential nutrients in the diets of animals?
2. True or False? Water accounts for 20 percent of total body weight.
3. Protein consists of nitrogen containing building blocks called ________________.
4. Give two examples of simple carbohydrates.
5. List the fat soluble vitamins.
6. Which vitamin helps to regulate blood clotting?

7. Are chromium and cobalt macro or microminerals?
8. Food moves through the esophagus by involuntary muscle contractions called ___________________.
9. Which part of the poultry digestive tract grinds food?
10. Name three different forages.
11. What are the four parts of a ruminant stomach?
12. What organism digests cellulose in a ruminant stomach?
13. Describe a symbiotic relationship.
14. Compare the function of a horse's cecum to a cow's rumen.
15. Contrast the use of microbial protein in cattle and horses.

ACTIVITIES

1. Make a classroom display of animal feedstuffs. Classify samples using the essential nutrients. Identify each sample with a flipcard. Write a number on one side of the flipcard and the correct name and nutrient classification on the other. See who can identify the most samples.
2. Collect labels of commercially prepared livestock rations. Discuss the similarities and differences of each. How does this information compare with what you learned in this chapter?
3. Contact a college or university animal science department or feed company. Ask if they will share information on recent digestion research. How can this information be used by a real-life livestock producer?

LABORATORY ACTIVITY

COMPARING AND CONTRASTING MONOGASTRIC AND RUMINANT STOMACHS

Purpose

To compare and contrast monogastric and ruminant stomachs

Materials

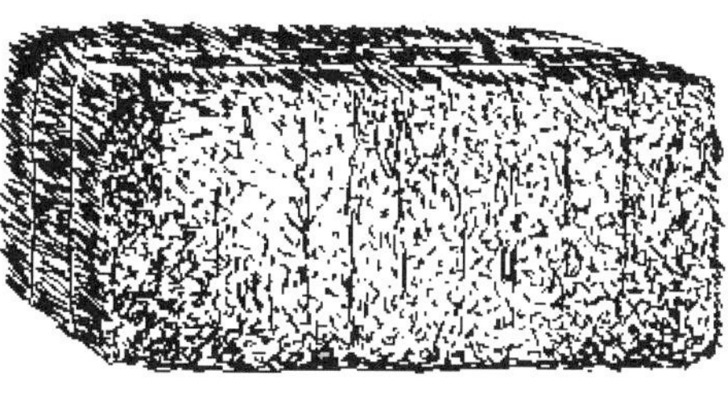

digestive tract from a pig
digestive tract from a cow or sheep
dissecting knife
plastic gloves for all participants
plastic aprons for all participants
tarp on which to perform dissection
well-ventilated area

Safety Precaution

Your local slaughterhouse may be able to provide you with the tracts and also dispose of them. If the tracts are secured from another source, be sure to arrange for appropriate disposal. The individual performing the dissection should be especially careful with the knife, which can become slippery from digestive fluids. Furthermore, plastic gloves and aprons can help keep participants clean. However, all participants should carefully wash their hands after handling digestive tissues and organs. *E. coli* and other potentially harmful contaminants can be present in digestive materials.

Procedure

1. Arrange the tracts side by side on a large plastic tarp.
2. Identify the following similar parts: esophagus, true stomach, small intestine, large intestine.
3. Cut open these four parts on each tract. Observe the appearance and contents of each. Record your findings.
4. Cut open the large rumen. Observe its appearance and contents. Record your findings.
5. Locate the reticulum on the side of the rumen. Note and record the texture and any unusual contents.
6. Cut open the ball-shaped omasum. Note and record the internal structure of the omasum.
7. Dispose of the dissection material properly.

Analysis

1. Make a chart listing the appearance and contents of each part of both digestive tracts.
2. Write a paragraph detailing the similarities and differences. Describe in detail any unusual findings.

Application of Laboratory Activity

This laboratory activity familiarizes you with the organs located in the digestive systems of monogastric and ruminant animals. After completing the lab, you should be able to compare and contrast the two different digestive tracts. An understanding of the systems will give you insight into the development of appropriate feeding rations for monogastric and ruminant animals.

In addition, veterinarians and producers sometimes dissect digestive tracts from animals that have died of unknown causes. Dissection findings, such as inflammation of the intestinal lining or blood found in the tract, can provide clues to the cause of death.

Chapter 5

BIOLOGY OF REPRODUCTION

Getting in the Family Way!

INTRODUCTION

Efficient reproduction may be the single most important factor in the profitability of livestock production. Without baby pigs to sell, swine producers have no income. Beef cows must each raise a calf every year, or the business of beef production is unprofitable. Likewise, dairy cows must have a calf in order to produce salable milk. The same theory holds true for sheep and, to some extent, horses. Although farm animals look very different, the biology of reproduction for all species is very similar.

Figure 5-1.

OBJECTIVES

1. Identify and explain the functions of male and female reproductive tract anatomy
2. Analyze the physiological changes during fertilization, gestation, parturition, and lactation
3. Describe practical reproductive differences among swine, cattle, sheep and horses, such as signs of heat, length of estrus, and gestation period

TERMS

capacitation
cervix
colostrum
corpus hemorrhagicum (CH)
corpus luteum (CL)
efferent ducts
epididymis
estrus cycle
follicle
free martin
infundibulum
Leydig cells
lordosis
ovary
oviduct
penis
rete testis
seminiferous tubules
testis
urethra
uterus
vagina
vas deferens
Vitelline Block
vulva
zona pellucida
zygote

ANATOMY OF REPRODUCTIVE TRACTS

FEMALE

The primary reproductive organ of females is the ***ovary.*** It is within the ovary that meiosis takes place and fertile eggs develop. Along with producing eggs, the ovary produces hormones at various times during the reproductive cycle.

Different structures are found on the ovary at different times during the reproductive cycle. A ***follicle*** is a developing egg that has not yet been released. Developing follicles produce a hormone called estrogen. Estrogen serves to further develop the follicle. When the follicle is mature, it ruptures (much like acne) and the egg is released. The release of the egg causes estrogen levels to fall. The blood clot remaining after the egg is released is called a ***corpus hemorrhagicum (CH)*** or bloody body. The CH lasts for a few days after which a yellow body begins to form at the site of the egg release. This new structure is called a ***corpus luteum (CL).*** The CL produces a hormone called progesterone, which sustains pregnancy. Progesterone also prevents a new follicle from developing. If pregnancy occurs, the CL will be present until just before the fetus is delivered. At that time, the CL will regress. If pregnancy does not occur, another hormone secreted by the uterus called prostaglandin causes the CL to regress and allows a new follicle to begin developing. Then, the cycle repeats. See Figure 5-2.

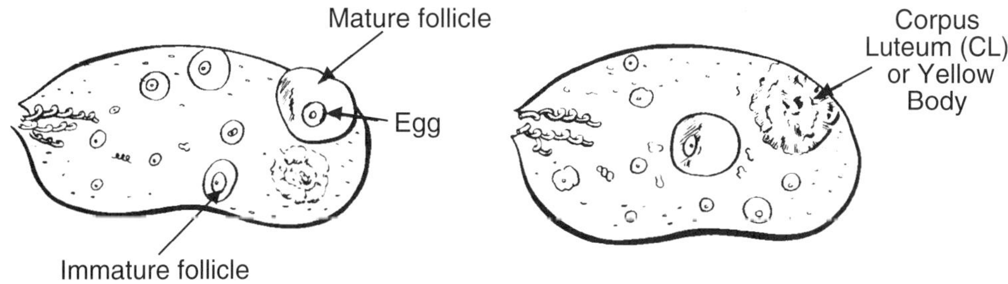

Figure 5-2. The ovary on the right contains a corpus luteum (CL) while the ovary on the left contains a mature follicle.

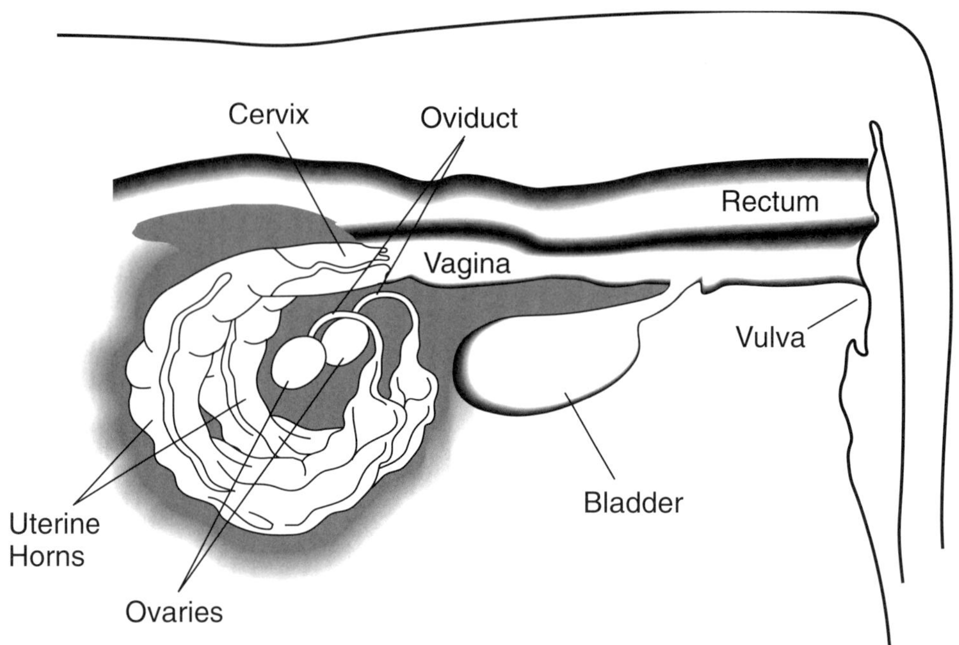

Figure 5-3. The reproductive tract of a cow.

This cycle of structures on the ovary corresponds with what is called the ***estrous cycle.*** The period during which a follicle is developing is called proestrus. The time when the egg is released is called estrus. It is during estrus when the female is receptive to be mated by a male. This period is also known as standing heat. The next period is called metestrus. During this period, a CL is actively producing progesterone whether the female is pregnant or not. If the female is not pregnant, she goes into the next phase called diestrus during which the CL recedes and a new follicle begins development. If the female is pregnant, the CL remains active until the end of the pregnancy.

During, or around estrus or standing heat, the egg is released into a funnel called the ***infundibulum,*** which moves the egg into a tube called the oviduct. Fertilization by sperm takes place within the upper portions of the ***oviduct.*** After fertilization, the egg continues moving toward the ***uterus*** or womb where it attaches and develops until birth.

The upper portion of the uterus consists of two horns, one leading to each oviduct and ovary. The lower portions of the uterine horns attach to form the body of the uterus. The size and length of the uterine horns vary among species. Embryos can attach in both horns in swine, but normally only attach in the horn closest to the ovary of origin in cattle, sheep, and horses. In species where single births are common, the ovaries alternate egg release.

First in the path leading from the uterus to the outside of the animal is the ***cervix.*** In some species, sperm is deposited in the cervix. The cervix also seals off the uterus after pregnancy has been established to guard against bacteria infecting the developing fetus. The next structure toward the outside of the female is the ***vagina,*** which serves as a site of insemination in some species. The visible part of the female reproductive tract is called the ***vulva.*** See Figure 5-3.

MALE

The primary male reproductive organ is the ***testis.*** See Figure 5-4. Meiosis takes place in ***seminiferous tubules*** buried deep within the testis. Surrounding the seminiferous tubules are ***Leydig cells,*** which produce the hormone testosterone. Testosterone is critical to the normal development

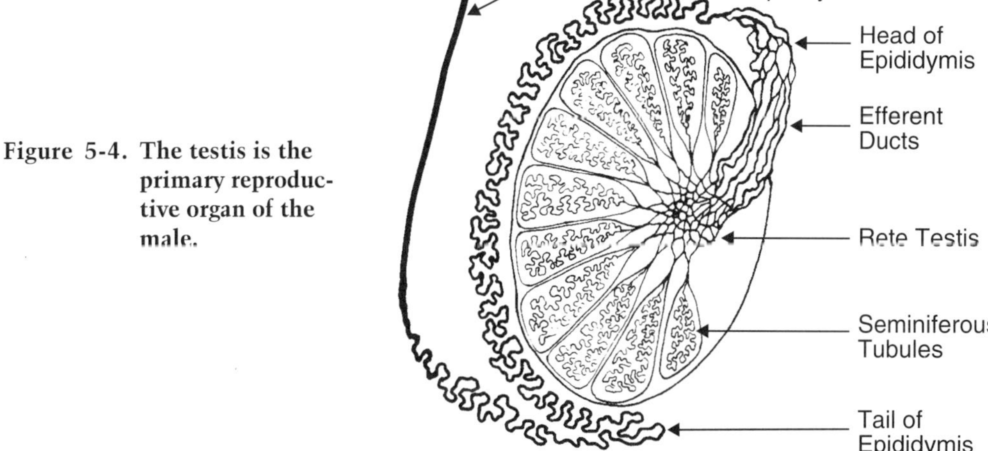

Figure 5-4. The testis is the primary reproductive organ of the male.

of sperm cells. Immature sperm collect in ducts called ***rete testis,*** and then travel to a bigger storage area in the ***efferent ducts.*** The final sperm collection site is the ***epididymis,*** which has a head, a body, and a tail. During ejaculation, the sperm travel from the epididymis, through the ***vas deferens,*** to the ***urethra*** and ***penis*** before being deposited in the female. During this journey, several sets of glands (Cowper's gland, prostate, and

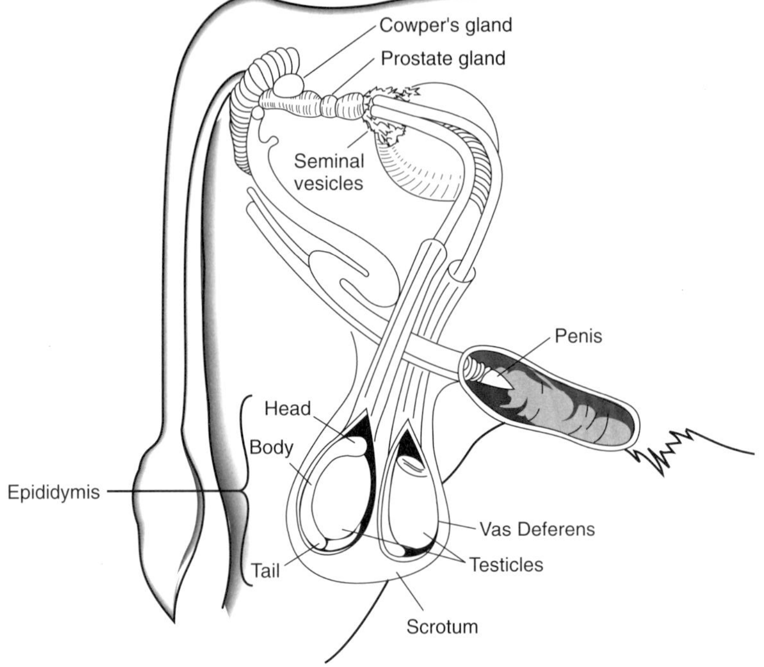

Figure 5-5. The reproductive system of a bull.

seminal vesicles) add fluids to the sperm. These fluids supply nutrients for the sperm to live after entering the female reproductive tract. Sperm cells mixed with these fluids are called semen. See Figure 5-5.

THE REPRODUCTIVE CYCLE

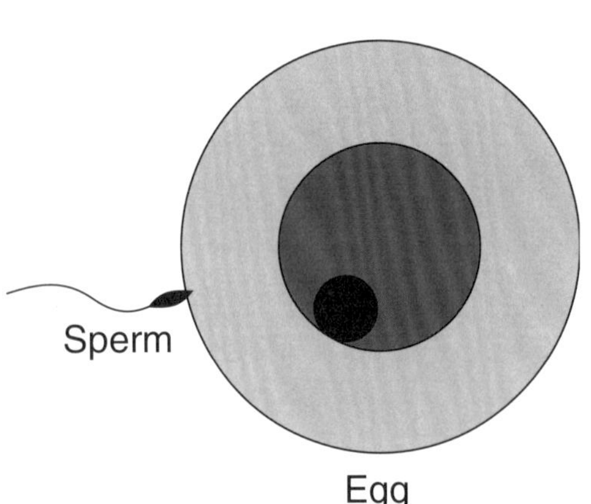

Figure 5-6. Only one sperm can enter an egg due to a process called the Vitelline Block. Drawing done to proportion.

FERTILIZATION

During mating, sperm are deposited and "swim" up the female reproductive tract while undergoing a process called ***capacitation.*** During capacitation, fluids present in the female reproductive tract "wash" the sperm and change its chemical composition, making it capable of fertilization. When sperm reach the upper portions of the oviduct, where fertilization takes place, they meet with the recently shed egg. The egg is surrounded by a barrier called the ***zona pellucida,*** which must be penetrated by a single sperm. Once a single sperm has penetrated the zona pellucida, the chemical composition changes, and no more sperm can enter. The technical name for this process is the ***Vitelline Block.*** See Figure 5-6. After the egg has accepted a sperm, the nuclei of both gametes mix, and the diploid fertilized egg is known as a ***zygote.***

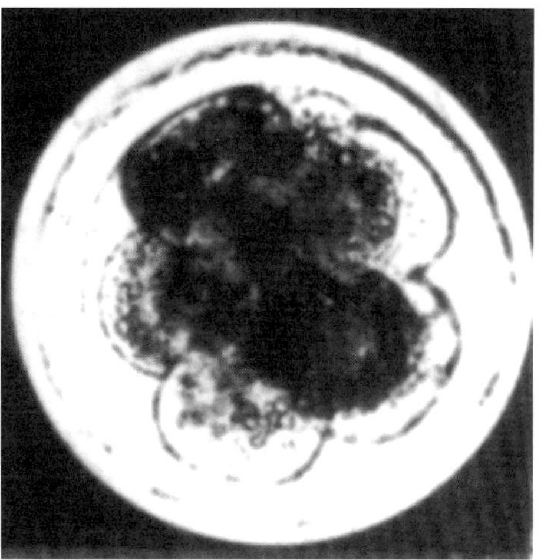

Figure 5-7. The embryo in early formation.

The zygote travels the remaining distance to the uterus where it is nourished by uterine fluids. The single cell zygote divides to 2, 4, 8, then 16 cells. During these early stages of development, the zygote becomes an embryo. The embryo soon attaches to the side of the uterus and develops an attachment to the female's blood supply through the umbilical cord. The unborn young is known as an embryo during early cell division until the internal organs form. After organ formation, the embryo is known as a fetus. See Figure 5-7.

PREGNANCY

During the period the embryo or fetus is developing in the uterus, no new follicles develop on the ovary. The cervix develops a thick plug of mucous to keep bacteria out of the uterus.

PARTURITION (birth)

When the pregnancy nears full term, the fetus signals to the mother when parturition will take place. The fetus starts to secrete a hormone, which causes the mother to release prostaglandin. In turn, prostaglandin ruptures the CL, ending the pregnancy, and causing the fetus to be born. The hormone released by the fetus also causes the mother to release two other hormones. The first is called relaxin, which relaxes the ligaments around the birth canal and allows the young to pass through during delivery. The second is called oxytocin, which causes milk let-down from the mammary gland and uterine contractions to eject the fetus.

LACTATION (giving milk)

Milk let-down is controlled by the hormone oxytocin. Milk formation in the mammary gland is controlled by the hormone prolactin. The first

ANIMAL SCIENCE FACTS

Using Correct Livestock Terminology

"Talk the Talk"

Swine

parturition–farrowing
gilt–young female prior to farrowing
sow–older female after farrowing
boar–intact male
barrow–castrated male

Cattle

parturition–calving or freshening
heifer–young female (usually prior to calving)
cow–older female (usually after calving)
bull–intact male
steer–castrated male

Sheep

parturition–lambing
ewe lamb–young female up to one year of age
ewe–female over one year of age
ram lamb–male up to one year of age
ram–male over one year of age
wether–castrated male

Horses

parturition–foaling
foal–young animal
filly–young female three years of age or less (Thoroughbreds—fillies extend to four years)
mare–female at least four years old (Thoroughbreds—five)
colt–young male three years of age or less (Thoroughbreds—colts extend to four years)
mare–older female after giving birth
stallion or stud–intact mature male at least four years old (Thoroughbreds—five)
gelding–castrated male

milk available from the female to newborn young is called ***colostrum.*** Colostrum is thicker and richer than the milk that will be produced later in lactation. Colostrum contains antibodies to any diseases the mother has contacted and protects the newborn. Newborn animals need colostrum within 24 hours after birth because after that time antibodies are much less readily absorbed.

Later lactation is characterized by frequent milk let-down. The amount of milk produced by the female changes as the newborn grows. Colostrum is produced in small amounts. As the young animal grows, more and more milk is produced to compensate for its increasing appetite. At some point, the female is producing as much milk as her genetics, nutrition, and mammary gland will allow. This point is called the "peak of lactation." After the lactation peak, milk production gradually drops off, and eventually stops.

In the real world of livestock production, producers usually stop lactation by weaning or removing the young animal from the mother. If young animals or milking machines do not remove milk from the mammary gland, milk production will naturally cease.

The mammary gland is amazing in its ability to capture nutrients from the bloodstream and synthesize large quantities of milk. High-producing dairy cows routinely produce over 100 pounds or 200 half-pints of milk per day. To produce this much milk, cows have to eat a lot. Lactating dairy cows eat about 3.5 percent of their body weight each day, not including water. That's about 50 pounds of dry feed. Many females cannot physically eat enough to maintain body weight during peak lactation.

REPRODUCTION IN SWINE

Swine display standing heat at approximately 21 day intervals. Heat can be determined by observing a swollen vulva or by placing weight on the female's back while rubbing the flank. If the sow or gilt does not try to run away, she is in standing heat. See Figure 5-8. The ears may "bob" up and down when pressure is placed on the back, further confirming standing heat. This response is called ***lordosis.*** The presence of a boar during heat-checking results in a stronger expression of standing heat. Gilts usually display first estrus at five to six months of age. Female swine will also display estrus throughout the year because they are not seasonal breeders.

Swine should be mated at least twice, with the first service timed 12 hours after the beginning of standing heat. Subsequent matings should be

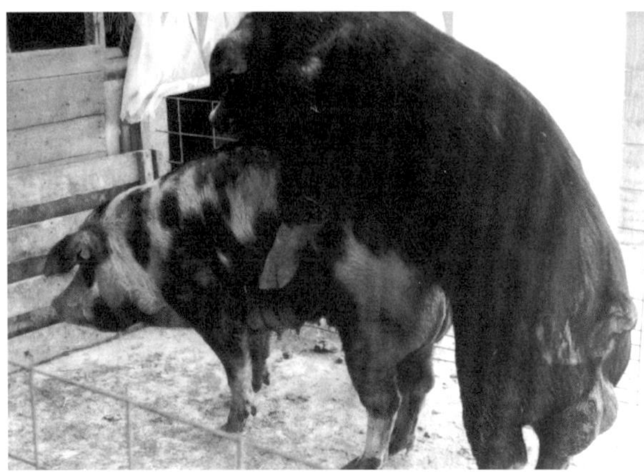

Figure 5-8. Sows are receptive to boars throughout the year.

spaced at 12 hour intervals until the sow or gilt will no longer stand for mating. Estrus normally lasts 24–72 hours. Swine can be bred artificially using either frozen or fresh semen.

Pregnancy in swine can be checked ultrasonically 30 days after breeding, or by watching for the absence of signs of heat 21 days after breeding. Gestation length for swine is normally between 112–118 days, averaging about 114 days.

Swine normally ovulate 8 to 20 ova per estrus. Most ova are fertilized, but several normally die before attaching to the uterus. The number of eggs that attach to the uterus is somewhat dependent on the length of the uterus; the longer the uterus, the more eggs that can attach. Therefore; older, larger sows normally have larger litters because more embryos can attach.

Swine give birth to or farrow litters averaging nine or ten pigs. Sows nurse their young until removal at weaning — normally at three to six weeks of age in most management situations. If not weaned earlier, sows will stop lactating after eight to nine weeks. Sows will usually return to estrus four to seven days after weaning.

REPRODUCTION IN CATTLE

The normal estrous cycle in cattle is 21 days. In other words, barring pregnancy, an egg is released into the oviduct every three weeks. Heifers reach puberty (first estrus) at about 9–10 months of age and should be bred to have their first calf at two years of age. Cattle will display estrus during all seasons of the year.

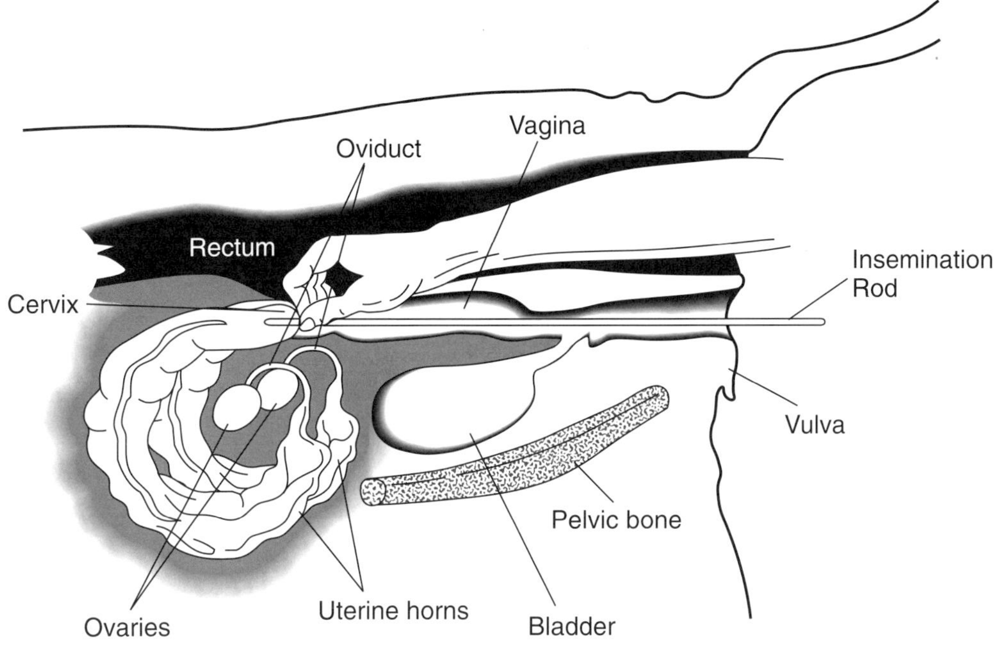

Figure 5-9. Cattle have been bred artificially since the 1930's.

Shortly before the release of the egg, cattle display "standing heat," which means they will stand still if another cow (or bull) tries to mount them. Standing heat is the only positive sign that a cow is in heat. However, many cows will "bawl", try to ride other cows, and act restless when nearing standing heat. Producers should routinely check for signs of heat twice daily. Early morning and evening are good times to check for heat since cattle are not as active during mid-day heat.

Figure 5-10. Cows can become pregnant while lactating.

ANIMAL SCIENCE FACTS

Embryo transfer allows a producer to get several progeny (offspring) from one female at the same time. Hormones are given to the animal to induce the ovulation of several eggs. Artificial insemination of the animal follows. The fertilized eggs are then flushed from the donor, evaluated for quality and implanted into a recipient female. This practice was first done commercially in the cattle industry in the 1970s.

Today many cattle are bred using artificial insemination with frozen semen. See Figure 5-9. This allows producers to select sires (fathers) of superior genetic quality without the cost of maintaining bulls. The ideal time for insemination in cattle is during the last half of standing heat which lasts eight to 18 hours. Cattle are usually inseminated only once during or shortly after the standing heat period. They will display heat about 40–50 days after calving, and normally become pregnant while still lactating. See Figure 5-10.

ANIMAL SCIENCE FACTS

15273

INDIANA CO. UNIT

RECEIPT

*Approved Artificial Breeding Association

Date of Breeding Mar 3 1948 Breed* of Cow Guernsey

Name of Cow Judy

Was Cow Identified by Checking Registration Papers

Registration No. Ear Tag or Tattoo

Owner Thomas Mikesell

Address Indiana Pa

Bull Used Antietam Admiral

Reg. No. Fees { Breeding $5.00 Extra Ass't.

Service No.

Previous Service: Date Bull

Name of Organization Western Pennsylvania Artificial Breeding Cooperative

Signed by
TECHNICIAN

This receipt is given as evidence of payment of fee for services rendered and also as certification of date of service and identity of semen used for service of animal identified hereon.

I hereby certify the above statements to be correct to the best of my knowledge.

Owner or Agent

(This certificate must accompany application for registry or birth report and must be given to member at time of service.)

*One approved and supervised by the State College of Agriculture as designated by the Purebred Dairy Cattle Association.

MOORE BUSINESS FORMS, INC., ELMIRA, N. Y.

The author's grandfather, Thomas Mikesell, was a dairy farmer in Western Pennsylvania. His farm records revealed a bill for artificial insemination done in 1948.

ANIMAL SCIENCE FACTS

Hormone List

Female

- estrogen–produced by follicle and further develops follicle
- progesterone–produced by CL, sustains pregnancy and inhibits new follicle development
- prostaglandin–produced by the uterus and causes CL to regress and new follicle to develop
- relaxin–produced by the CL and causes ligaments around birth canal to relax during parturition
- oxytocin–produced by the posterior pituitary and causes milk let-down

Male

- testosterone–produced by the Leydig cells and is critical to normal sperm development

Pregnancy in cattle is ideally confirmed by a rectal examination called palpation about 40–50 days after breeding. Otherwise, the absence of heat signs 21 days after breeding can indicate a pregnancy. The average gestation length for cattle is 280–285 days or about nine and a half months. Approaching birth in cattle, called "calving" or "freshening," can be signaled by restlessness, frequent urination, relaxation of ligaments around the tailhead, presence of milk, and loosening of the vulva.

Cattle normally give birth to a single offspring, although twins are fairly common. Triplets are rare. In a split set of twins (one bull and one heifer), the heifer will be sterile 80–90 percent of the time. This phenomenon occurs almost solely in cattle but has been observed in sheep. This condition is due to the passing of substances from the male to the female fetus, which inhibits the growth and development of the female reproductive tract. This sterile heifer is known as a ***freemartin.***

Dairy cows are normally "dried off" or left unmilked for about 60 days before the birth of a calf to allow them to gain body weight. Beef calves are normally weaned at four to eight months of age. After weaning, beef cows remain dry until the next calf is born.

Figure 5-11. A marking harness placed on a vasectomized ram can help detect ewes in heat.

REPRODUCTION IN SHEEP

Ewes have a somewhat shorter estrous cycle than cattle and swine, averaging 16 days instead of 21. Detection of estrus is very difficult in sheep. Standing for the ram to breed is the best indication of estrus. This difficulty limits the use of artificial insemination in sheep. Some producers use a vasectomized ram with a chalking device attached to his chest to mark ewes in heat. These ewes can immediately be bred artificially. See Figure 5-11.

Some breeds of sheep are known as seasonal breeders. In general, black-faced breeds, such as Hamphires or Suffolk, are the most seasonal, showing estrus only in the fall months as day length shortens. White-faced sheep are less seasonal, and may show estrus at any time during the year. Ewe lambs experience their first estrus at six to nine months of age. Furthermore, ewe lambs may be bred their first fall so that they will lamb when they are yearlings; however, they will give birth to a greater percentage of single lambs than older ewes do. Older ewes have a high incidence of twins and triplets, with white faced sheep averaging more lambs per year than black-faced sheep.

Pregnancy can be determined ultrasonically or by palpation at 40–60 days after mating. Producers normally check pregnancy by observing for estrus 16 days after mating. If the ram does not breed the ewe again, she is determined to be pregnant. The average gestation length for sheep is about 150 days, or 5 months.

Ewes normally give birth to one to three lambs. Four lambs are fairly common with very prolific breeds, such as Finnsheep.

Lambs may be weaned anytime after they start eating feed on their own. Most management strategies call for only one lamb crop per year. However, using white-faced sheep that will breed "out of season," it is possible to produce a lamb crop every eight months.

REPRODUCTION IN HORSES

Estrus detection in horses requires an intact male (stallion), and is known as "teasing." When a teaser stallion is given visual contact to a mare in heat, the mare will assume a squatting position, urinate frequently, and display "winking" of the vulva. Heat detection can also be aided by palpation of the ovaries for mature follicles.

Figure 5-12. A foal is usually weaned in the fall of the year. (Courtesy, Illinois Department of Agriculture)

Horses are naturally seasonal breeders, showing estrus every 21 days during the period from mid-April to mid-September. Controlling the lighting in a box stall can fool the mare into displaying estrus at other times of the year.

Young female horses (fillies) normally show their first estrus at 12–15 months of age, but breeding is normally delayed until the filly is two to three years old. Estrus may last from three to seven days, and ovulation occurs one to two days before the end of estrus.

Horses can be mated either naturally with a stallion or artificially. However, some breed associations either will not allow registry or severely restrict the registration of horses sired by artificial insemination.

Pregnancy can be determined by the absence of the next heat period, rectal palpation, ultrasound examination, or blood test. The normal gestation period for a horse is about 330 days or 11 months.

Mares normally show heat 7 to 10 days after foaling. This heat is known as "foal heat." Mares may be re-bred during foal heat or during the next heat period 20–25 days later. Mares normally rebreed while lactating. Foals are normally weaned in the fall of the year. See Figure 5-12.

SUMMARY

The primary reproductive organ of the female is the ovary. Ovaries produce eggs as well as hormones that allow and sustain pregnancy. Eggs are fertilized in the oviduct then travel to the uterus where attachment occurs. Fertilization can only occur during the time of ovulation at a certain point in the estrous cycle called estrus or standing heat. The primary male reproductive organ is the testis. Sperm are manufactured in the testis and are transferred through a system of ducts to the epididymis. During mating, sperm are mixed with other fluids and ejaculated into the female reproductive tract through the penis. Sperm and egg meet at fertilization. Only one sperm can fertilize an egg. All others are blocked from entrance. The fetus signals the female when parturition or birth will occur. The first milk is called colostrum. Later in lactation, milk production peaks then drops until weaning. There are minor differences in the reproductive systems and reproductive performance of swine, cattle, sheep, and horses.

CHAPTER SELF-CHECK

___ ovary	1. first milk from female to the newborn
___ follicle	2. yellow body at site of egg release
___ corpus hemorrhagicum	3. site of transported sperm from rete testis
___ corpus luteum	4. seals off the uterus after pregnancy
___ estrus cycle	5. primary male reproductive organ
___ infidibulum	6. produce testosterone
___ oviduct	7. transports sperm to exterior of testis
___ uterus	8. sterile heifer twin born with a bull
___ cervix	9. female organ which produces eggs
___ vagina	10. funnel to oviduct
___ vulva	11. final collection site for sperm
___ testis	12. male mating organ
___ seminiferous tubules	13. sign of heat in swine
___ Leydig cells	14. developing egg not yet released
___ rete testis	15. washing of sperm by vaginal fluid
___ efferent ducts	16. bars second sperm from egg
___ epididymis	17. between cervix and vulva
___ vas deferens	18. barrier surrounding egg
___ urethra	19. corresponds with ovarian structures
___ penis	20. blood clot remaining after egg is released
___ capacitation	21. tube leading from ovary to uterus
___ zona pellucida	22. visible part of the female reproductive tract
___ vitelline block	23. meiosis in males takes place here
___ colostrum	24. tube from epididymis to urethra
___ lordosis	25. first collection site of immature sperm
___ free martin	26. womb

QUESTIONS AND PROBLEMS FOR DISCUSSION

1. Which hormone is produced by developing follicles on the ovaries?
2. Describe the four stages of estrus.

3. List the organs of the female reproductive tract.
4. Where in the male reproductive tract is testosterone produced?
5. Explain why no more than one sperm can penetrate a single egg.
6. Follow the path of a fertilized egg. Where does fertilization and implantation occur?
7. Which hormone triggers milk let-down?
8. Explain the signs of standing heat in cattle.
9. Which two species discussed in this chapter can be seasonal breeders?
10. How long is the estrus cycle for swine, cattle, sheep, and horses?
11. How long is the gestation period for swine, cattle, sheep, and horses?
12. At what age do swine, cattle, sheep, and horses experience their first estrus period?
13. Why are dairy cows "dried off" two months prior to calving?
14. Do older or younger ewes tend to have more incidents of multiple births?
15. A stallion used to detect mares in heat is called a ___________.

ACTIVITIES

1. Contact a local artificial insemination technician. Ask her or him to speak to the class. Perhaps he or she could give a demonstration and bring materials on heat detection and sire selection.
2. Divide the class into species interest groups: swine, cattle, sheep, and horses. Interview producers and do additional research on problems associated with reproduction in that species. Follow with reports to the class.
3. Complete the following simulations.
   a. You want a sow, cow, ewe, and a horse to give birth on February 15. When should they be bred?
   b. Your county fair market hog show is July 28. When should you breed your sow in order to exhibit her pigs at the show? Figure six months from birth to market weight for swine.

LABORATORY ACTIVITY

REPRODUCTIVE TRACT IDENTIFICATION

Purpose

To identify anatomy of a female reproductive tract

Materials

female reproductive tract from a pig, sheep, or cow
dissecting knife
pins, paper flags and marking utensil for labeling
plastic gloves for all participants
aprons for all participants
tarp or newspaper on which to perform dissection

Safety Precaution

A local slaughterhouse may be able to provide you with the tract and assist you in proper disposal. Be sure to arrange for proper disposal if tracts are secured from another source. The individuals using the knife should be especially careful. Be sure to thoroughly wash hands after handling the reproductive tract.

Procedure

1. Arrange the tract on the tarp.
2. Identify the following: ovary, infundibulum, oviduct, uterine horns, cervix, vagina, and vulva.
3. Identify any structures on the ovaries.
4. Carefully cut open the tract starting at the vulva and ending at the uterus. Observe the interior of the cervix. A cattle or sheep cervix will have a ring type appearance while a swine cervix will appear folded.
5. Slice any ovarian structures in half. Observe the color.

Analysis

1. Using flagged pins marked a, b, c, and d, label the following:
   a. fertilization site
   b. implantation site
   c. site of mucous plug during pregnancy
   d. site of only visible part from exterior

2. Label all anatomical parts.

Application of Laboratory Activity

Labeling the parts of the reproductive tract will allow you to better understand the physiology of ovulation, insemination, and pregnancy. Artificial insemination technicians must be able to locate the parts of the reproductive tract and deposit the semen in the correct location.

Reproductive physiology advances in animal science often precede similar developments made with humans. Artificial insemination was routinely used by dairy farmers over 50 years ago, while this same procedure in humans is just beginning to become common place. Furthermore, newly developed reproductive technologies raise a host of ethical questions for debate.

Chapter 6

GENETICS

A Chip Off the Old Chromosome

INTRODUCTION

Animal selection based on ***genetics*** (the study of heredity) can be traced to two insightful individuals. The first was Gregor Mendel, an Austrian monk who discovered that there are factors that pass from one generation to the next that influence physical characteristics. These factors are known as genes. The second individual was Robert Bakewell, an Englishman who is known as "the father of modern animal breeding." Bakewell demonstrated that if a male and female that excelled in a certain trait were mated, the resulting offspring also excelled in the same trait. The science of genetics has broadened greatly since the primitive efforts of these two forefathers; however, the basic premises remain the same.

Figure 6-1.

OBJECTIVES

1. Describe how the gender of offspring is determined
2. Explain how genotype and phenotype are different
3. Differentiate between qualitative and quantitative inheritance
4. Identify breeding systems commonly used in animal science
5. Explain why heritability, selection intensity, and generation interval are important to genetic progress

TERMS

alleles
codominance
complete dominance
contemporary group
crossbreeding
generation interval
genetics
genotype
heritability
heterosis
heterozygous
homozygous
inbreeding
incomplete dominance
linebreeding
outbreeding
phenotype
qualitative traits
quantitative traits
selection intensity

ANIMAL SCIENCE FACTS

Lethal genes are rare recessive genes that cause death if present in the homozygous form. Parents can carry a lethal gene in the heterozygous form and be normal in appearance and performance. However, if two parents pass the recessive lethal gene to an offspring, death occurs prior to birth or shortly thereafter. Examples of lethal genes are bulldog syndrome in cattle (calf appears with shortened legs and stout head), hydrocephalus (water on the brain), and aretesia coli (closed colon in horses).

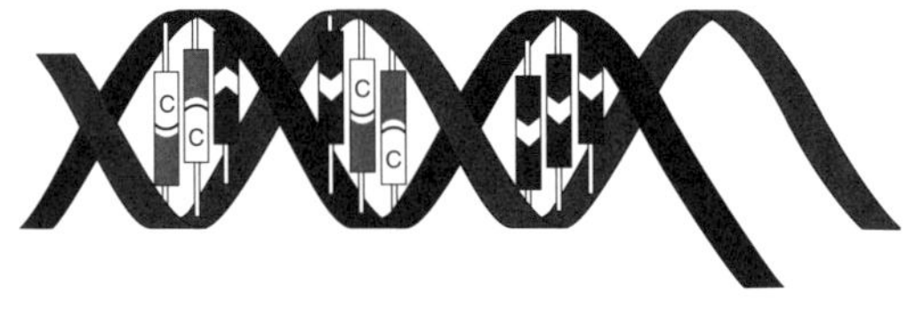

MENDELIAN GENETICS

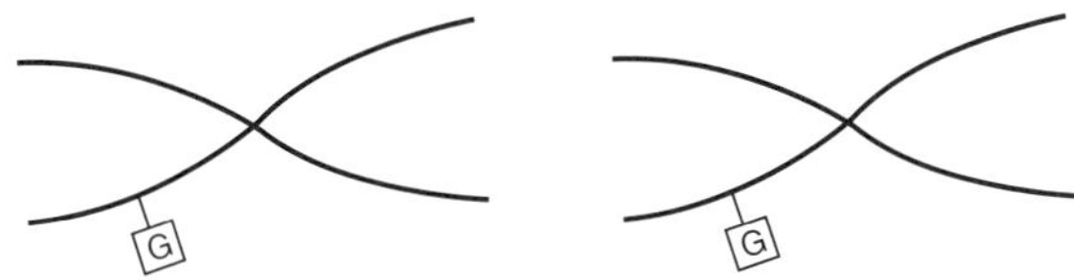

Figure 6-2. Chromosomes are X-shaped structures occurring in pairs. Genes coding for the same trait occupy analogous space on each chromosome.

Gregor Mendel was the first to discover that traits were inherited through units called genes that are present on chromosomes. He also deduced that genes are present in pairs with half of each pair being inherited from each parent. Each gene could then be independently transmitted, unchanged, to the next generation. Pairs of genes have come to be known as ***alleles***. See Figure 6-2 for a representation of a pair of alleles on homologous chromosomes. Remember, chromosomes are made of DNA and are found in the nucleus of cells. Genes are present on chromosomes.

DETERMINATION OF GENDER

As discussed in Chapter 1, the gametes, sperm and egg, meet at fertilization. At this meeting, the genetic material from the sperm mix with the genetic material of the egg. The resulting fertilized zygote thus contains genetic material (genes) from both the mother and father, and will display characteristics from both. The determination of the sex of the zygote depends on a pair of chromosomes called the sex chromosomes. Male sex chromosomes are either X or Y. A zygote that gets an X chromosome from the sperm will be female. A zygote that receives a Y chromosome from the sperm will be male. All eggs from the female receive an X chromosome. So, a female zygote will have two X chromosomes (XX) and a male zygote will have an X and a Y chromosome (XY).

GENOTYPE AND PHENOTYPE

The ***genotype*** of an animal is its actual genetic code. The animal's genotype controls physical traits, such as color, as well as potential performance traits, such as average daily gain. Genotype is the actual genetic code that cannot be changed by environmental factors. ***Phenotype*** is the part of the genotype the animal expresses. In some cases, such as coat color, phenotype cannot be changed by the environment. In other cases, it can. For example, a lamb may have the genotype to gain 1 pound per day. However, because it receives an inferior diet, the lamb only gains 1/2 pound per day. The lamb's phenotype for daily gain is 1/2 pound per day. In this case, an environmental factor (the inferior diet) is masking the true genetic potential (genotype) of the lamb.

QUALITATIVE TRAITS

Qualitative traits are traits controlled by only a single pair of genes and cannot be altered by the environment. They are sometimes called "either, or" traits because their phenotype is either one thing or the other. Coat color is a good example of a qualitative trait. If a cow is genetically programmed to be black, there is no environmental factor that can change the fact that the cow is black. Other examples of qualitative traits include the polled or horned conditions of cattle and sheep and blood type of all species.

Qualitative traits most easily show how genes are inherited. Coat color in Shorthorn cattle is a classic example. Shorthorns have three phenotypes: white, red, and roan (a mixture of white and red hairs). The two possible genes are red (designated R) and white (designated r). A red Shorthorn has two copies of the red gene (genotype designated by RR). A white Shorthorn has two copies of the white gene (genotype designated by rr). A roan animal has one copy of each gene (genotype designated by Rr).

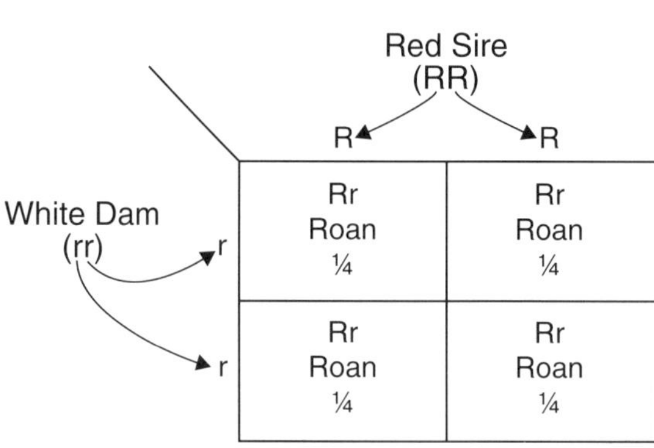

Figure 6-3. Punnet square of RR (red) × rr (white) mating in Shorthorn cattle. Each box represents a possible genotype of the offspring. All offspring are Rr (roan).

Coat color in Shorthorn cattle is an example of a qualitative trait with ***codominance***. This means that both the R and the r gene are expressed. Neither gene completely masks the other's presence. Both are expressed to some degree (both red and white hairs are present with the Rr genotype). On the other hand, when ***incomplete dominance*** occurs, a blending of the allele pair is expressed. If a red

Shorthorn bull (RR) is mated to a white Shorthorn cow (rr), all resulting offspring will receive one copy of each gene from each parent and will be roan (Rr). Punnet squares aid in determining the genotype of resulting offspring. See Figure 6-3. If a roan bull is mated to a white cow, half of the resulting offspring will be white and half will be roan. See Figure 6-4.

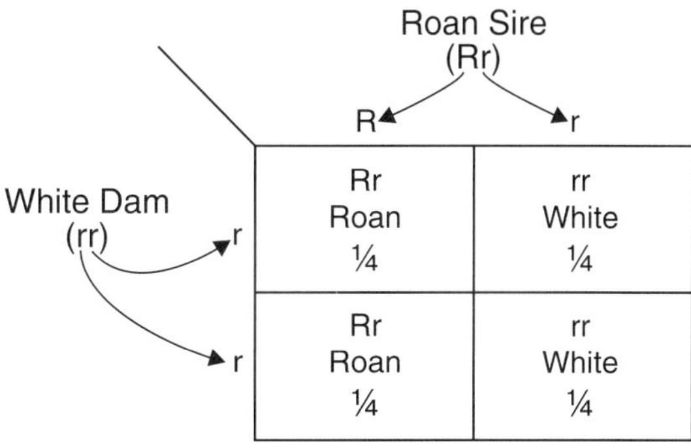

Figure 6-4. Punnet square of Rr (roan) × rr (white) mating in Shorthorn cattle. On average, half the offspring will be roan (Rr) and half will be white. (rr).

A qualitative trait with ***complete dominance*** can be illustrated by the coat color in Angus cattle. When dominance is complete, one gene completely masks the presence of the other. The two possible genes for coat color in Angus cattle are black (B) and red (b). If an animal has either one (Bb) or both (BB) copies of the black gene, phenotypically it will be black. Only if an animal has both copies of the red gene (bb) will the animal be red. See Figure 6-5 displaying the inheritance of coat color in Angus cattle.

Animals that have two copies of the same gene are called ***homozygous*** for that trait. An Angus cow with the genotype BB for coat color would be referred to as homozygous black. Since black is dominant to red in Angus cattle, genotype BB could be called homozygous dominant. Genotype bb would then be called homozygous recessive. Animals that have one

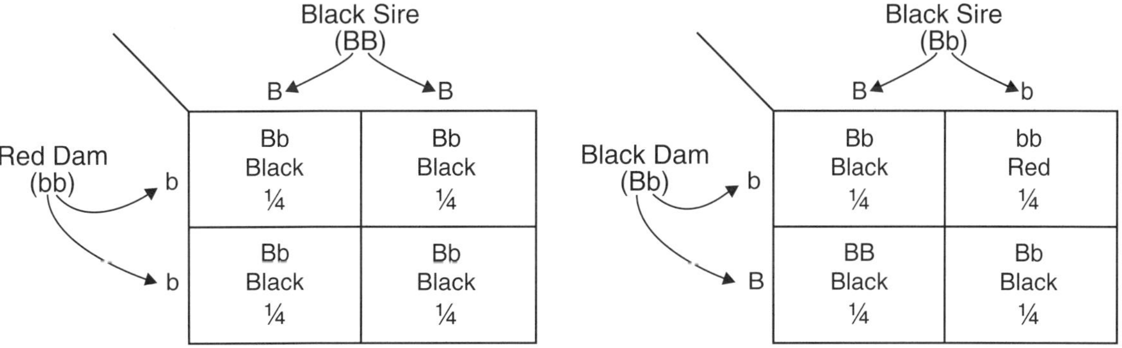

Figure 6-5. Punnet squares of the inheritance of coat color in Angus cattle. The black gene (B) is dominant over the red gene (b).

copy of each gene are called ***heterozygous*** for that trait. An Angus cow with the genotype Bb would be called heterozygous black. Animals that are homozygous have a 100 percent chance of passing that gene to their offspring. Therefore, if an Angus bull is homozygous black (BB), all its offspring will be black. However, if an Angus bull is heterozygous black (Bb) and is mated to a heterozygous cow, one-fourth of the offspring (on average) will be red. In other words, in a mating of two heterozygous Angus cattle, the chance of getting a red calf is one in four.

QUANTITATIVE TRAITS

Quantitative traits are controlled by several pairs of genes. Most economically important livestock traits, such as growth rate, backfat depth, speed, wool production, etc. are quantitative traits. The expression of quantitative traits can be influenced by environmental factors. Instead of the "either, or" inheritance of qualitative traits, quantitative traits are expressed across a range. For instance, average daily gain is a quantitative trait that, in pigs, may range from 0 to 3 pounds per day. The expression of average daily gain is a combination of genes and environment. Even though these traits are controlled by many pairs of genes, some animals have more "good" genes for a trait than other animals. Animals that are homozygous for these "good" genes have a better chance of passing those traits to their offspring. See Figures 6-6 and 6-7.

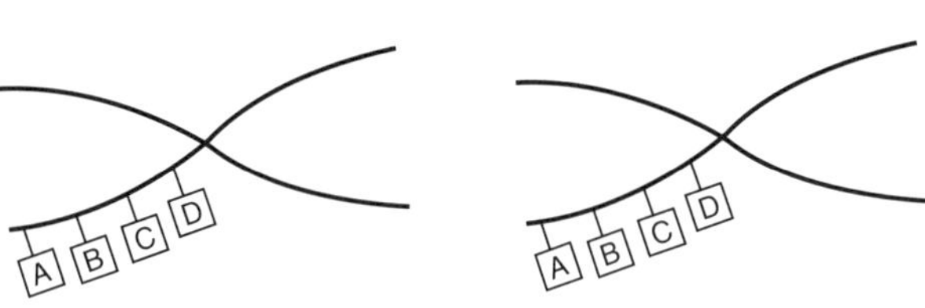

Figure 6-6. Chromosome representation of an animal that is homozygous for a quantitative (multiple gene) trait.

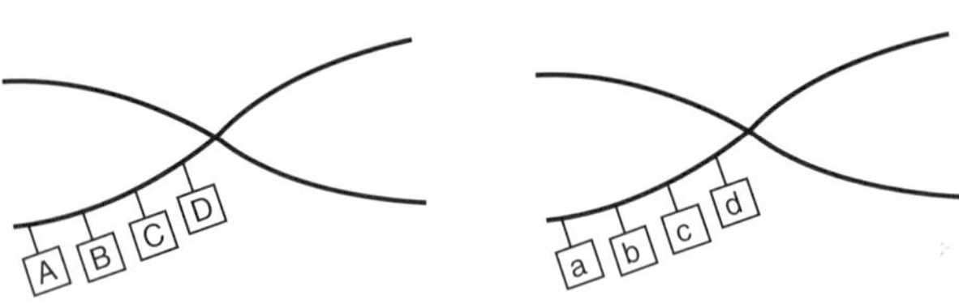

Figure 6-7. Chromosome representation of an animal that is heterozygous for a quantitative (multiple gene) trait.

To unmask the true genetic potential of an animal, it must be given an ideal environment. This means the elimination of any obstacle that could cause the animal to express less than its genotype. Some of these environmental factors include nutrition, disease, temperature, and competition from other animals. Unfortunately, total elimination of these factors is nearly impossible. Therefore, livestock producers identify animals with superior genes for quantitative traits by comparing expression of those traits among animals that are fed and managed the same. These similarly fed and managed groups of animals are called ***contemporary groups.***

BREEDING SYSTEMS

For many years, livestock producers have manipulated genes through inbreeding and crossbreeding in an effort to improve quantitative traits of livestock.

INBREEDING

Inbreeding is the mating of closely related animals, such as parent–offspring or brother–sister. The concentration of genes resulting from inbreeding serves to make animals more homozygous for all traits, both qualitative and quantitative. For instance, mating a sire with outstanding performance to his daughters increases the probability that the resulting offspring will possess a greater proportion of the original sire's genes. One drawback with inbreeding is the concentration of undesirable genes along with the desirable ones. Any genetic flaw in the original sire would also be magnified in the offspring of a father-daughter mating. ***Linebreeding*** is the mating of related animals that are not immediate family members.

CROSSBREEDING (OUTBREEDING)

Crossbreeding or ***outbreeding*** is the mating of animals that are not related. Crossbred animals have a higher proportion of heterozygous gene pairs than non-crossbred animals. Resulting offspring of a crossbred mating receive genes from the sire and dam that are very different. The main advantage of outbreeding is the increase in a phenomenon called ***heterosis*** in crossbred animals. Heterosis (sometimes called hybrid vigor) is the increase in a performance trait that exceeds the average of the parents. For instance, cattle breed A may have an average weaning weight of 450 pounds and cattle breed B may have an average weaning weight of 500 pounds. The crossbred progeny of breeds A and B may display an average weaning weight of 550 pounds. It would make sense that since half the genes for weaning weight come from the sire and half from the dam, the average weaning weight of the crossbred progeny would be 475 pounds. The difference between the actual weaning weight (550 pounds) and the expected weaning weight (475 pounds) is a result of heterosis.

HERITABILITY

As previously discussed, the true genetic merit of an animal can be altered or masked by many environmental factors. Geneticists have developed a system for separating genetics from environment, following this simple equation:

genetics + environment = genetic expression

The "genetic expression" is a quantitative trait that we can measure, such as butterfat percentage, backfat, or wool production. The "genetics" and "environment" parts of the equation must be estimated by genetic breeding experiments. Geneticists have estimated the proportion of many different traits that results from genetics and the proportion that results from the environment. The proportion that results from genetic differences is called ***heritability*** and is expressed as a percentage.

For example, the heritability of reproductive traits, such as pigs born per litter, is quite low — only about 15 percent. This means that only 15 percent of the difference seen in pigs born per litter is a result of genetic differences. The other 85 percent is a result of environmental factors, such as nutrition, air quality, number of breedings per sow, or disease. In contrast, the heritability of carcass traits, such as ribeye area in beef cattle, is 50–60 percent. Thus, only 40–50 percent of the observed differences in ribeye area are a re-

ANIMAL SCIENCE FACTS

The following traits are listed with the approximate corresponding heritability (percentage of the trait that results from genetics).

	Trait	Heritability
a.	Number of Live Pigs Farrowed	15%
b.	Backfat Thickness in Swine	45%
c.	Birth Weight in Beef Cattle	40%
d.	Carcass Tenderness in Beef Cattle	60%
e.	Mastitis Susceptibility in Dairy Cattle	25%
f.	Milking Ability in Dairy Cattle	30%
g.	Twinning in Sheep	10%
h.	Weaning Weight in Sheep	30%
i.	Pulling Power in Horses	25%
j.	Jumping Ability in Horses	20%

sult of environmental influences. Growth and feed efficiency traits are intermediate in heritability, averaging 25–40 percent.

Traits that have high heritability are most responsive to genetic selection. For instance, mating a bull and a cow that are both heavily muscled will increase the chance that the offspring also will be heavily muscled. However, mating a boar and a sow from big litters does little to guarantee the offspring will also farrow big litters.

Interestingly, low heritability traits show high levels of heterosis, while high heritability traits show low levels of heterosis. In our example, we would expect less heterosis in a crossbred animal for ribeye size (a high heritability trait) than for pigs born per litter (a low heritability trait).

SELECTION

Using genetics to improve livestock in economically important areas has long been the objective of producers. The first rule in selective breeding is, "mate the best to the best." Identifying the best animals for a certain trait or set of traits requires good production records. For example, if a dairy farmer wants to keep replacement heifers from the highest producing cows, the farmer must have records of which cows give the most milk. Likewise, if a cow-calf producer is selecting for growth rate, the producer must keep weight records of possible replacement heifers and prospective herd bulls to know which ones gained weight the fastest. Producers then make selection decisions based on contemporary groups. A fall-born beef calf cannot be genetically compared to a calf born in April because they were raised under different environmental conditions.

Selection intensity is the percentage of animals at the top of the contemporary group that are used to produce the next generation. A population of animals will show a normal distribution for any quantitative trait (see Figure 6-8). The fewer top performing individuals that can be used as parents for the next generation, the faster the genetic improvement for that trait will be. In this

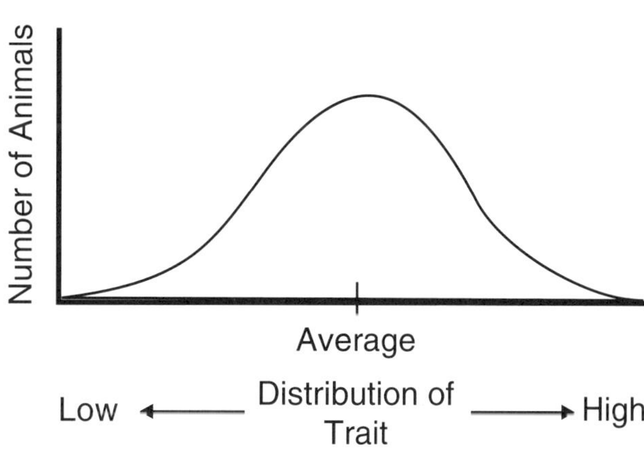

Figure 6-8. Normal distribution curve for any quantitative trait.

example of selection intensity, 100 bulls were fed together in a contemporary group. Average daily gain was calculated for each. If the five bulls with the highest average daily gain are kept to sire the next generation, the selection intensity for average daily gain would be 5 percent. Artificial insemination (AI) is used by livestock producers to increase selection intensity for sires. Semen can be collected from proven top sires, diluted and distributed to all parts of the country. Therefore, all producers have a chance to sample some of the very best genetics available for a given trait. The selection intensity using AI may be less than 1/2 percent for sires. Because males can sire offspring by many females, the selection intensity for males is usually higher than that for females. Superovulation coupled with embryo transfer is one way to increase selection intensity in females.

Figure 6-9 illustrates the genetic change for a highly heritable trait using relatively high selection intensities for both males and females. Notice that the average trait value of each generation keeps increasing. Low heritability traits would respond similarly, but by a lesser amount per generation. Figure 6-10 shows the genetic change over three generations using low selection intensities for both males and females.

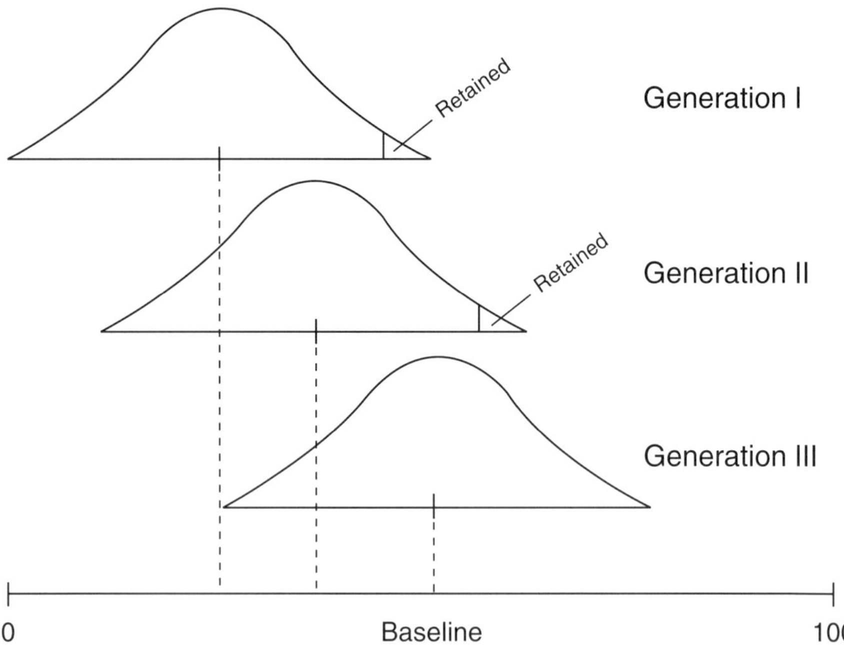

Figure 6-9. Change in baseline averages for a highly heritable quantitative trait over three generations with high selection intensity.

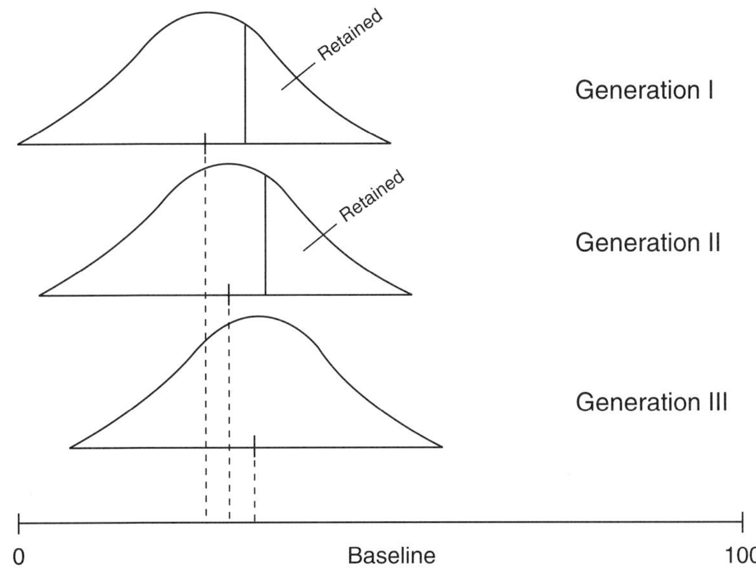

Figure 6-10. Change in baseline averages for a highly heritable quantitative trait over three generations with low selection intensity.

Obviously, making genetic changes relies on replacing the old generation with a new, genetically better generation. Therefore, the faster generations can be turned over, the faster genetic improvements can be made. The amount of time it takes to replace each generation is called the ***generation interval.*** Generation interval for cattle is about five years, swine one year, sheep one year, and horses about four to five years. Species with shorter generation intervals can make more rapid genetic changes.

Selection for a single trait will cause faster improvement than selecting for many traits at the same time. For instance, if a sheep producer selects replacements based solely on wool yield, wool weights will increase relatively quickly, but fertility, longevity, and structural soundness may suffer. However, a swine producer selecting replacements based on three traits (backfat, feed efficiency, and pigs born per litter) may not see great improvement in any one area for several generations.

SUMMARY

Genes from both parents are mixed at fertilization. As a result of genetic mixing, offspring display traits from both parents. Genotype is the actual genetic code of the animal, while phenotype is the part of that genetic code that is expressed. Qualitative traits like coat color are controlled by a single set of genes. Quantitative traits, such as average daily gain, are controlled by many sets of genes and are economically important to producers. Inbreeding and linebreeding are breeding systems where related animals are mated. This serves to concentrate genes from a common ancestor. On the other hand, crossbreeding results in heterosis, an improvement in performance over the average of the parents. Heritability is the portion of a quantitative trait that is controlled by genetics rather than environment. Selection intensity and generation interval must be used in conjunction to realize rapid genetic change. Single trait selection results in more rapid genetic change than multiple trait selection.

CHAPTER SELF-CHECK

___ genetics	1. time needed to replace each generation
___ alleles	2. controlled by single pair of genes
___ genotype	3. mating of unrelated animals
___ phenotype	4. expression of genetic code
___ qualitative traits	5. animals similarly fed and managed
___ codominance	6. top percent of animals used for production of next generation
___ incomplete dominance	7. a blending of genes in pair is expressed
___ complete dominance	8. animal's genetic code
___ homozygous	9. allele that has one copy each of the two possible genes
___ heterozygous	10. allele with two copies of identical genes
___ quantitative traits	11. study of heredity
___ contemporary group	12. one gene masks presence of other
___ inbreeding	13. two genes that have the same location on a pair of chromosomes
___ linebreeding	14. controlled by several pairs of genes
___ crossbreeding	15. mating of immediate family
___ outbreeding	16. percentage of trait that results from genetics
___ heterosis	17. increase in performance that exceeds parents' average
___ heritability	18. mating of distant relatives
___ selection intensity	19. same as crossbreeding
___ generational interval	20. both genes in pair are expressed

QUESTIONS AND PROBLEMS FOR DISCUSSION

1. Name two scientists credited with the early development of the study of genetics.
2. Explain why males determine the gender of offspring.
3. Write the chromosome pairings for males and females.
4. Differentiate between genotype and phenotype.
5. Give a specific example of a qualitative trait.
6. Give a specific example of a quantitative trait.
7. List one breed of beef cattle that exhibits codominance in coat color.
8. Animals with two copies of the same gene for a certain trait are said to be ________________.
9. Write a heterozygous gene code for a Black Angus cow.
10. Discuss the importance of feeding and managing all animals alike in a contemporary group.
11. List one disadvantage of inbreeding.
12. Describe the difference between inbreeding and linebreeding.
13. The increase in performance above the parents' average is ______________.
14. Give one example of a low, a medium, and a high heritability trait.
15. Explain how heritability and heterosis are related.

ACTIVITIES

1. Contact a teaching hospital or breeding stock company for information on new breakthroughs in genetics. Report findings to the class.
2. Divide the class into two groups. Debate the ethical issues involved in genetic research in humans. Choose a specific topic for consideration.

Research the topic in your school library. Use the point/counterpoint approach.

3. Contact local livestock producers to speak to the class about genetics. Ask them to discuss how they choose sires and replacement females.

LABORATORY ACTIVITY

PRACTICAL USES OF GENETICS

Purpose

To determine the effects of heterosis, calculate genotype and phenotype in matings, and evaluate the effects of different selection intensities

Materials

pencil
paper
calculator (optional)

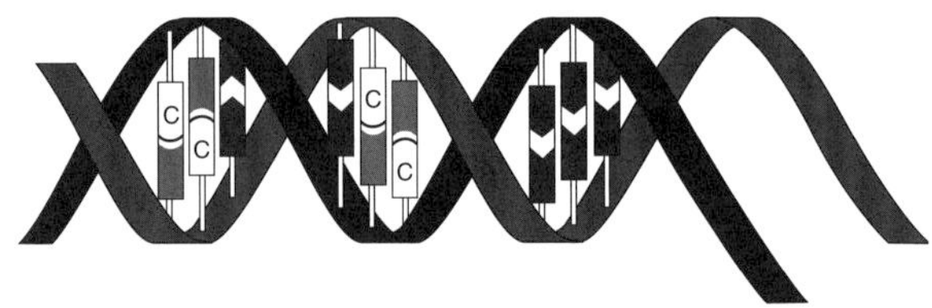

Procedure

Part one: Heterosis
Calculate heterosis for the following scenarios:

a) Average lambs born in Suffolk sheep = 1.5
Average lambs born in Dorset sheep = 2.1
Average lambs born in Suffolk × Dorset crossbreds = 2.2

b) Average ribeye area in Suffolk sheep = 3.4 square inches
Average ribeye area in Dorset sheep = 2.9 square inches
Average ribeye area in Suffolk × Dorset crossbreds = 3.2 square inches

Questions:

1. Use the following formula to calculate percent heterosis for a and b.

$$\frac{\text{Offspring Average} - \text{Parental Average}}{\text{Parental Average}} \times 100$$

2. Which is higher?
3. Using what you know about the relationship between heritability and heterosis, which (lambs born or ribeye area) would most likely have a higher heritability?

Part two: Genotype and Phenotype

A qualitative trait controlled by a single gene pair is the spotted or not spotted condition in Maine-Anjou cattle. The not-spotted gene (S) is completely dominant over spotted gene (s). Following are nine possible matings of animals homozygous dominant (SS), heterozygous (Ss), or homozygous recessive (ss). Use punnet squares to calculate the genotypic and phenotypic percentages expected in the offspring of each mating.

			Offspring	
Male	**x**	**Female**	**Genotype**	**Phenotype**
SS		ss	All Ss	All not-spotted
Ss		ss		
ss		ss		
SS		SS		
Ss		Ss		
Ss		SS		
ss		Ss		
ss		SS		
SS		Ss		

Part three: Selection Intensity

Fifty boars were fed together in a contemporary group. The top 5% of the boars had an average daily gain of 2.75 pounds per day. The top 10% of the boars had an average daily gain of 2.50 pounds per day. At the same time, 200 gilts were also fed together in a contemporary group. The top 25% of the gilts averaged 2.10 pounds of gain per day. The top 50% of the gilts had an average daily gain of 1.80 pounds of gain per day.

Calculate the expected average daily gain of the progeny of these boars and gilts if the following percentages were selected to be parents of the next generation. Assume no heterosis.

a. Top 5% of the boars and top 25% of the gilts

b. Top 10% of the boars and top 25% of the gilts

c. Top 5% of the boars and top 50% of the gilts

d. Top 10% of the boars and top 50% of the gilts

Which of the above would you recommend to farmer Buford Hamhock if he were interested in raising the fastest gaining pigs possible? Why? Which would you recommend if Mr. Hamhock told you he needed to keep 100 of the gilts?

Application of Laboratory Activity

Calculating heterosis and selecting animals for herd breeding stock illustrate practical agricultural uses of genetics. Differing farm scenerios, such as a need for more efficient gain or even the desires for a certain color, require special genetic considerations. Producers need to understand the basics of genetics in order to run the most profitable farm possible.

On the other hand, misinformation and a lack of genetic understanding often cost producers countless dollars in lost profits. Agricultural educators and extension agents can assist producers in making sound selections as well as understanding the genetics behind the decision.

Chapter 7

ETHOLOGY: ANIMAL BEHAVIOR AND WELFARE

Monkey See, Monkey Do

INTRODUCTION

The science of animal behavior related to the environment has existed for ages, but was not recognized as a legitimate field of study, called ***ethology,*** until relatively recently. Farmers in the dark ages undoubtedly noticed how animals acted when they were sick or in need of nourishment. More recently, animal behavior has been studied in an effort to better understand the physical needs of animals.

Figure 7-1.

OBJECTIVES

1. Give specific examples of maintenance, social, and learned behaviors in animals
2. Identify behavioral traits that will respond to selection
3. Apply the principles of animal behavior to defecating, movement, and facilities
4. Debate the difference between animal welfare and animal rights

TERMS

animal rightists
animal welfarists
conditioning
eliminative
ethology
Flehman response
ingestive
learned behaviors
lip prehenders
maintenance behaviors
negative reinforcement
positive reinforcement
social behaviors
tongue prehenders

ANIMAL SCIENCE FACTS

Researchers have spent many hours patterning the ingestive behaviors of pigs. Video tapes of various pigs eating have been analyzed to determine the most common motion of a pig's head and mouth as it eats. This information was used to design a sow feeder that will almost eliminate spilled feed. The feeder is J-shaped with the short end of the J facing the sow. Several bars are installed horizontally inside of the feeder up the long side of the J. When a sow takes a bite of feed from the bottom of the trough, she naturally lifts her head and bumps her nose on the horizontal bars, knocking any loose feed in her mouth back into the feeder. The feeder is wide enough so that, as the sow chews, any feed that spills out of her mouth also falls back into the trough. This new feeder, designed using the sow's own ingestive behavior as a guide, reduces feed wastage by over 50 percent.

ANIMAL BEHAVIORS

Animal behaviors can be divided into three broad categories. The first is ***maintenance behaviors,*** such as eating, defecating, and sleeping. The second category is ***social behaviors,*** which deal with the interactions of two or more animals. The third category is ***learned behaviors.*** A horse adapting to carry a saddle and rider is an example of a learned behavior.

The study of animal behavior gives producers an understanding of how to design facilities to make moving and training easier for both livestock and their handlers.

INDIVIDUAL MAINTENANCE BEHAVIORS

Individual maintenance behaviors include those behaviors basic to the sustenance of life. This category would include ***ingestive*** (eating), ***eliminative*** (defecating), and sleeping behaviors.

Ingestive behaviors vary among species. Ruminants (cattle and sheep) do not have incisor teeth on their upper jaw with which to bite off their food. Cattle grasp standing forages with the tongue and upper palate. Cattle are therefore known as ***tongue prehenders*** because they use their tongue to gather forages. See Figure 7-3. Sheep are known as ***lip prehenders*** because they grasp forages with their lips. See Figure 7-4. Sheep tend to nibble their feed and can graze closer to ground level than cattle. Horses and swine have incisor teeth on both jaws and can bite off chunks of feed the same as humans.

All livestock species except swine excrete feces at random. Swine are particularly sensitive to where they defecate. Pigs are clean animals by nature, and usually choose a corner of the pen to use for dunging. There

Figure 7-2. Eating is a maintenance behavior essential to life.

Figure 7-3. Cattle grasp forages with their tongues and are referred to as tongue prehenders.

Figure 7-4. Sheep nibble forages with their lips and are referred to as lip prehenders.

are patterns, however, to exactly where this dunging occurs. For instance, pigs are more likely to defecate in an area that is the most draft-prone. Areas where pigs have nose-to-nose contact with pigs from another pen are also attractive dunging areas.

Sleep or deep rest is common in all livestock species. Pigs, like humans, sleep for extended periods. Judging from the jerking of pigs' legs during sleep, it appears that pigs dream. Sheep and cattle sleep lying down, but their sleep is sporadic and very short. Adult horses sleep in the standing position. A system of ligaments allows the horse to sleep while resting muscles. While sleeping, horses rest weight on only one hind leg with the other cocked. During sleep, horses will switch the weight-bearing leg.

SOCIAL BEHAVIORS

Social behaviors of animals are very interesting and informative to those managing livestock. Behaviors that determine social structure or hierarchy among groups of animals have been studied extensively. In most animal groups, a pecking order is established. There is a "boss" animal that has asserted itself through bullying or fighting. The hierarchy of a group may not be completely linear. In a group of animals, Animal A may be dominant over B, B over C, and C over D. However, sometimes there can be an inversion where D is dominant over B. See Figure 7-5.

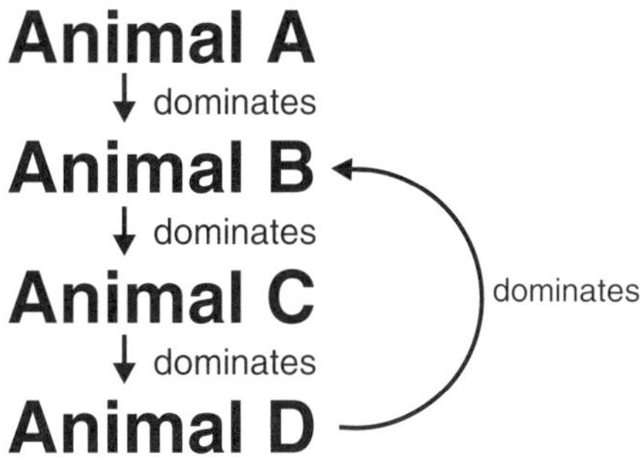

Figure 7-5. Animal dominance can be linear, where each animal dominates over a lower individual (A over B, B over C, C over D), or inverted, where a lower animal rises to dominance over a higher ranking individual not directly above it (A over B, B over C, C over D, D over B).

Swine have a nearly complete hierarchical social structure. In horses, groups of mares normally contain a dominant mare, but if a mature stallion is included in the group, he usually dominates the mares. Cattle dominance is based on age, with older animals in a herd usually more dominant over younger animals. Social dominance does not seem to be a component in the flocking behavior of sheep. Sheep simply want to be members of a flock. They do not seem to care who is boss.

Sexual behaviors are another example of social behaviors. Knowledge of sexual behaviors is extremely important to producers using artificial insemination, because they must recognize behaviors associated with estrus. Sows nearing estrus will be restless, urinate frequently, and attempt to mount other sows. See Figure 7-7. Cows exhibit the same signs. During heat, both sows and cows will stand to be mounted. Ewes show few signs of estrus except for the acceptance of the male. Mares in heat also urinate frequently, seek the companionship of other horses, and may lift their tail sideways.

Figure 7-6. Sheep want to flock with others and have little regard for a social hierarchy.

Figure 7-7. Sows show impending estrus by trying to mount other animals.

In natural mating, sexual behaviors include both the male and female component. The male component in all species involves courtship. In swine, the boar will root at the flank and grunt before attempting to mount the sow. A ram's courtship patterns consist of walking with a stiff-legged gait and biting the wool on the ewe's flank. The ram may also smell the urine of the ewe and curl his upper lip in what is called the ***Flehman response.*** Bulls often smell or lick the cow's genitalia then exhibit the Flehman response. Bulls will follow a receptive cow around the pasture, head resting on her rump. The stallion will smell the genitalia of the mare and may also exhibit the Flehman response. In addition, stallions may nip the mare with his teeth.

Social behaviors also include the behaviors exhibited between mother and young. Sows aggressively protect their newborn pigs against intruders. When a sow thinks her pigs may be threatened, she will charge the intruder with mouth open. Another reason farrowing crates are used by swine producers is to allow piglets to be handled without fear of confrontations with the sow. Baby calves are often born in open areas, such as pastures. A mother cow separates herself from other cattle as much as possible before calving. After the calf is born, the cow will lick the calf dry, which stimulates the calf to nurse. The cow then spends the next day or two forming a strong bond with her calf before she and the calf rejoin the herd. Cows and calves

ANIMAL SCIENCE FACTS

The author studied the social dominance of barrows versus gilts for his master's degree research. No difference was found. Neither barrows nor gilts are dominant over the other group.

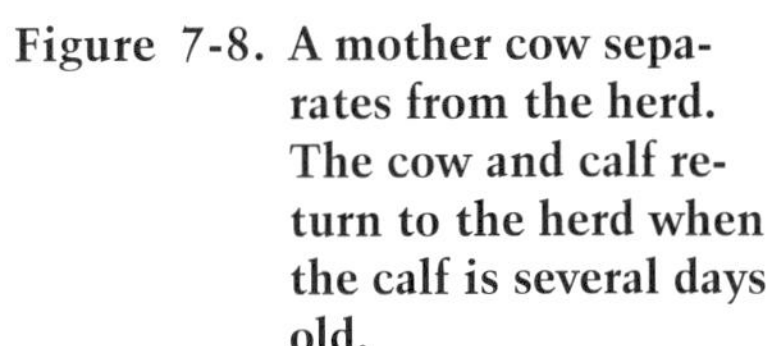

Figure 7-8. **A mother cow separates from the herd. The cow and calf return to the herd when the calf is several days old.**

can locate each other by smell, sound, or sight. Ewes, also, stimulate their young immediately after birth by licking them. The bond between ewes and their lambs is obvious. In a pasture of several hundred ewes and lambs, a ewe will nurse only her own lambs. Mares also exhibit a bond with their foals and will defend them against intruders. Voice recognition seems to be the most important means of communication between mares and foals.

LEARNED BEHAVIORS

Animals learn by three methods. The first is through ***positive reinforcement.*** Positive reinforcement is giving the animal something good (a treat, attention, feed) for doing something correctly. The second is ***negative reinforcement.*** Negative reinforcement is punishing the animal for doing something wrong. Animals can be trained in both ways but generally,

Figure 7-9. **Animals can be halter trained using positive reinforcement. (Courtesy, Jasper S. Lee)**

Figure 7-10. A cattle headgate can be associated with negative reinforcement. If every time cattle are run through a catch chute, they are dehorned, castrated, or prodded with needles, they come to regard the chute with fear.

positive reinforcement works best. A well-placed scratch, some special feed, or simply a pat and soothing words are examples of positive reinforcement. Animals can regard simple management practices as negative reinforcement. For instance, if every time cattle are run through a catch chute, they get prodded with vaccination needles, they come to regard the chute with fear. The needle acts as a negative reinforcement. Some positive reinforcement combined with the shots would make the cattle less fearful.

A third type of learned behavior is known as ***conditioning.*** Animals may become conditioned to associate a certain stimulus with a behavior. For example, cows entering a milking parlor know from past experience that milking time is near. They respond to the parlor stimulus with a conditioned response called milk let-down.

BEHAVIORAL GENETICS

Behaviors can also be inherited. Behaviors would be classified as quantitative traits, but they are usually more difficult to measure objectively than something like average daily gain. Few studies have investigated the heritability of behaviors, but the area will undoubtedly be further researched. Producers know that certain breeds of animals are easier to handle and manage than others. For instance, beef bulls are generally known to be calmer and less dangerous than dairy bulls. The natural gaits of horses are a good example of a behavior that is genetically controlled. Tennessee Walking Horse foals naturally assume their distinctive gait with no training, and thoroughbred foals gallop across a field with the ease of their parents.

One behavior that has received recent research attention is the sitting behavior in pigs. This trait is important because sows that spend considerable time sitting tend to crush more baby piglets than those that do

Figure 7-11. Beef bulls tend to be calmer than their dairy counterparts.

not. Researchers measured the amount of time sows from different families spent sitting down. Results proved that sitting behavior is genetically controlled. Pigs from some families spent considerably more time sitting than those from other families. Producers can use this information to select replacement gilts. They can, therefore, expect their sows not to sit and crush so many piglets.

BEHAVIOR AND LIVESTOCK MOVEMENT

The movement of animals from place to place requires a knowledge of animal behavior. Body position is the key to moving animals. Each animal has a zone surrounding it that, if invaded, causes the animal to move away. Body position within that zone determines the direction the animal will move. If the animal is approached to the rear of the point of the shoulder, it will most likely move forward, angling to the opposite side of the approacher. If an animal is approached to the front of the point of the shoulder, it will most likely turn away from the approacher and travel in the opposite direction. See Figure 7-12.

Animals will balk at physical obstacles. Steps up or down or a change in flooring type will cause animals to stop and look. If animals are given enough time to investigate the steps or change in flooring, they will normally continue to move. Rushing animals will only worsen the situation.

Walls of livestock-movement alleys and loading facilities should also be modified to make moving livestock easier. Animals are easily distracted by other animals outside a passageway. Walls should be solid to avoid this problem. Alleyways should be either wide enough that two animals can fit side by side, or, preferably, narrow enough that only one animal can fit, without room to turn around. Sharp turns, especially blind turns

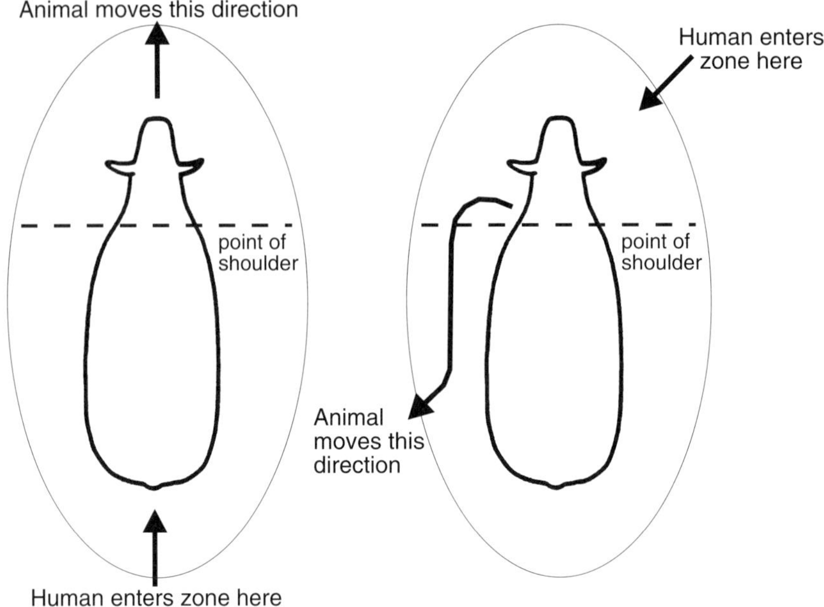

Figure 7-12. Animals move in the opposite direction of an approaching human.

should also be avoided. Turns should be circular. Cattle, especially, have a natural behavioral tendency to follow a curved passageway.

Air movement and light should also be considered when moving livestock. Strong drafts entering a doorway through which livestock are being moved can cause livestock to balk. Likewise, moving livestock from bright light into darkness can cause movement problems. Movement from darkness to light does not seem to be as much of a problem. However, any such obstacles should be eliminated before animal movement is attempted.

ANIMAL WELFARE AND ANIMAL RIGHTS

Many people mistakenly believe that animal welfare and animal rights are the same concept. Welfare is defined in the 1989 edition of *The New Lexicon Webster's Dictionary of the English Language* as being "healthy, happy, and free from want." The welfare of animals is, without question, the responsibility of the owner, producer, or caretaker. Most producers would consider themselves ***animal welfarists.*** On the other hand, those who advocate animal rights think that animals should have the same rights and privileges as humans. ***Animal rightists*** do not think animals should be used for any human benefit, such as food (milk, meat, eggs), clothing (leather, wool) or pleasure (horse racing, pleasure riding).

Animal welfarists contend that for animals to grow, reproduce, and perform at profitable levels, they must be well tended. Livestock under the care of competent producers have their basic needs met. These basic needs include feed, water, and protection from parasites, disease, and predators. Producers who are animal welfarists believe confining livestock in pens serves to keep the animals from many dangers, like foul weather or busy roadways. Along that same line, confining sows to crates during gestation eliminates competition from other sows for feed. Housing sows in farrowing crates also reduces the number of baby pigs crushed by the sow, and increases the welfare of the piglets.

ANIMAL SCIENCE FACTS

Cow tipping is an old college hoax designed to trick naive newcomers. The newcomers are told that cows sleep while standing and a group of people can tip or push the cows over. The tricksters then take the newcomers to a cow pasture and tell them to give cow tipping a try. In the meantime, the newcomers are abandoned. Hopefully, you won't fall for this old prank!

Animal rights advocates often have little connection with food and fiber production. However, they usually do have access to finances with which to lobby lawmakers. Animal rightists want lawmakers to pass legislation making it illegal to raise animals for human use. They label modern livestock production as "factory farming" where livestock are routinely mistreated and abused. The best way for animal welfarists to avoid these types of laws, is to support scientific study of animal behavior. Researching animal behavior is the best way for humans to objectively determine the needs and wants of animals.

SUMMARY

Maintenance behaviors are critical for life. Examples include eating, defecating, and sleeping. Social behaviors involve two or more animals and include hierarchy, sexual behaviors, and bonding between mother and young. Learned behaviors can be taught through both positive and negative reinforcement, or conditioning. Some behaviors are inherited as a quantitative trait. Animal movement can be eased through knowledge of animal behaviors and modification of facilities. Animal rightists assign the same privileges to animals that humans enjoy. Animal welfare involves making sure animals are happy, healthy, and free from want. Many management practices, although seemingly inhumane, actually increase animal welfare.

CHAPTER SELF-CHECK

___ ethology	1. the science of animal behavior related to the environment
___ maintenance behaviors	2. cattle
___ social behaviors	3. believe that animals should not be used for human benefit
___ learned behaviors	4. associate a behavior with a stimulus
___ animal welfarists	5. essential for life
___ animal rightists	6. sheep
___ ingestive	7. eating
___ eliminative	8. positive and negative reinforcement, conditioning
___ tongue prehenders	9. reward for good behavior
___ lip prehenders	10. punishment for bad behavior
___ Flehman response	11. interaction of animals
___ positive reinforcement	12. defecating, dunging
___ negative reinforcement	13. curling upper lip in courtship
___ conditioning	14. believe animals should be well tended

QUESTIONS AND PROBLEMS FOR DISCUSSION

1. List three examples of maintenance behaviors.
2. List three examples of social behaviors.
3. List three methods of learned behaviors.
4. Give two advantages of farrowing sows in crates. Can you name any disadvantages?
5. Do ruminants have upper incisor teeth?
6. Which species, cattle or sheep, graze forages closer to the soil? Why?
7. Which livestock species sleeps for extended periods of time in the prone position?
8. In which position do cattle sleep?

9. Social dominance in cattle is usually based on _____.
10. Is social dominance seen more in cattle or sheep?
11. Name three treats used in positive reinforcement of animals.
12. Milk let-down upon entry into a milking parlor is a learned behavior referred to as ________________.
13. Explain how behavior and genetics can be related.
14. Describe four physical obstacles that inhibit animal movement.
15. Contrast the beliefs of an animal welfarist and an animal rightist.

ACTIVITIES

1. Conduct a classroom debate using an animal rightists vs. animal welfarists format. Randomly select sides. Obtain information from the extension service, library, animal welfarist and rightist groups.
2. Attempt to train your pet or livestock using positive reinforcement.
3. Prepare a speech concerning animal behavior, welfare, or rights for an FFA public speaking contest.

LABORATORY ACTIVITY

OBSERVING ANIMAL BEHAVIOR

Purpose

To observe animal behavior

Materials

group of animals in normal setting
livestock pasture or pen would be great,
but fish tank or hamster cage will also work
paper and pencils to make observations

Safety Precaution

Maintain a safe distance from unfamiliar animals.

Procedure

1. Observe a group of animals for a set period of time. Each student watches only one animal. Try not to disturb the animals you are watching.
2. Record the amount of time out of the period that the animals spend standing up, lying down, or sitting.
3. Record the time spent eating, drinking, and sleeping.
4. Divide a diagram of the enclosure into quadrants. Record the amount of time the animal spends in each quadrant.
5. Record any social interactions between two animals including grooming behaviors, vocalizations, or sexual behaviors.

Analysis

1. Compile and tabulate the amount of time spent standing, sitting, and lying down. Calculate the percentages of total observation time animals spent in each of these behaviors.
2. Compile and tabulate the amount of time spent eating, drinking, and sleeping. Calculate the percentages of total observation time animals spent in each of these behaviors.
3. Compile and tabulate the amount of time animals spent in each quadrant. Was there a certain quadrant or quadrants that animals spent more time in than others? If so, why?
4. Did you notice any patterns in social behaviors between animals? Any grooming behaviors between mothers and their young? Any signs of females nearing estrus?
5. What other behaviors did you observe besides those you were watching for? Categorize them into maintenance, social, or learned behaviors.

Application of Laboratory Activity

Understanding animal behavior allows scientists and producers to make decisions which affect productivity. Animals raised in low stress circumstances will gain or produce more efficiently. Research findings in animal behavior influence decisions in building design as well as the number, age, and gender of animals housed together. Moreover, animal equipment like watering devices and round bale feeders has been designed as a result of behavioral research.

Recent animal rights claims have triggered increased behavioral research. The findings of such research allow logical conclusions to questions poised by activists.

UNIT II

Application

Chapter 8

SWINE MANAGEMENT

This Little Piggy Went to Market

INTRODUCTION

Pigs have paid for many farms in the United States, earning their title as the "mortgage lifter." The business of swine production is in the midst of a massive consolidation where more pigs are owned by fewer producers. No matter how the industry changes, pork will still be a favorite meat of American consumers. Pigs will continue to be raised by American producers and the principles of swine production will remain the same.

Figure 8-1.

OBJECTIVES

1. Compare the physical characteristics of the eight major swine breeds and classify each as a maternal or paternal breed
2. List and explain four breeding systems used in the swine industry
3. Identify six steps in processing baby pigs
4. Develop a feeding program for a market hog
5. Make a chart of a common vaccination and parasite control schedule for breeding and market swine
6. Discuss the housing requirements for swine
7. Calculate the cash value of a 250-pound market hog using the current cash market price

TERMS

all-in all-out production
atrophic rhinitis
backcrossing (crisscrossing)
biosecurity
cash price
cross-fostered
futures price
leptospirosis
maternal breeds
mycoplasmal pneumonia
PRRS (porcine reproduction and respiratory syndrome)
parvovirus
paternal breeds
pseudorabies
rotaterminal cross
split-sex feeding
swine erysipelas
terminal crossbreeding
three breed rotational cross

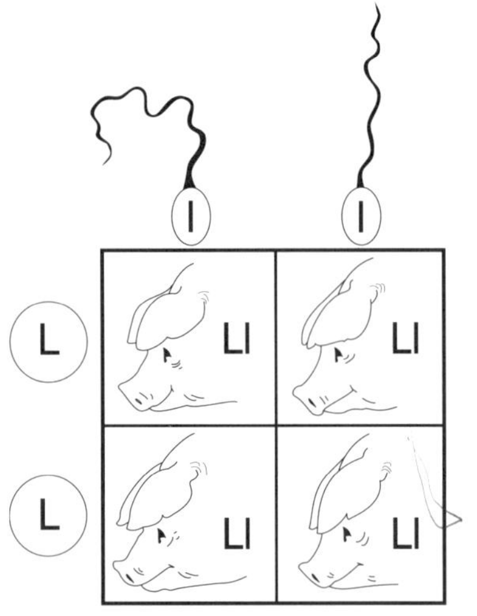

ANIMAL SCIENCE FACTS

Swine facts

Average body temperature = 102.5°F
Average pigs/litter = 8.2
Average market weight = 250–255 pounds
Average 10th rib backfat = 1.1 inches
Average loineye area = 5.0–5.5 square inches
Average daily feed intake (feed), feed per pound of gain (F/G), and average daily gain (ADG)

	feed(lbs)	F/G(lbs)	ADG(lbs)
40-pound pig	2.8	2.2	1.3
140-pound pig	4.6	2.9	1.6
240-pound pig	7.0	3.7	1.9
Average 40–240 pounds	4.8	2.9	1.6

BREEDS

There are eight traditional breeds of swine in the United States. They each have a place in the various crossbreeding schemes practiced by swine breeders. Some breeds excel in maternal traits, such as pigs born per litter or milking ability. These breeds are called ***maternal*** or dam breeds. ***Paternal*** or *sire breeds* are those that excel in growth rate, muscling, or leanness. Recently, swine breeders and companies that specialize in swine genetics have sought other breeds from around the world for either their maternal or paternal traits. In this section, we will discuss the eight major swine breeds, as well as other recently imported breeds from around the world.

Berkshires trace their ancestry to Berkshire, England, and were first imported to the United States in the 1820s. Berks, as they are sometimes called, can be distinguished by their black color, six white points (four feet, nose, and tail), and upright-ears. Berks used to be known for their short pug noses. However, since the early 1970s, breeders have concentrated on breeding pigs with longer noses. The longer nose facilitates the animal's ability to eat from automatic self-feeders. Berkshires can function as either a maternal or paternal breed because they are not dominant in either trait. For this reason, they may be used in crossbreeding systems where both maternal and paternal traits are important. Berks are known to have exceptional muscle quality in terms of color, texture, and flavor. There were 20,710 Berkshires registered in the United States in 1990. See Figure 8-2.

Chester Whites were developed in Chester County, Pennsylvania, from several other breeds of hogs in the early 1800s. Chester Whites are obviously white with rather small, droopy ears. This breed is known to be

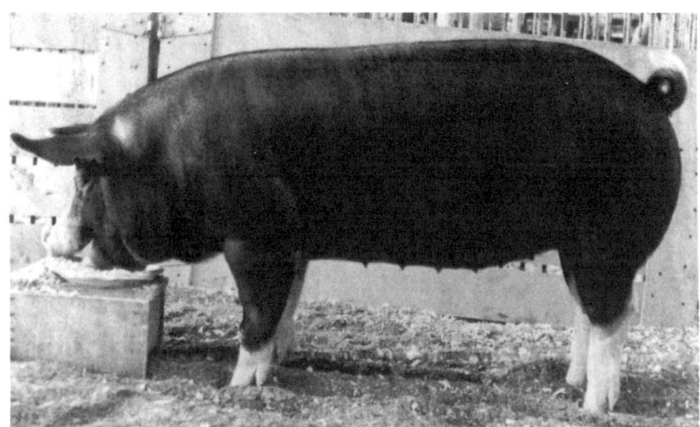

Figure 8-2. Berkshires are black with six white points and upright ears. (Courtesy, American Berkshire Association)

Figure 8-3. Chester Whites are a white maternal breed with small droopy ears. (Courtesy, Chester White Swine Record)

exceptional in mothering ability and would be classified as maternal. Registrations totalled 55,440 in 1990. See Figure 8-3.

Durocs originated from a breed of hogs called "Jersey Reds" from New Jersey. These were crossed over a period of years with another red breed from New York called Durocs. The resulting breed established in the 1860s was originally called "Duroc-Jerseys." They are now simply known as Durocs. Durocs can be a variety of red shades with small, drooping ears. Durocs are best known for their growth rate and carcass traits and would be classified as a paternal breed. They have been exported to other countries where breeders have spent many years selecting for extreme leanness and muscling. American breeders are now importing some of these extremely lean Durocs into the United States. Imported Durocs serve to further improve American Durocs as a paternal breed. Most of the imports are from Canada or Denmark and are called Canadian Durocs or Danish Durocs. Durocs are the most popular breed based on registrations which totalled 221,790 in 1990. See Figure 8-4.

Figure 8-4. Durocs are a red paternal breed with droopy ears. (Courtesy, United Duroc Swine Registry)

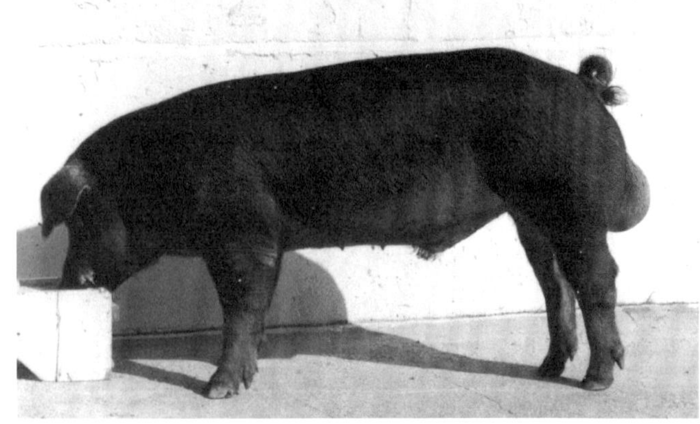

Hampshires are black with a white belt circling the shoulders. They originated in Boone County, Kentucky, where the breed association was formed in 1893. Hamps, as they are known, are often used as both a maternal and paternal breed. They are known for their muscling and leanness, but Hamp crossbred females are just as likely to be used as commercial sows. Like Durocs, Hamps were exported and are now being reimported, mainly from Sweden and Canada. Hampshires are the third most popular breed with 189,250 pigs registered in 1990. See Figure 8-5.

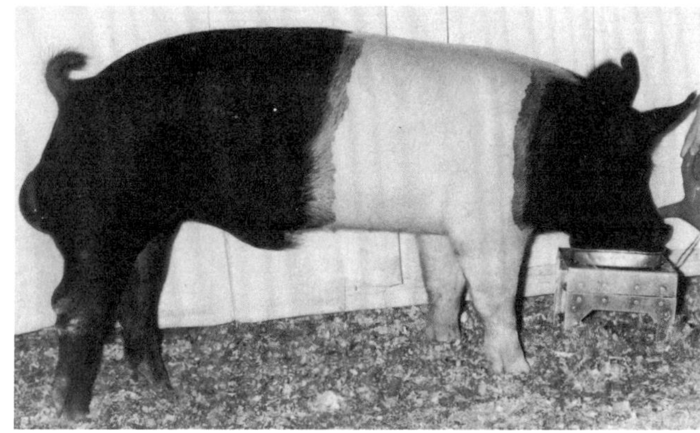

Figure 8-5. Hampshires are black, belted hogs that can function as both a paternal and maternal breed. (Courtesy, Hampshire Swine Registry)

The Landrace breed was first imported from Denmark in the 1930s. Since then, imports have arrived from both Norway and Sweden. These pigs are white and extremely long-bodied with large, drooping ears. They are used almost exclusively as a maternal breed because of their large litters and their exceptional milking ability. There were 43,650 Landrace registered in 1990. See Figure 8-6.

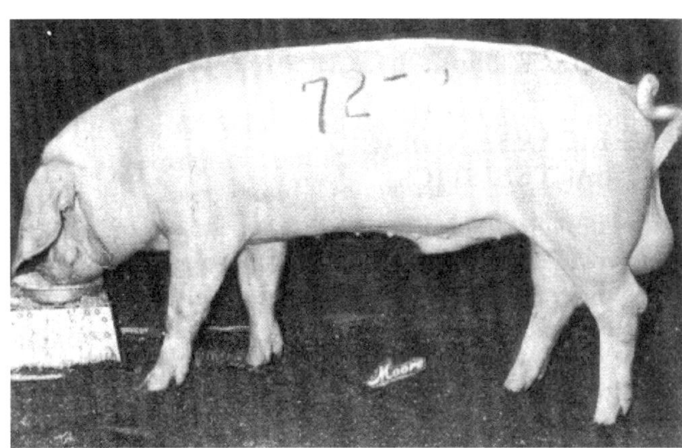

Figure 8-6. Landrace is a white maternal breed with large droopy ears. (Courtesy, American Landrace Association, Inc.)

Poland Chinas, or Polands, were first bred in Butler and Warren counties of Southwest Ohio, in the late 1800s. Their appearance is black with white points, much the same as Berkshires, except Polands have drooping ears. Polands are known for their growth rate, muscling, and hardiness. Therefore, their main use is as a paternal breed. A total of 18,480 Polands were registered in 1990. See Figure 8.7.

Figure 8-7. Poland Chinas are a black paternal breed with droopy ears and white on the nose, tail, and feet. (Courtesy, Poland China Record Association)

Spotted Swine, or Spots, were developed in the state of Indiana from predominantly the same base stock as Polands. The Spotted Swine Association was formed in 1914. As indicated by the name, Spotted Swine are spotted black and white and have drooping ears. Spots are also used as a paternal breed in crossbreeding systems. Registrations of Spotted Swine numbered 64,430 in 1990. See Figure 8-8.

Figure 8-8. Spotted Swine is a black and white paternal breed. (Courtesy, National Spotted Swine Record)

Yorkshires, or Yorks, are a maternal breed of great renown. White with upright ears, they play an important role in many crossbreeding systems because of their ability to farrow and wean large, heavy litters. Yorks originated in and around the county of Yorkshire, England, and were imported in the early 1800s. European Yorkshires are known as Large Whites. Large Whites are frequently imported in an attempt to improve American Yorkshires. Registrations barely trailed those of Durocs in 1990 with 206,000. See Figure 8-9.

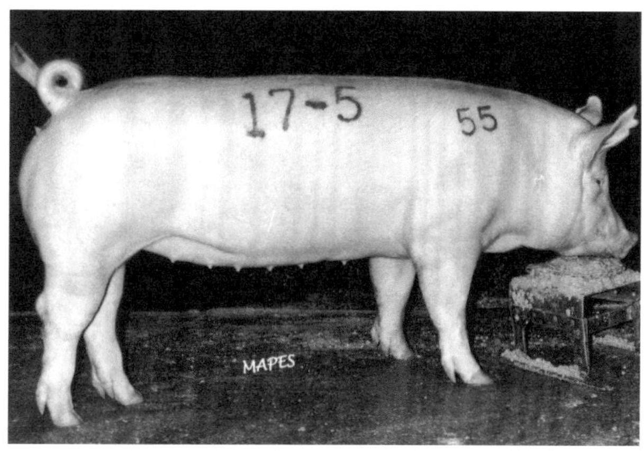

Figure 8-9. Yorkshires are a white maternal breed with upright ears. (Courtesy, American Yorkshire Club, Inc.)

Other breeds less commonly used by American swine breeders include the Tamworths (red with upright ears). Tamworths were originally developed in Ireland and imported in the early 1800s. Herefords (red with white face, legs, and underline) are a relatively minor breed, developed in the United States in the early 1900s There were 2,490 Tamworths and 400 Herefords registered in 1990.

A more recent addition to America's swine breeds is the Pietrain from Germany and Belgium. Most Pietrains are black and white spotted. They are extremely lean and heavily muscled. Many Pietrains carry two copies of the stress gene, which can adversely affect muscle quality. PSE (pale, soft, and exudative) muscle is common from Pietrain and Pietrain crossbred pigs. Their main use is to be crossed with other paternal breeds to produce extremely lean, heavily muscled, crossbred terminal sires with zero or one copy of the stress gene. There is no breed association for Pietrains in the United States. See Figure 8-10.

Chinese strains of swine were imported in the 1980s by universities to study their amazing maternal abilities. Chinese breeds routinely farrow 15–20 pigs per litter, but are extremely fat and light muscled. The goal

Figure 8-10. Extremely lean, heavily muscled Pietrain gilts. (Courtesy, White Oak Mills)

of researchers is to isolate the genes responsible for these large litters, and transfer them to leaner, heavier-muscled breeds. Chinese strains have not yet been released from research facilities to American producers. See Figure 8-11.

Figure 8-11. The amazing maternal abilities of Chinese pigs are being studied by university scientists. (Courtesy, Keith Bryan)

ANIMAL SCIENCE FACTS

Before the advent of the pickup truck or semi-trailer, hogs were driven on foot (sometimes hundreds of miles) from the farm to the major packing cities. Pork packing cities included Chicago and Cincinnati, which was known as "Porkopolis."

BREEDING SYSTEMS

There are many ways to combine the breeds of pigs in crossbreeding systems to maximize heterosis and use the relative maternal and paternal advantages of each. Discussed here are four crossbreeding systems.

The first crossbreeding system is called ***backcrossing*** or crisscrossing. See Figure 8-12. In a backcrossing system, only two breeds are used—usually one maternal and one paternal. At the beginning of this system, breed A is bred to breed B. Replacement gilts are kept and bred to another boar of breed A. Replacement gilts from the third generation are retained and bred back to a boar of breed B, and so on. This system does little to maximize heterosis after the first generation. This is because the dam will always contain some of the same genes as the boar to which she is bred.

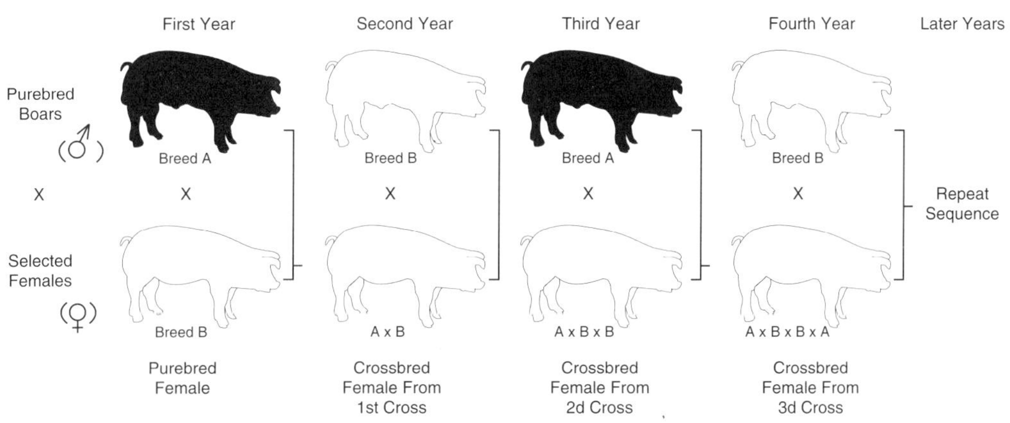

Figure 8-12. A backcrossing or crisscrossing breeding system involves just two breeds.

The second crossbreeding system is called a ***three-breed rotational cross.*** See Figure 8-13. Three-breed systems normally contain one maternal, one paternal, and one combination breed. This system is similar to backcrossing, except a third breed is introduced into the program. Breeds A and B are crossed; retained A × B gilts are mated to boars of breed C. Gilts from this generation are mated back to breed boars of A, and so on. The three-breed rotational cross maximizes heterosis better than the backcrossing system. However, after three generations, dams contain some of the same genes (although fewer than in the backcrossing system) as the boars to which they are bred.

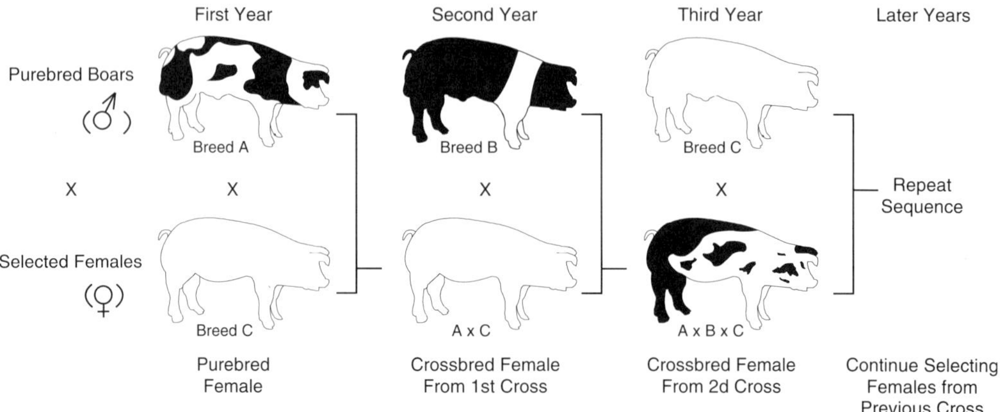

Figure 8-13. A three-breed rotational crossbreeding system adds a third breed to the criss-crossing system.

The third system is called a ***rotaterminal crossbreeding*** system. See Figure 8-14. Three maternal breeds are required for this system. These three maternal breeds are treated like a three-breed rotational cross. The three-way cross gilts from this rotational program are bred to paternal

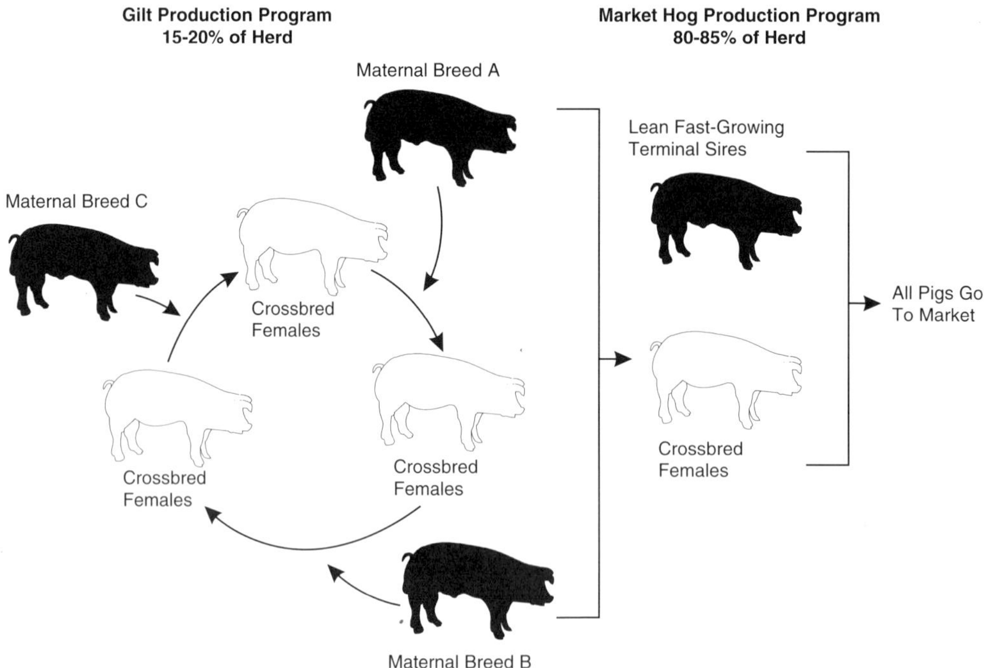

Figure 8-14. The rotaterminal crossbreeding system utilizes three maternal breeds and a terminal sire breed.

boars that are extremely lean and heavily muscled. These boars are referred to as "terminal sires" because no gilts from their litters are kept for replacements. The resulting pigs are all destined for slaughter. This system maximizes heterosis in the terminal pigs headed for slaughter, but not in the sow herd. A disadvantage of this system is that it requires the maintenance of at least four breeds of boars—a difficult task for small producers.

The fourth and most effective system for maximizing heterosis is a ***terminal crossbreeding*** program. In a terminal program, producers buy crossbred gilts of two or more maternal breeds from reputable producers or breeding stock companies. These gilts are bred to terminal sire boars, which may also be crossbred, but are extremely thick and heavily muscled. The resulting pigs are strictly fed for slaughter and no gilts are retained. See Figure 8-15.

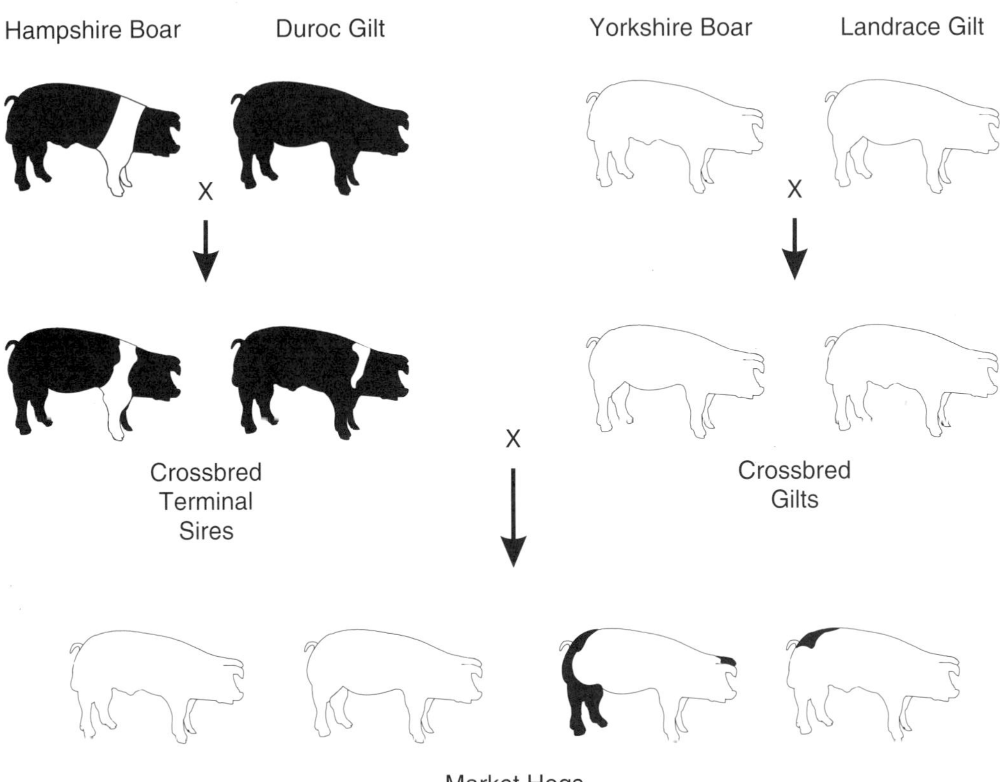

Figure 8-15. The terminal crossbreeding system maximizes heterosis by using a planned crossing system where no two breeds are used twice. All resulting pigs are destined for slaughter.

PIGLET AND NURSERY MANAGEMENT

Newly-born piglets require some special management procedures the first few days of life. These procedures are collectively called "baby pig processing."

Shortly after birth, umbilical cords should be cut about four inches from the navel. The remaining cord should then be dipped in or sprayed with iodine.

The next step to processing baby pigs is cutting needle teeth. Piglets are born with eight sharp teeth, four on the top jaw and four on the bottom. The tips of these teeth should be cut with small, side-cutting pliers when the pig is one day of age. If not removed, the needle teeth may cut the noses of other piglets during fights over teats. Piglets can also bite the sow's udder. This could lead to infection or the sow's refusal to nurse.

Piglets' tails are often docked to half their original length at the same time as the needle teeth are cut. Older pigs, if fed in confinement conditions will often bite other pigs' tails. Docking tails at a young age reduces the amount of tail biting that occurs later. Ears should also be notched for identification of individual pigs. The notch in the left ear indicates the pig's number within the litter. The notch in the right ear indicates the litter number. Traditionally, producers use number one for the first litter born after January 1. Subsequent litters are numbered two, three, etc. Large swine operations have adopted their own notching systems. See Figure 8-16. The notching system is more permanent than ear tags which may be lost.

Piglets should be given an appropriate dose of long-lasting antibiotic at one day of age to help ward off any infection. Sow's milk is low in iron; therefore, to prevent anemia, piglets should be given injectable iron at about seven days of age. Boars may be castrated at three to seven days of age to reduce the stress of this procedure. All injections should be given on the side of the neck about one inch behind the ear.

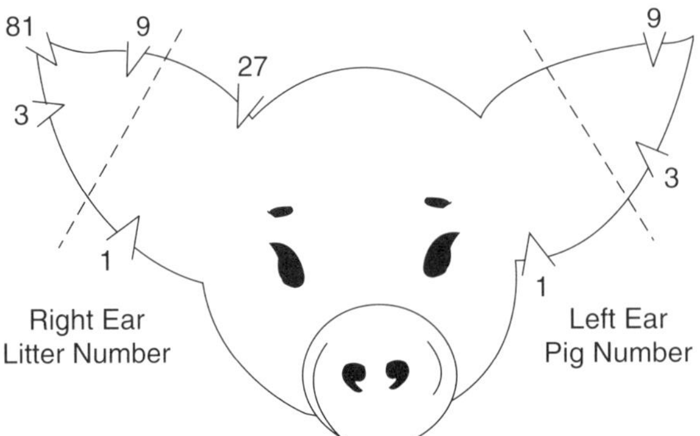

Figure 8-16. The Universal ear notching system identifies litters of pigs and individual pigs within a litter.

If a group of sows is farrowing about the same time, piglets from one sow, which has too many piglets and too few functioning nipples, can be ***cross-fostered*** to other sows with fewer piglets and more functioning nipples. Along the same line, smaller, weaker piglets can be cross-fostered onto heavier milking sows. This gives them a better chance to catch up with the average size of the group.

Baby pigs should be kept warm. The temperature requirement for newly born pigs is 90–95°F. Sows reduce feed intake when the temperature gets much above 70°F. Therefore, the room temperature should be set where the sow is comfortable, and supplemental heat via a heat lamp or mat should be provided. Supplemental heat can be gradually decreased to 75–80°F at three to four weeks of age.

In most large swine production facilities, piglets are weaned at 21 days of age or about 12 pounds. Ideally, they should be weaned by size, with big and small piglets (runts) kept separate. Small pigs are often transferred to a "nurse" sow so that they may continue to grow. Piglets should be weaned into a nursery room that has been washed and disinfected. They should stay in the nursery until they are 9 to 10 weeks of age, and then moved to a finishing facility. At this time, piglets should weigh about 50 pounds. Management, where pigs are kept in separate rooms by age, is called ***all-in all-out production.*** This type of system serves to keep disease problems to a minimum.

FEEDING PROGRAMS

Various sizes and ages of pigs require different feeds. For example, the nutrient requirements of mature, pregnant sows are drastically different from those of nursery pigs. Diets can be classified as follows: nursery, grower, developer, finisher, gestation, and lactation.

Nursery pigs, especially those weaned at 21 days of age, are perhaps the most challenging group to feed. This is because their digestive system is not fully developed and cannot fully digest feedstuffs that are fed to older pigs, such as corn and soybean meal. Young pigs are used to drinking sow's milk and often need a few days to adjust to eating feed of any kind. For this reason, nursery feeds usually contain some milk-based feedstuffs along with plant feeds that are easily digested. See Figure 8-17 for the formula of a basic nursery feed for 21 day old pigs. Nursery feeds should be high in protein and energy to meet the requirements of young, rapidly growing pigs. As pigs get older, their digestive systems become more mature

	Nursery Diet	Grower Diet	Developer Diet	Finisher Diet	Gestation Diet	Lactation Diet
Shelled Corn or Milo	820 lbs	1440 lbs	1540 lbs	1640 lbs	1100 lbs	1455 lbs
Soybean Meal (48%)*	500 lbs	500 lbs	400 lbs	300 lbs	200 lbs	375 lbs
Rolled Oat Groats	200 lbs					
Dried Skim Milk	200 lbs					
Dried Whey	200 lbs					
Dried Fat	20 lbs					100 lbs
Oats					620 lbs	
Vitamins, Minerals & Salt**	60 lbs	60 lbs	60 lbs	60 lbs	80 lbs	70 lbs
	2000 lbs	2000 lbs	2000 lbs	2000 lbs	2000 lbs	2000 lbs
% Protein	21.7	18.5	16.5	14.6	13.5	16.1

* Soybean meal can purchased in combination with salt, vitamins, and minerals. This mixture is called a protein supplement.

** Vitamins, minerals, and salt are sold as a vitamin/mineral premix.

Figure 8-17. Formulas for typical nursery, grower, developer, finisher, gestation and lactation diets.

and the expensive, milk-based feedstuffs gradually can be replaced by cheaper grains. Most nursery feeds contain some kind of antibiotic to help ward off diseases and infections. Nursery pigs convert feed to gain at about 1.5 pounds of feed per pound of gain. Feed intake will be somewhere around 1 pound per day.

Grower diets are normally fed to pigs weighing between 50 and 110 pounds. Although lower in protein than nursery diets, grower diets provide plenty of amino acids during a time when pigs are depositing large amounts of muscle protein.

Developer diets should be fed to pigs weighing between 110 and 180 pounds. Again, the protein percentage in the diet should decrease. When pigs weigh 180 pounds, protein requirements again drop. Pigs should be fed a finisher diet at this time. See Figure 8-17 for grower, developer, and finisher diets. Feed efficiency for pigs from grower to finisher is normally between about 3 pounds of feed per pound of gain. Feed intake will increase from about 3 pounds per day for a 50-pound-pig to about 7 to 8 pounds per day for a 250-pound pig.

Pigs may be fed differently according to sex. Since gilts inherently deposit more lean and less fat than barrows, their protein requirements are higher. Therefore, if gilts are penned separately from barrows, they can use higher protein grower and developer diets than their male counterparts. This practice is called ***split-sex feeding.***

The nutrient requirements of gestating sows are very low since there are few nutritional demands placed on these sows. The developing fetuses require few nutrients over those already required for sow maintenance. Gestating sows are normally fed 4 to 6 pounds of feed per day of a low energy, low protein diet. However, the diet must be higher in vitamins and minerals than normal since the sow's intake is low. See Figure 8-17 for a gestation diet that, if fed at 4 to 6 pounds per day, will maintain or slightly increase sow condition. Boars are normally fed in the same manner of gestating sows, 4 to 6 pounds of the gestation diet.

Lactating sows, on the other hand, have tremendous nutrient demands placed upon them by a large litter of nursing piglets. Lactating sows should be fed as much high protein, high energy feed as they will consume (usually 9 to 12 pounds). If lactating sows do not eat enough high quality feed to maintain body weight, they will be too thin to rebreed. See Figure 8-17 for a typical lactation diet.

PARASITES, DISEASES AND PREVENTION

Swine parasites are many, but those of major economic importance are limited to worms, mange, and lice. A heavy worm load in the digestive tract can cause reduced growth rate and organ damage. A variety of worm species affect different organs, but all are easily controlled by feed or water treated with a commercially available wormer. See Figure 8-18.

Mange mites burrow under the skin and are a common problem in all areas of the United States. Scratching and hair loss are common signs of mange infestation. Even minor mange infestation will reduce growth rate and worsen feed efficiency. Mange is difficult to eradicate since mites can stay alive in bedding for several days. Mange can be controlled by insecticidal sprays that kill adult mites, but do not damage unhatched eggs.

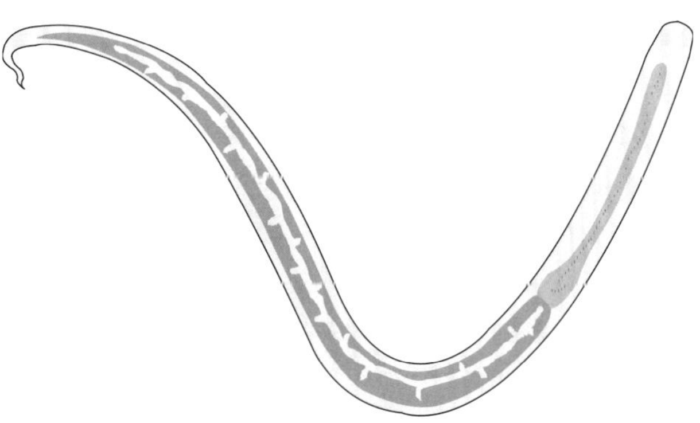

Figure 8-18. Roundworms are the most common internal parasite found in pigs. They are easily controlled through proper medication.

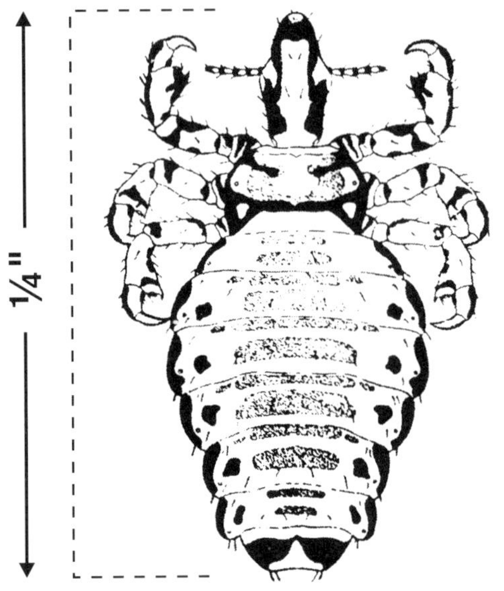

Figure 8-19. Lice cause severe itching in hogs. They can be easily seen on infected areas.

Lice are small insects that spend their entire life on the skin of their host, in this case, the pig. See Figure 8-19. Lice usually congregate around the ears and under the flanks where they cause severe itching. Like worms and mange, lice can reduce animal performance. Insecticidal sprays can also be used to control lice. Ivermectin is a commonly used drug that kills parasites of all types.

A parasite control program is essential to hog producers. Sows should be treated when moved to the farrowing room. This practice frees the sow of any parasites and reduces the chances of spreading them to the piglets. Piglets should be treated for parasites at about 50 pounds, and wormed again at about 130 pounds, if worms are a problem.

Many diseases affect swine to one degree or another throughout the United States. Entire books have been written on the subject. The more economically important diseases include atrophic rhinitis, erysipelas, leptospirosis, mycoplasmal pneumonia, parvovirus, PRRS, and pseudorabies.

Atrophic rhinitis is an infection of the nose causing the degeneration of the turbinate bones inside the snout. The result is a twisted snout, which makes eating difficult. The degeneration of the turbinate bones (which function to filter incoming air) also allows bacteria easier access to the lungs. Aside from twisted snouts, sneezing is a primary sign. Rhinitis can be controlled using all-in, all-out production in combination with vaccine administered to pregnant sows and baby pigs. See Figure 8-20.

Classic ***swine erysipelas*** is readily diagnosed by observation of raised diamond-shaped patches on the skin. Pigs may also have swollen ears, nose, and legs. Erysipelas is caused by a bacterium. Vaccination of gilts upon arrival and sows after weaning gives adequate protection.

Leptospirosis causes severe reproductive failure manifested by abortions or dead or weak pigs at birth. Leptospirosis is easily prevented by vacci-

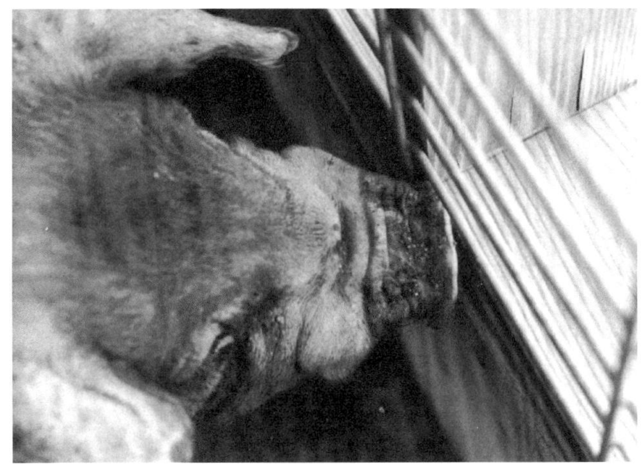

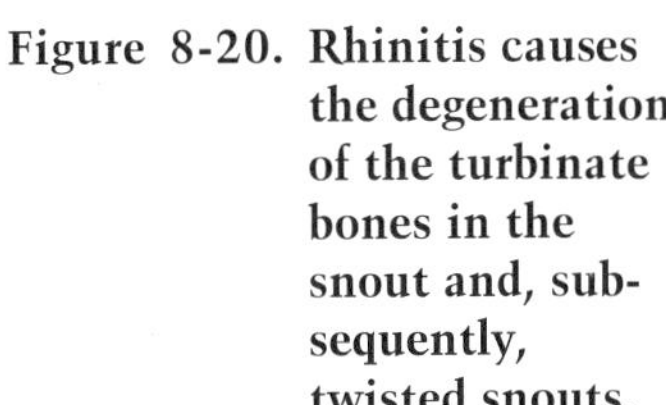

Figure 8-20. Rhinitis causes the degeneration of the turbinate bones in the snout and, subsequently, twisted snouts.

nation of sows and gilts before breeding. All sows should be routinely vaccinated for leptospirosis as an insurance measure.

Mycoplasmal pneumonia is a long-lasting disease that can exist with very few outward signs. Coughing in the morning is the most common sign that mycoplasmal pneumonia is present in the herd. Mycoplasma does cause reduced growth rate and poor feed conversion even though pigs do not look sick. Mycoplasmal pneumonia makes pigs more susceptible

ANIMAL SCIENCE FACTS

American Berkshire Association
P.O. Box 2436
West Lafayette, IN 47906
(317) 497-3618

Chester White Swine Record
Box 9758
Peoria, IL 61612
(309) 691-0151

United Duroc Swine Registry
P.O. Box 2339, 1769 U.S. 52 W.
West Lafayette, IN 47906-0339
(317) 497-4628

Hampshire Swine Registry
P.O. Box 2339, 1769 U.S. 52 W.
West Lafayette, IN 47906-0339
(317) 497-4628

American Landrace Association
P.O. Box 2340
West Lafayette, IN 47906
(317) 497-3718

Poland China Record Association
Box 9758
Peoria, IL 61612
(309) 691-6301

National Spotted Swine Record
1803 W. Detweiller Drive
Peoria, IL 61615
(309) 693-1804

American Yorkshire Club, Inc.
P.O. Box 2339, 1769 U.S. 52 W.
West Lafayette, IN 47906-0339
(317) 497-4628

to other, more harmful respiratory diseases. If mycoplasmal pneumonia is diagnosed as a problem, vaccination of feeder pigs, combined with good management can successfully combat it.

Parvovirus is characterized by failure of sows to come into heat, farrowing of small litters, and farrowing mummified pigs. Prevention is easily accomplished by vaccination of all sows and gilts before breeding. Sows should be vaccinated for parvovirus as insurance whether or not disease signs have been observed.

PRRS (porcine reproductive and respiratory syndrome) is a recently characterized viral disease that has two components affecting different stages of swine production. The first component is the reproductive component. Sows do not come into heat, have reduced conception rates, small litter sizes, and small, weak pigs. The second component is the respiratory component that makes the pigs' respiratory system extremely susceptible to other diseases. Normally, a herd will experience an outbreak of PRRS, then develop an immunity that lasts one to three years. Subsequent outbreaks are usually less severe. The PRRS virus can be isolated from most swine herds. A PRRS vaccine has been approved for baby pigs that shows promise in controlling the respiratory component.

Pseudorabies is caused by an easily spread virus and is usually most prevalent in areas where large numbers of hogs are raised. Symptoms include the death of large numbers of baby pigs under three weeks of age. Abortion is also a common symptom. Some infected herds, however, may show almost no signs at all. A pseudorabies vaccine is now available for use in areas that are at high-risk for an outbreak. Steps to control the spread of pseudorabies include 30-day isolation and blood testing of all new breeding stock. The practice of good ***biosecurity*** (keeping visitors, animals, birds, and rodents away from the herd) helps control the spread of all swine diseases.

HOUSING

Pigs can be raised successfully under a wide range of conditions. Many pigs in the temperate climates are raised outside. In the northern parts of the country, pigs must be raised with access to shelter during the coldest part of the year.

Shelter from inclement weather is very important if pigs are raised outside. They can tolerate cold weather to a point, but the amount of feed

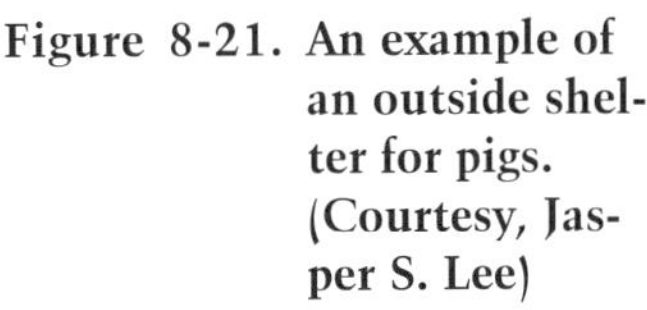

Figure 8-21. An example of an outside shelter for pigs. (Courtesy, Jasper S. Lee)

must be increased to maintain body temperature. The growth rate of pigs slows if the temperature drops below 65°F for growing pigs and 60°F for finishing pigs. Pigs must have a warm, dry, well-bedded place to lie down and keep warm during cold weather especially if no supplemental heat is available. See Figure 8-21. Sows can farrow in outside huts in cold weather if heat lamps are provided. Most pigs are farrowed in environmentally controlled buildings that are heated to a constant temperature.

Heat is harsher on pigs than cold. Pigs reduce feed intake when the temperature climbs over 80 or 85°F, thus reducing growth rate. This reduction in feed consumption is particularly hard on lactating sows, which often lose too much body condition to rebreed after weaning. Boars can become infertile if their body temperature gets too high. Pigs do not have sweat glands or any natural way to cool off, so a cooling mechanism must be provided. Indoor confinement buildings normally have a sprinkler system

Figure 8-22. Sprinkler systems are installed in environmentally controlled buildings to keep pigs cool during hot weather.

installed that drips water onto pigs when the temperature reaches 85°F. See Figure 8-22. Outdoor facilities should have shade and access to a wet place for pigs to lie down and keep cool during the hottest part of the day.

MARKETING

The sale of market hogs is the culmination of the swine production process. To receive top price, producers must provide what the buyers (packers and consumers) want. This means lean, high quality pigs of uniform size.

Leanness can be traced to genetics and feed. Muscle quality is also a function of genetics. Size is determined by producers. Packers want pigs to fall within a given weight range (for instance 240–270 pounds) so that all the hams, loins, and other cuts of meat are of similar size. Therefore, packers penalize producers for sending pigs outside the accepted weight range. This penalty is called a "sort penalty." Moreover, lean pigs within this weight range may receive a bonus and fat pigs may be penalized even though they are of the correct weight.

Producers use certain strategies to receive more bonuses and fewer discounts. The first is to weigh all pigs going to slaughter. It is very difficult to correctly guess the weights of pigs. An accurate scale removes the guesswork. The second strategy is to sell barrows at the lighter end of the weight range, since barrows get fatter at lighter weights than gilts. For example, if a barrow is sold at 240 pounds, it still may be lean enough to garner a premium. However, if the same barrow were allowed to reach 270 pounds, it would probably be penalized for being too fat. Gilts are more likely to stay lean at heavier weights and can be sold somewhat heavier than barrows.

Cash price on a given day is the price paid by packers for market hogs. The cash price is determined by supply of, and demand for, pigs on a given day. The cash price is a base price. An individual pig may be worth more or less than the base price depending on bonuses (for leanness) or discounts (for pigs outside the weight range or fat pigs). For example, a cash price may be quoted at 40 cents per pound. An individual pig weighing 250 pounds would be worth \$100 (.40 × 250). If the pig is leaner than average, it may receive a two cent per pound bonus. In that case, the pig would be worth \$105 (.42 × 250). If the pig was fatter than average, it may receive a three cent per pound discount. This fat version of a 250-pound pig would be worth only \$92.50 (.37 × 250). The lean

250-pound pig is worth $12.50 more than the fat 250-pound pig. It pays to produce lean pigs!

Futures prices are the expected cash prices at a certain point in the future. They are no guarantee of what the cash market will be on that date, but producers can use them as an insurance policy. If a producer has pigs to sell in three months, he or she can "lock in" a price for that time according to the futures market. This means that when three months pass, the producer will receive the price set three months ago, regardless of the cash price on the day of sale. Producers use futures as a way to reduce the risk of wild market fluctuations. Futures and current cash prices can be found in local farm papers, on the radio or television, or through various data services.

SUMMARY

Breeds of swine are many and varied, but all have their place in the crossbreeding schemes practiced by American pork producers. Most are known for either their maternal or paternal traits. The breeds are used in several basic crossbreeding systems: backcrossing, three-breed rotational cross, rotaterminal cross, and terminal cross.

Baby pigs should undergo some special management procedures shortly after birth. These practices are collectively known as baby pig processing. Nursery pigs require different feeds from grower and finisher pigs. All ages of pigs have different nutrient requirements and require different diets. Worms, mange, lice, and many diseases can infest and infect swine herds. Most can be prevented or controlled by good management, a solid vaccination program, and good biosecurity. Pigs require relief from extremes in temperature, but can be raised either indoors or outdoors. Receiving top price for the sale of market hogs is dependent on selling lean pigs at the correct market weight. Futures can be used by producers to reduce risk on market price.

Most registration numbers and addresses given in this chapter were courtesy of The American Livestock Breeds Conservancy via the following publication:

Bixby, D. E., Christman, C. J., Ehrman, C. J., & Sponenberg, D. P. (1994). *Taking stock: The North American livestock census.* Blacksburg, VA: The McDonald & Woodward Publishing Company.

CHAPTER SELF-CHECK

___ maternal breeds	1. piglets moved to other sows with fewer piglets
___ paternal breeds	2. expected price
___ backcrossing (crisscrossing)	3. causes twisted snouts
___ three breed rotational cross	4. uses only two breeds
___ rototerminal cross	5. causes diamond-shaped skin patches
___ terminal crossbreeding	6. excel in pigs born and milking ability
___ cross-fostered	7. isolation of herd
___ all-in all-out production	8. disease with reproductive and respiratory components
___ split-sex feeding	9. requires three maternal breeds
___ atrophic rhinitis	10. barrows and gilts fed separately
___ swine erysipelas	11. chronic disease with mild coughing
___ leptospirosis	12. excel in growth rate, muscling, and leanness
___ mycoplasmal pneumonia	13. uses one maternal, one paternal, and one combination breed
___ parvovirus	14. mummified piglets farrowed
___ PRRS	15. crossbred gilts used with heavily muscled boars
___ pseudorabies	16. actual price on given day
___ biosecurity	17. easily spread virus prevalent where large numbers of pigs are raised
___ cash price	18. causes severe reproductive failures
___ futures price	19. pigs kept in separate rooms by age

QUESTIONS AND PROBLEMS FOR DISCUSSION

1. Classify the eight major breeds of swine as being maternal, paternal, or combinations breeds.
2. Which three breeds of swine are predominantly black?

3. What extremely lean breed of pig has recently been introduced into the United States from Germany and Belgium?

4. Only one maternal and one paternal breed are used in the ______________ or ______________ system.

5. Why is the three-breed rotational cross superior to the backcrossing or crisscrossing system?

6. How many maternal breeds are required for the rototerminal crossbreeding system?

7. Which breeding system retains no gilts for replacements?

8. What are the six steps involved in processing baby pigs?

9. Differentiate between nursery, grower, developer, finisher, gestation, and lactation diets.

10. List two external swine parasites.

11. Why is it essential to provide pigs with a cooling mechanism if the temperature rises above 85°F?

12. Why do packers penalize producers with a sort penalty for pigs not weighing within a specified range (240–270 pounds)?

13. Do barrows or gilts remain leaner at heavier weights?

14. Packers often give bonuses for ______________.

15. What is the current cash price for a market hog?

ACTIVITIES

1. Have each student develop a herd health program for a fictitious swine herd.

2. Pretend to notch the ears of a litter of pigs using precut paper ears.

3. Divide class into groups representing the major breeds of pigs. Write to breed associations for information. Have groups prepare a one minute TV commercial promoting their breed. Videotape the presentations.

LABORATORY ACTIVITY

USING SWINE MANAGEMENT SKILLS

Purpose

To use swine management skills learned in this chapter

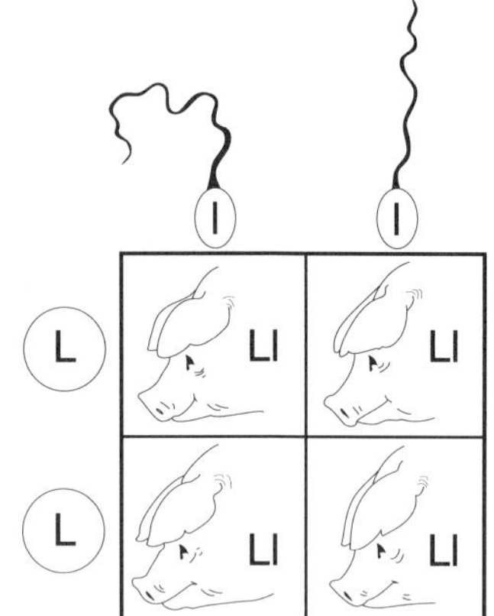

Materials

pencils
paper
text as reference

Procedure

1. a. Develop a rototerminal crossbreeding program.
   b. Select three maternal breeds along with terminal sire breeds for the above program.
   c. Diagram your breeding program on paper.
   d. Write an essay explaining why you selected the specified breeds in your program.
2. Calculate the profitability of raising a market pig from 40 to 240 pounds using the figures listed. Show all your work.

   Consult chapter for feed conversion averages.

   Cost of feeder pig—$1.00/pound

   Average cost of feed—$7.00/cwt (cwt means per 100 pounds)

   Overhead cost (use of buildings, etc.)—$10.00

   Market price at four figures—.30/lb, .35/lb, .40/lb, .45/lb

Application of Laboratory Activity

In order to be successful, swine producers must be able to develop efficient breeding programs and calculate the profitability of raising hogs. Good management skills are crucial in maximizing product quality and minimizing associated costs.

This laboratory activity helps to demonstrate the practical management skills producers use in developing a breeding program and calculating profitability. After carrying out this activity you should have an understanding of the kinds of decisions swine producers must regularly make in order to maximize profits and minimize costs.

Chapter 9

BEEF CATTLE MANAGEMENT

Home on the Range

INTRODUCTION

Beef cattle are different from hogs in their ability to convert forages, nutritionally of little use to humans, into valuable, high quality protein. Much of the land pastured by beef cattle is too steep, rocky or wet to be cultivated by conventional methods. Beef cattle make excellent use of this land. Beef cattle can be successfully wintered outdoors while consuming poorer quality forages than those required by dairy cattle. Extremely hardy creatures, beef cows can survive and produce a calf with a minimum of management, making them ideal for part-time producers. Beef cattle are indeed "at home on the range."

Figure 9-1.

OBJECTIVES

1. Describe the physical characteristics of the major beef breeds
2. Explain the relationship between the seasons and beef production
3. Develop feeding programs for beef cattle
4. Make a chart of common parasites and diseases that infest and infect beef cattle
5. Discuss the housing requirements for beef cattle
6. List three factors that affect the cost of feeder cattle

TERMS

backgrounding
baldies
commercial cattle
compensatory gain
continental breed
dressing percentage
dystocia
eared cattle
feeder calf
finishing cattle
intensive rotational grazing
lactating
marbled
performance testing
polled
progeny
replacement heifers
settle
shipping fever
terminal sire
weaning

ANIMAL SCIENCE FACTS

Beef facts

Average body temperature = 101.5°F

Average market weight = 1,100–1,350 pounds

Average 12th rib backfat = 0.6 inches

Average ribeye area = 10–14 square inches (depends on carcass weight)

Average birth weight = 75–100 lbs.

Average daily gain (full feed) = 2.5–3.5 lbs. per day

Average feed conversion = 6:1–7:1

BREEDS

The first American breed of beef cattle was the Texas Longhorn, introduced by the Spanish to the Southwest in the 1500s. Until the mid-1800s, the Longhorn was the only beef breed available to the majority of American cattle producers. Longhorns were hardy creatures—a necessity at that time—but lacked a beefy appearance and were difficult to fatten. See Figure 9-2. The importation of Shorthorn cattle, and subsequently Herefords and Angus, (all from the British Isles) to cross with the native Longhorns gave ***commercial cattle*** (crossbreds) heavier muscled carcasses. Other advantages of these three British breeds, including increased growth rate, improved mothering ability, and shorter horns, drove Longhorns to the brink of extinction. However, a recent increase in the popularity of Longhorns due to their lean carcasses and for use as roping stock has boosted registrations to 12,373 in 1990, from 500 in 1970.

For the better part of 100 years, Shorthorns, Herefords, and Angus ruled the American cattle industry.

Shorthorns were the first breed imported from Britain to the United States. This importation occurred in the late 1700s. The Shorthorn name derives from the breed's ability to shorten the Longhorn's lengthy horns. Shorthorns can be red, white, roan, or red and white spotted. Because they are earlier maturing and growthier than the Longhorns, Shorthorns can be given credit for stamping a beefy appearance on American cattle. Shorthorns are still relatively popular, with registrations being 8,002 in 1990. See Figure 9-3.

Figure 9-2. Longhorns were the first cattle of the American West. (Courtesy, Texas Longhorn Breeders Association)

Figure 9-3. This high capacity, correct Shorthorn cow probably looks much different than the original Shorthorns imported to the United States. (Courtesy, American Shorthorn Association)

During the early 1800s, Herefords became the second British breed to be imported to the United States. Herefords are red with white faces, bellies, and tail switches. White markings on the back of the neck are also common. Hereford crossbreds retain the characteristic white face and are often called ***"baldies."*** Hereford × Angus crosses are known as "black baldies" exhibiting black bodies and white faces. Herefords are extremely hardy animals able to tolerate a wide range of environmental conditions. In 1990, Herefords were the second most popular beef breed with 97,424 registrations recorded. See Figure 9-4.

In the early 1900s, Warren Gammon located ten registered Hereford cows and one registered Hereford bull that were naturally ***polled*** or horn-

Figure 9-4. Herefords have left their trademark white face on many of America's crossbred cattle. (Courtesy, American Hereford Association)

less. These eleven cattle were the foundation of today's Polled Hereford breed. Polled Herefords are used much the same as Herefords by American cattle producers, but the polled gene eliminates the task of dehorning calves. In 1990, 72,113 animals were registered as Polled Herefords. See Figure 9-5.

Figure 9-5. Modern Polled Herefords carry on the tradition of bringing a heavy calf to weaning. (Courtesy, American Polled Hereford Association)

Aberdeen Angus, or more simply Angus, were first imported from Scotland in the late 1800s. Angus are the most popular breed of beef cattle with registrations totaling 143,520 head in 1990. Angus are usually black in color and all Angus are naturally polled. Angus breeders have paved the way for industry-wide ***performance testing,*** which details growth rate data. See Figure 9-6. Some Angus possess a recessive gene for red coloration. When two Angus carrying a single copy of the red gene are mated, a 25 percent chance of producing a red calf exists. If two red Angus are mated, all of the calves will be red. In the 1950s, a separate breed association was formed for the registration of Red Angus cattle. Since then, Red Angus have been a distinct breed. In 1990, registrations of Red Angus totalled 15,353. See Figure 9-7.

Until the mid-1900s, these three breeds comprised most of the American cattle industry. There were two new breeds that arrived during the mid-1800s and early 1900s, which were not given much attention at the time.

Zebu or Brahman cattle were imported in the mid-1800s, and the Charolais in the 1930s. Neither was seriously considered as a threat to the three established breeds.

Brahman cattle are actually a different species of cattle (*Bos indicus* rather than *Bos taurus*) than the other beef breeds. Their native home is India from where they were first exported to the United States. This

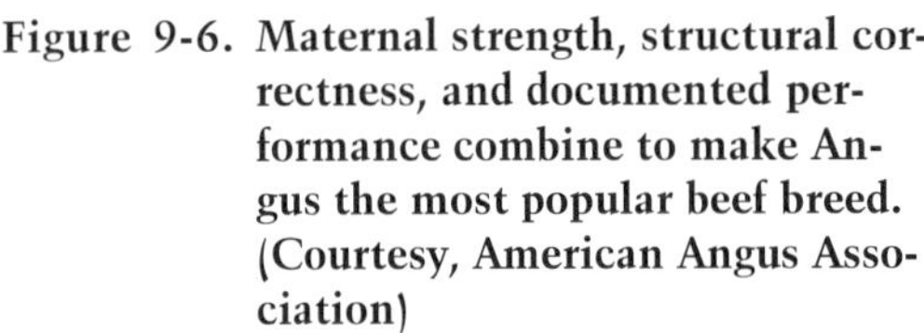

Figure 9-6. Maternal strength, structural correctness, and documented performance combine to make Angus the most popular beef breed. (Courtesy, American Angus Association)

Figure 9-7. Red Angus were developed by isolating the recessive red gene from Black Angus. (Courtesy, Red Angus Association)

Figure 9-8. Loose skin increases the surface area of Brahman cattle, allowing heat to escape. This feature makes Brahmans well suited for hot climates. (Courtesy, American Brahman Breeders Association)

exportation occurred in several waves from the mid-1800s to the early 1900s. Purebred Brahmans have loose hides and long ears, which continue to be evident in their crossbred offspring. For this reason, crossbred Brahman cattle are sometimes called ***"eared cattle."*** A hump over the shoulders and a relatively long face, as compared to other breeds, are also trademarks of Brahmans. Purebred Brahmans are usually gray, but all colors can be found. Brahmans are usually found in hot climates because their loose hides dissipate heat efficiently. Brahmans are also more tolerant to diseases than many other breeds of cattle. They have been used as a foundation breed for the development of several new breeds of cattle. Registrations of purebred Brahmans numbered 16,454 in 1990. See Figure 9-8.

Several distinctly American breeds of cattle rely on a high percentage of Brahman blood. Brangus ($^5/_8$ Angus, $^3/_8$ Brahman), Beefmaster ($^1/_2$ Brahman, $^1/_4$ Hereford, $^1/_4$ Shorthorn), and Santa Gertrudis ($^5/_8$ Shorthorn, $^3/_8$ Brahman) are very popular breeds in the South and Southwestern ranges where their hardiness and maternal abilities are well-known. Registrations for these breeds numbered 32,070, 36,458, and 6,050, respectively, in 1990.

Charolais (shar-lay) cattle originated in France as draft animals. Entirely white in color, Charolais are noted for their rapid growth rate and muscling. The first imports came from Mexico in the 1930s, but the breed association was not established in the United States until the early 1950s. The Charolais was the first ***continental breed*** (so-called because it came from the continent of Europe, not the British Isles) to be imported. Cattle breeders were amazed with the growth rate and muscling of Charolais crossbreds. However, the early imports tended to sire large calves that were born with difficulty. Many other continental breeds have had the

Figure 9-9. Heavily muscled Charolais bulls such as this one are used to sire thick, growthy calves. (Courtesy, American International Charolais Association)

same problem. American breeders have judiciously selected breeding stock for low birth weights, therefore, genetically solving most of the birth weight problems. Registrations of Charolais cattle numbered 44,725 in 1990. See Figure 9-9.

In the 1970s, imports of "new" continental breeds mushroomed. The availability of these new breeds was often limited to small numbers of bulls available only through artificial insemination. In order to rapidly increase registration numbers, breed associations allowed producers to "breed up" to purebred status. This involved breeding purebred bulls to crossbred (or purebreds of another breed) cows and registering the ***progeny*** or offspring as half-bloods. The half-blood progeny were then bred to another purebred bull, and the resulting three-fourths bloods were also registered. Many continental breed registries consider 7/8 blood females and 15/16 blood bulls to be purebreds. Depending on the breed of the original cows in the breeding up process, black and/or polled status could be maintained in a purebred animal from a continental breed that was originally red and horned. Three examples, Limousin, Maine-Anjou, and Simmental will be discussed later.

American cattle producers recognized the usefulness of crossbreeding and the advantages of heterosis. Newly imported breeds arrived in large numbers and breeders snapped them up, using them in crossbreeding programs. Presently, cattle producers have a vast array of genetics from which to choose including the "old" British breeds, as well as many newer breeds. A brief description of the more popular new continental breeds used in the United States follows. This list is not comprehensive.

Chianina (key-a-nee-na) are the largest breed of beef cattle with mature bulls standing six feet high and weighing up to two tons. Originating in Italy as a draft animal, purebred Chianinas are white to gray in color with black pigmented noses, tails and hooves. The white color is recessive to any other color in a Chianina crossbred animal. Chianinas were first introduced to the United States through semen imports in the early 1970s. They are noted for their growth rate and lean carcasses. Registrations of Chianina cattle totalled 8,022 in 1990. See Figure 9-10.

Gelbvieh (gelb-vee) cattle originated in southern Germany and were first introduced to the United States in the early 1970s. Gelbviehs are solid red in color. The American Gelbvieh Association requires individual performance data for birthweight and growth before animals can be registered. In 1990, 21,509 animals were registered. See Figure 9-11.

Figure 9-10. Large framed Chianinas are known for their lean carcasses and rate of gain. (Courtesy, Terry Atchison, Platte City, MO)

Figure 9-11. Functional, performance-oriented Gelbvieh bulls such as this one are extremely useful to commercial producers. (Courtesy, American Gelbvieh Association)

Limousins are a French continental breed first introduced to the United States through semen imports in the late 1960s and early 1970s. Originally red in color and horned, American breeders have developed black and polled lines. Limousins are moderately framed cattle known for their heavily muscled carcasses and rapid growth rates. They have proven to be an excellent choice for crossbreeding with large framed or light muscled cows. The popularity of Limousins is rising with registrations totaling 59,074 in 1990, up from 44,484, in 1985. See Figure 9-12.

Maine-Anjou (main an-ju) cattle were imported from France in the early 1970s. Known for their maternal abilities and docile temperament, Maines are the largest framed and heaviest of the French breeds. They were originally red and white, but as with Limousins, American breeders have developed black and polled lines. Maines are also popular in terminal crossbreeding programs and as show steer sires. Registrations of Maine-Anjou cattle reached 8,000 in 1990. See Figure 9-13.

Salers (sa-lair) are a third French breed imported shortly after Maines. Salers are dark red in color and may have white tail switches and belly spots. Salers are gaining a reputation as a performance oriented breed. In 1990, 21,464 Salers were registered. See Figure 9-14.

Simmental cattle were developed as a true multipurpose animal in their native Switzerland. Large and heavily muscled, they could work as draft animals, produce large amounts of milk, and have acceptable carcasses when slaughtered. The first Simmentals came to America in 1971, but semen was available several years prior. Simmentals were originally red or yellow and white with a prominent white face that passes to crossbred progeny. Black and polled lines are available. Simmentals can be successfully used as a maternal breed or as a ***terminal sire*** (father in a crossbreeding system where all progeny are marketed). The versatility of the Simmental explains its popularity with 90,516 registrations posted in 1990. See Figure 9-15.

BREEDING SYSTEMS

Breeding systems are not as well defined in beef cattle as in swine. Many commercial producers have essentially purebred herds. Others may use a loose rotational crossbreeding program. In the future, beef producers must identify specific crossbred cows that wean heavy calves, rebreed easily, and live a long time. These cows should be bred to heavily muscled, growthy bulls from early maturing bloodlines. Many producers may look to buy ***replacement heifers*** of specific crosses rather than raise their own.

BEEF PRODUCTION BY SEASONS

The beef production cycle is a year-long affair. The cycle begins at calving time. In most parts of the country, calving starts just before the grass begins to turn green in the spring (as early as January in southern states to March or April in northern states). However, some producers keep fall calving herds to take advantage of the higher priced ***feeder calf*** (young cattle just weaned) market brought on by a shortage of feeder calves in the spring.

Management at calving is crucial to reproductive success. A live calf is the only product beef producers have to sell. Therefore, they make a

Figure 9-12. Limousin cattle originated in France and are known for their exceptional muscling and carcass quality. (Courtesy, North American Limousin Foundation)

gure 9-13. A typical, high quality Maine-Anjou brood cow. (Courtesy, American Maine-Anjou Association)

Figure 9-14. Salers bulls are developing a reputation for performance. (Courtesy, American Salers Association)

Figure 9-15. Strong maternal characteristics and high milk production lead to heavy weaning weights typical of Simmentals. (Courtesy, American Simmental Association)

special effort to help ensure all calves are born live and get a good start. Calving should occur in clean pastures with enough room for the cow to get away from the herd. See Figure 9-16. Producers check cows frequently during the calving season to catch cows that may be having calving problems ***(dystocia).*** Heifers calving for the first time, small cows with big calves, and abnormal positions of the calf are the most common causes of dystocia. Experienced managers know when to assist the cow. Calves should be standing and nursing within one or two hours of birth. Disinfecting the naval with iodine, tagging or tattooing for identification, and castration are all management practices performed within twenty-four hours of calving. See Figure 9-17.

Figure 9-16. A cow will separate from the herd prior to calving, and return a few days after the birth.

National FFA Organization

Compliments of the National FFA Organization

Science and Evaluation

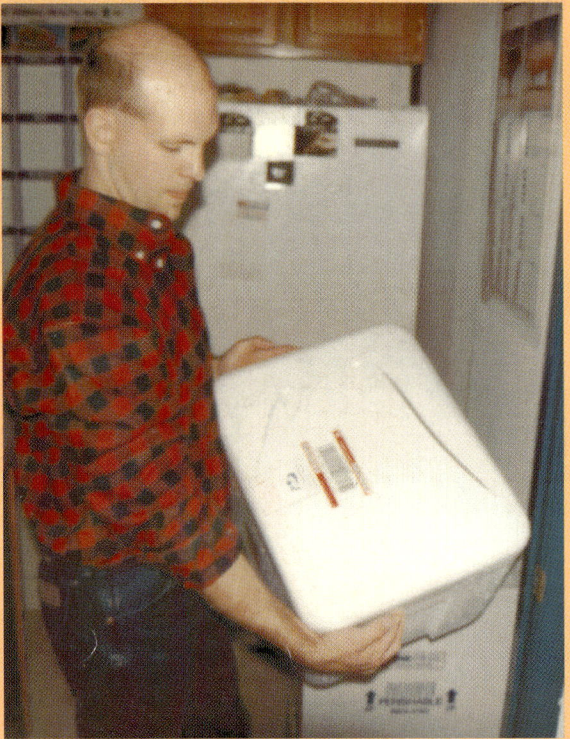

Transporting boar semen

Judging market barrow

Sow with litter in farrowing crate

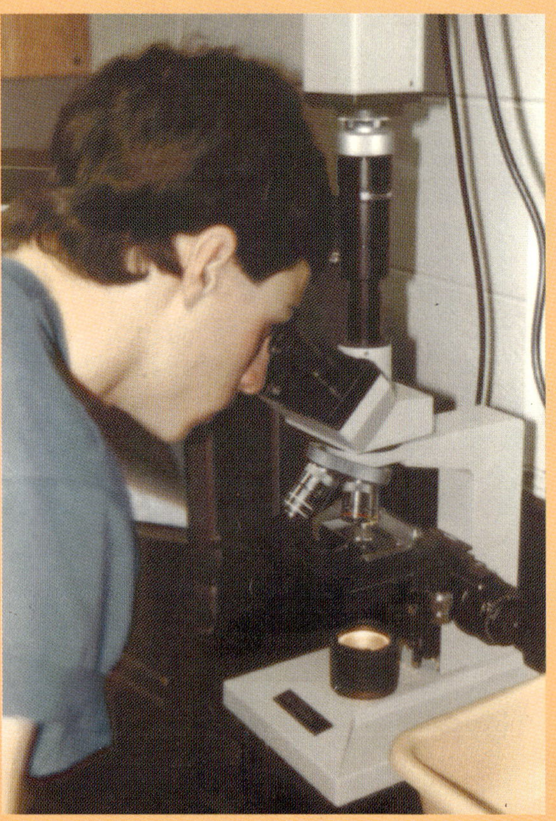

Evaluating morphology

Industry

Horse barn in Kentucky

A laboratory is often used in industry

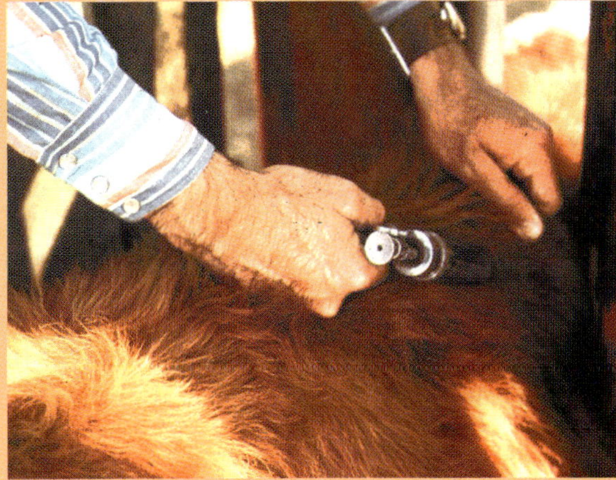

Implanting a beef animal with a growth hormone

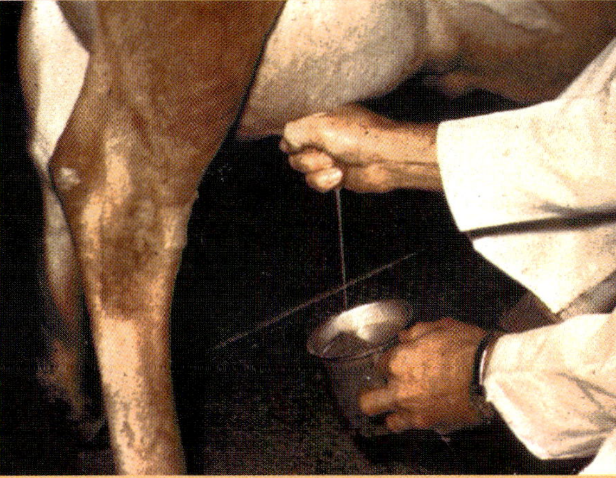

Obtaining a milk sample for testing

Filling a meat case in a supermarket

Side of ribs on a grill

Beef Retail Cuts

Beef

• RETAIL CUTS •
WHERE THEY COME FROM
HOW TO COOK THEM

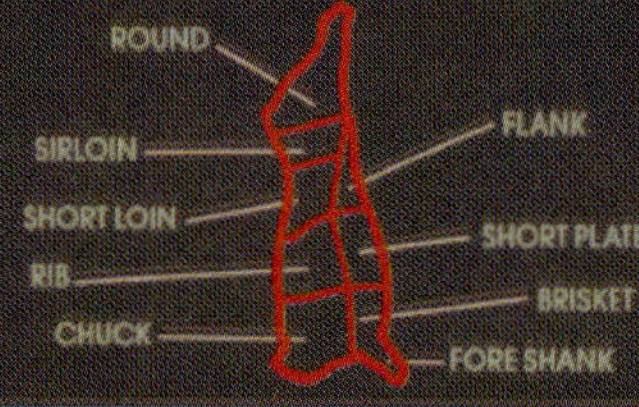

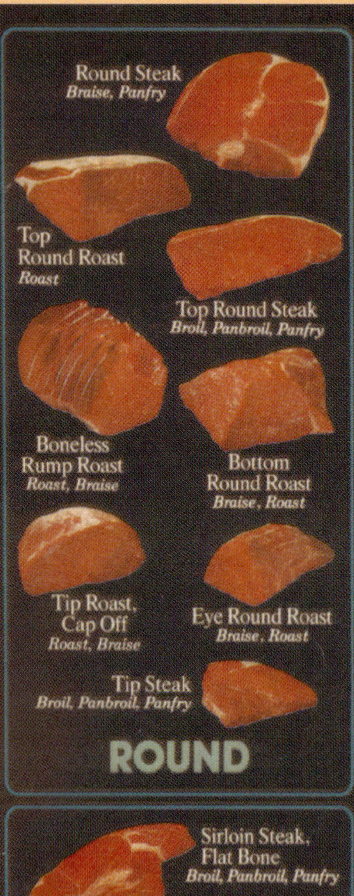

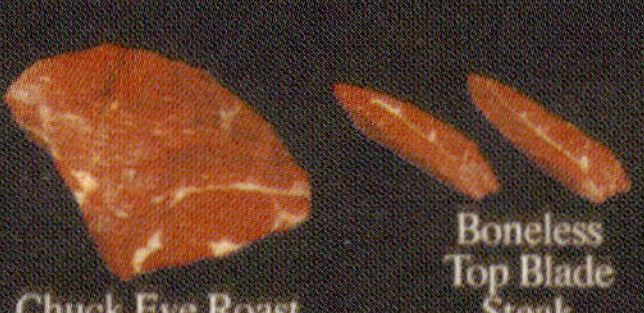

CHUCK

THIS CHART APPROVED BY
NATIONAL LIVE STOCK & MEAT BOARD

Pork Retail Cuts

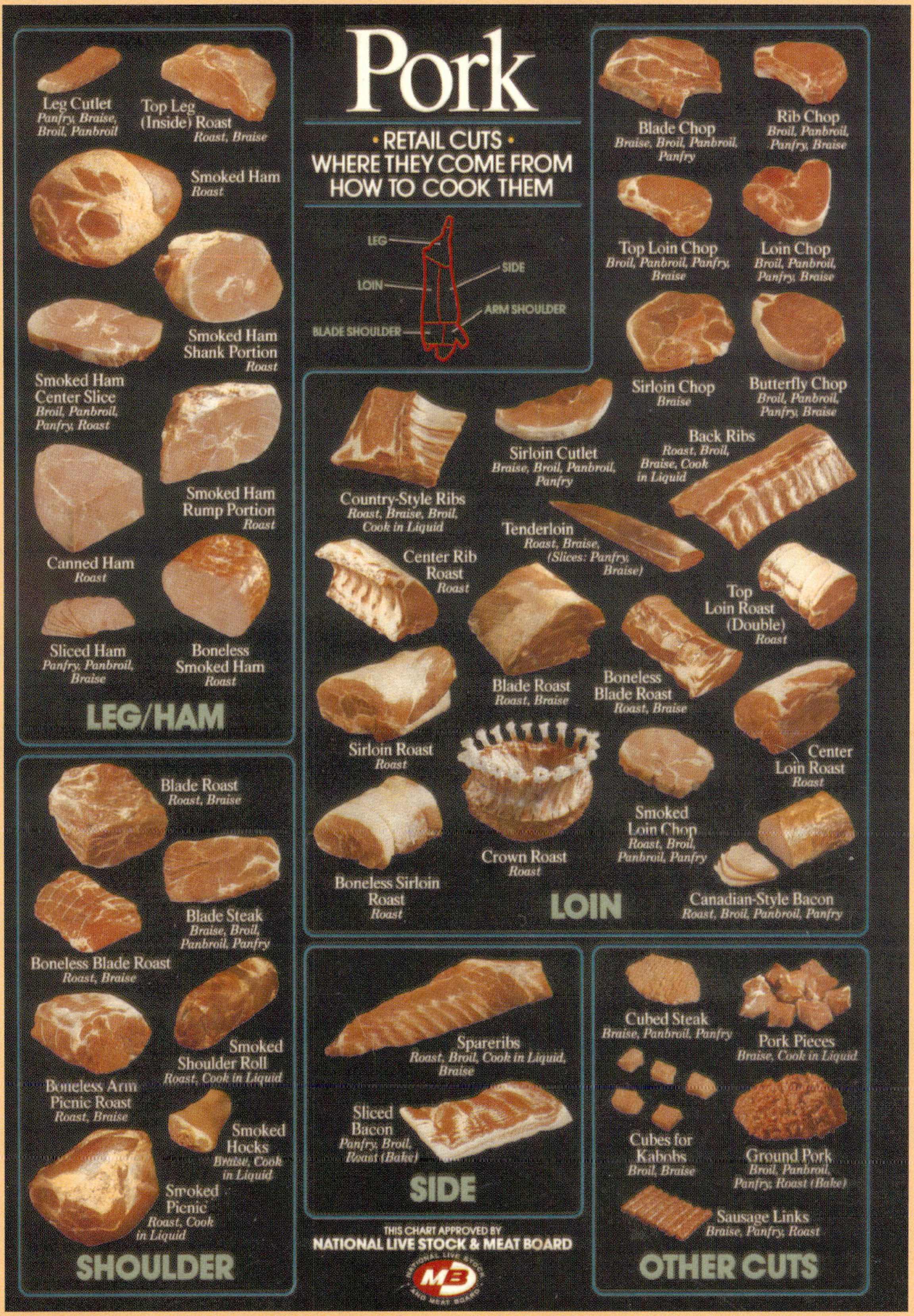

Pork

• RETAIL CUTS •
WHERE THEY COME FROM
HOW TO COOK THEM

LEG
LOIN
SIDE
ARM SHOULDER
BLADE SHOULDER

Blade Chop
Braise, Broil, Panbroil, Panfry

Rib Chop
Broil, Panbroil, Panfry, Braise

Top Loin Chop
Broil, Panbroil, Panfry, Braise

Loin Chop
Broil, Panbroil, Panfry, Braise

Sirloin Chop
Braise

Butterfly Chop
Broil, Panbroil, Panfry, Braise

Country-Style Ribs
Roast, Braise, Broil, Cook in Liquid

Sirloin Cutlet
Braise, Broil, Panbroil, Panfry

Back Ribs
Roast, Broil, Braise, Cook in Liquid

Tenderloin
Roast, Braise, (Slices: Panfry, Braise)

Center Rib Roast
Roast

Top Loin Roast (Double)
Roast

Blade Roast
Roast, Braise

Boneless Blade Roast
Roast, Braise

Sirloin Roast
Roast

Center Loin Roast
Roast

Smoked Loin Chop
Roast, Broil, Panbroil, Panfry

Crown Roast
Roast

Boneless Sirloin Roast
Roast

Canadian-Style Bacon
Roast, Broil, Panbroil, Panfry

LOIN

Boneless Blade Roast
Roast, Braise

Smoked Shoulder Roll
Roast, Cook in Liquid

Boneless Arm Picnic Roast
Roast, Braise

Smoked Hocks
Braise, Cook in Liquid

Smoked Picnic
Roast, Cook in Liquid

SHOULDER

THIS CHART APPROVED BY
NATIONAL LIVE STOCK & MEAT BOARD

MB
NATIONAL LIVE STOCK AND MEAT BOARD

Lamb Retail Cuts

Lamb

RETAIL CUTS
WHERE THEY COME FROM
HOW TO COOK THEM

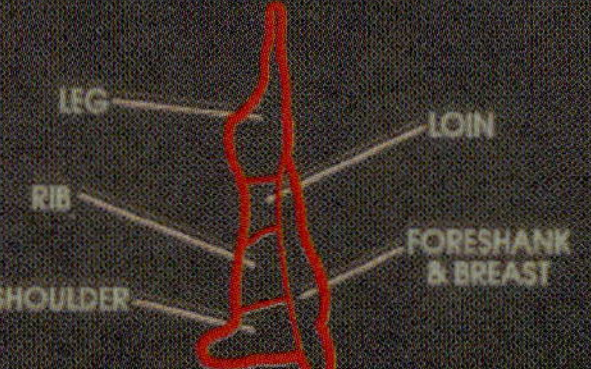

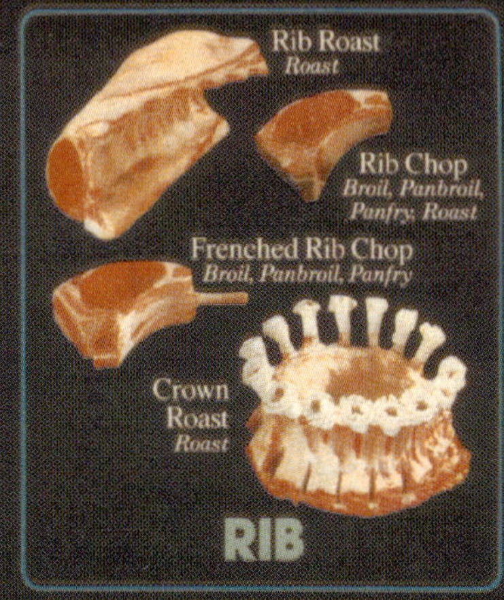

THIS CHART APPROVED BY
NATIONAL LIVE STOCK & MEAT BOARD

Veal Retail Cuts

Veal

• RETAIL CUTS •
WHERE THEY COME FROM
HOW TO COOK THEM

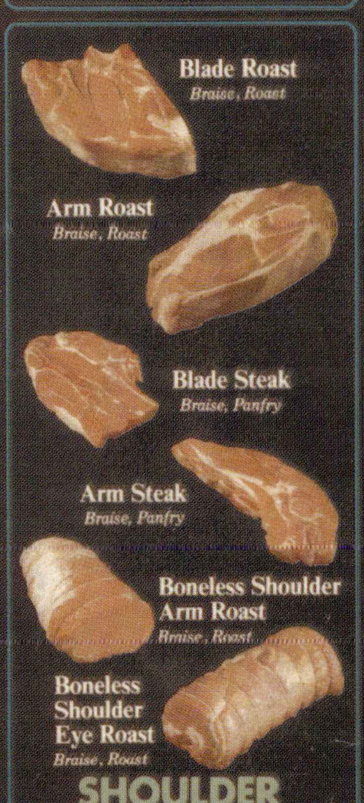

LEG (ROUND)
SIRLOIN
LOIN
RIB
SHOULDER
FORESHANK & BREAST

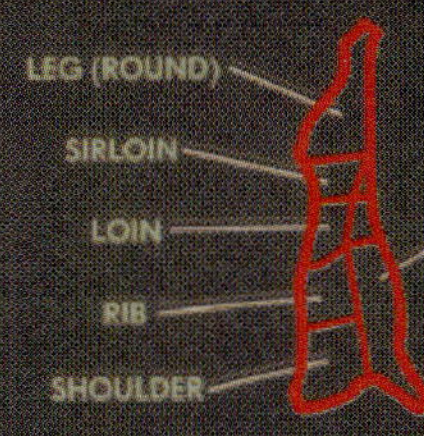

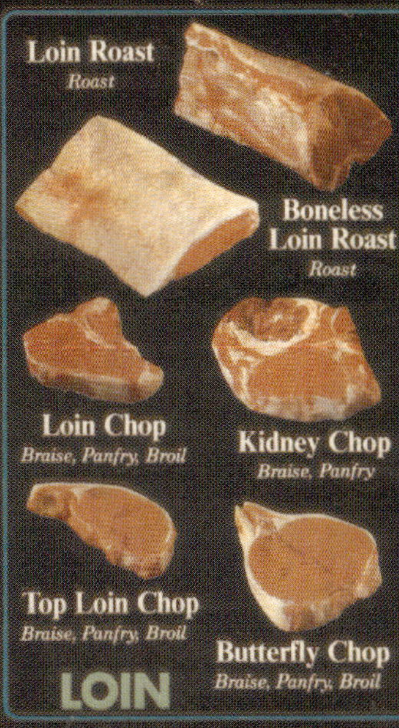

THIS CHART APPROVED BY
NATIONAL LIVE STOCK & MEAT BOARD

MB

Official USDA Marbling Photographs

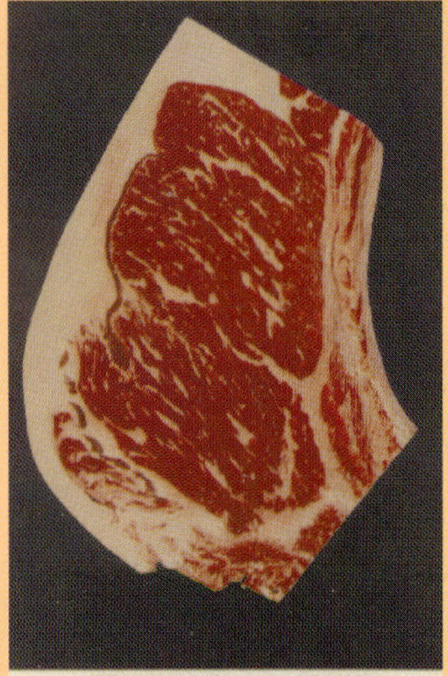
Moderately Abundant (MdA⁰)

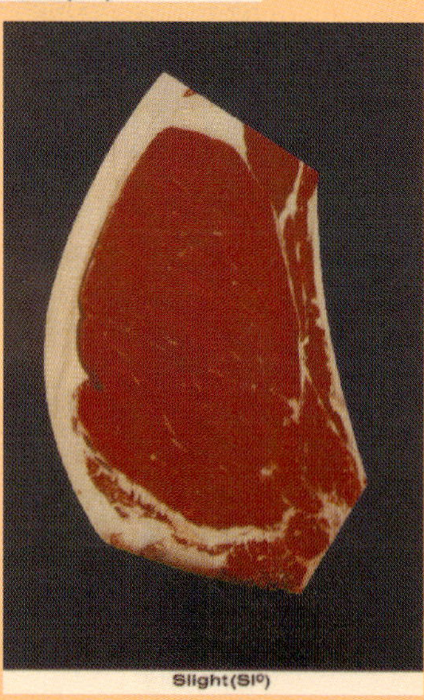

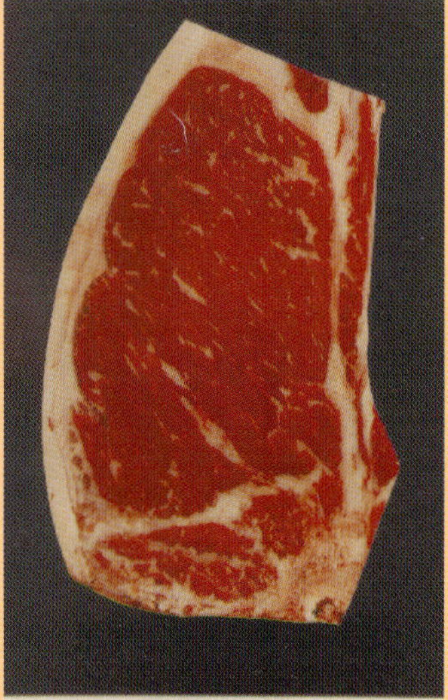
Moderate (Md⁰)

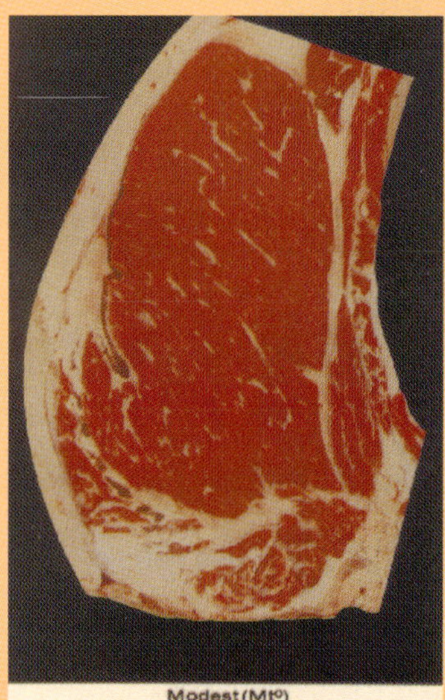
Modest (Mt⁰)

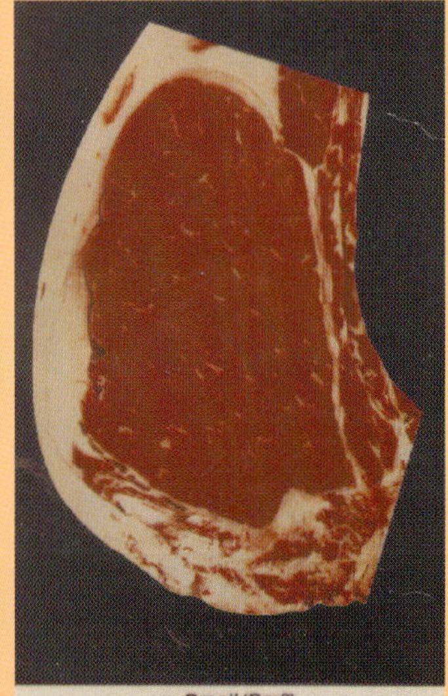
Small (Sm⁰)

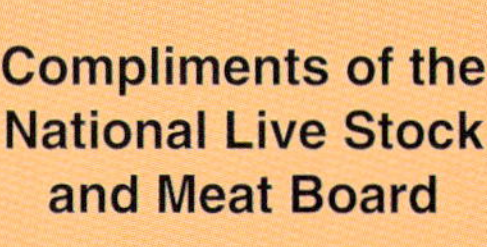
Compliments of the National Live Stock and Meat Board

Slightly Abundant (SlA⁰)

Slight (Sl⁰)

Lactating cows (those giving milk) need large amounts of high quality forage. Spring growth of most pasture and range grasses is high in energy and protein, which allows cows to give milk, build body condition, and subsequently be rebred for the next year. Since lactation demands are low when the calf is first born, and peak about two to three months into lactation, calving is often timed about a month before cows can be turned onto grass. At that time, the peak of lactation occurs when the highest levels of quality forage nutrients are available.

Figure 9-17. Ear tag number 214 identifies this calf.

The breeding season should begin 60–75 days after calving and should last about 60 days. This season normally occurs in early to midsummer depending on the part of the country. Heifers should be bred about 30 days before mature cows. There are two reasons for this practice. Heifers will deliver the first calves of the season. Therefore, they can receive more attention at calving time. Also, this schedule allows first calf heifers an extra month to rebreed. This is necessary since first calf heifers are often the most difficult cows to ***settle*** (become pregnant) because of the added nutritional demand of their own growth in addition to lactation. Gestation length for cows is approximately 280 days. Next year's calf should be born at roughly the same time as this year's calf [gestation (280 days) + prebreeding period (60 days) + average time before breeding (30 days) = 370 days].

It is very important to restrict the calving season by introducing and removing bulls in a timely manner. A calving season of 60 days can be accomplished by introducing the bull(s) and removing him (them) 60 days later. Pregnancy checks can be performed 50–60 days after removal of the bull (late summer to early autumn). Any cows not pregnant or open should be sold. A 60-day calving season helps keep groups of calves uniform in size, an economically important factor at marketing time. A short calving season also concentrates the period of intense labor.

Figure 9-18. Stored round bales can provide winter forage for beef cows.

Weaning (removal from the dam) normally takes place in autumn when grass production ceases. At this time, calves are about 205 days old. After weaning, pregnant cows are "wintered" on a diet of lower energy forage, such as poorer quality hay, until the month prior to the next calving season. See Figure 9-18. At that time, energy levels should be increased by feeding a higher energy ration so cows are in good flesh before calving.

Bull calves not castrated at birth are usually castrated at weaning. Also, all horned calves should be dehorned. Calves not retained for breeding stock are often implanted with a growth promotant.

Weaned calves take one of several possible routes. Some heifer calves may be retained for breeding the following summer. Steer or heifer calves may be placed in a feedlot where they are fed high energy diets in preparation for slaughter. See Figure 9-19. Calves may also be placed on a low energy, high forage, economical diet for their first winter, then placed in a feedlot the following spring. This practice is called ***"backgrounding."***

Figure 9-19. This feedlot in the high plains of Texas has a 30,000 head capacity.

FEEDING PROGRAMS

Beef cows depend heavily on forages for nutrition. More specifically, they depend on the nutrients (especially energy) contained within those forages. As discussed previously, the beef cow needs the most nutritious diet (good forages) when her demands are the highest (during late gestation and lactation). Dry, gestating mature cows require relatively little energy and protein.

Forages come in two types: pasture or range grasses and stored forages, such as hay or silages. During the growing season, pasture or range is the obvious choice for feeding. Pasture and range management varies tremendously in different parts of the country. For instance, in the East, some pastures are intensively managed to produce as much forage as possible. In an ***intensive rotational grazing*** system, cattle are allowed access to a paddock of grass for as little as a day or two. Under such a system, 1 to 1.5 acres can support a cow/calf pair for an entire year. In drier western states, however, the lack of moisture and alkaline soils will not support such intensive management. Many beef cattle are raised in areas that require up to 110 or more acres per cow/calf pair.

Stored forages are normally fed to dry, pregnant cows in the winter when range and pasture grasses are dormant. Hay, fed free-choice, is perhaps the most commonly stored forage. Depending on body size, dry, pregnant cows will consume approximately 25–30 pounds of hay per day. Pregnant heifers should be fed better quality hay than mature cows since they have not yet reached mature body weight.

Bulls can be fed stored forages when not breeding. They may require some grain before, during, and after the breeding season to maintain body condition. Bulls, cows, and bred heifers should have access to free-choice salt and minerals at all times.

During their first winter, heifer calves are often fed a grain ration in addition to stored forage. Five to seven pounds of a grain mix, similar to that shown in Figure 9-1, coupled with good quality, free-choice hay will ensure heifers are heavy enough (at least 750 pounds) for breeding at 11 to 12 months of age.

Finishing cattle (feedlot cattle) are normally fed a high grain diet containing very little forage. This results in rapid gains and increased carcass quality. See Figure 9-20. The amount of grain calves receive when they first enter a feedlot is relatively low only three pounds per day, with the rest of the diet made up of forage. The amount is gradually increased until 80–90 percent of the diet is grain mix. Young feedlot cattle require

Feedstuffs	Heifers	Low Energy	High Energy
Ground Ear Corn	95		
Shelled Corn		55	99
Oats		35	
Soybean Meal		7	
Salt, Vitamins, Minerals	5	3	1
Total	100	100	100
Amount to feed/hd/day	3-6 lb.	6-18 lb.	18-25 lb.
Amount of good quality hay	Free Choice	6 lb.	4 lb.

1. Total feed consumption should be 2.5-3.0% of live weight.
2. Salt, vitamins, and mineral recommendations are approximate. Consult your instructor or follow directions of a commercial premix.
3. The amount of grain should be increased gradually, starting at 2-3 pounds per day and increasing amount by 1 lb. every 2 days.
4. Heifer diet is based on 500-800 pound calves. Low energy diet is for 500-900 pound calves. High energy diet for finishing cattle 900 pounds and above. High energy diet can be fed to large framed lighter calves for the entire feeding period at lesser amounts and with higher levels of vitamins and minerals

Figure 9-20. Rations for heifer growing, and low and high energy diets for finishing cattle.

some supplemental protein, but as they reach finished weight, the supplements can be abolished. Monensin and Lasalocid are two feed additives that increase weight gain and improve feed efficiency in finishing cattle as well as developing heifers. Frame size affects how finishing cattle should be fed. Because high energy grains fatten cattle quickly, small framed cattle should be fed a low energy "grower" grain mix for a period of time. Ninety to 120 days of high energy grain at the end of the finishing period sufficiently fattens average and small framed cattle. Silages are often substituted for grain in low energy finishing diets. Large framed cattle require high energy grain mixes for the entire finishing period.

Growing heifers and finishing cattle should have salt, minerals, and ionophore (Monensin or Lasalocid) added to the grain mix or available free-choice.

PARASITES, DISEASES, AND PREVENTION

As with all livestock, cattle are subject to a wide variety of parasites and diseases. A discussion of the symptoms and prevention of the more common or severe parasites and diseases follows.

Flies are the most common parasites of beef cattle. Face flies irritate cattle by circling the eyes and feeding from the discharge. In addition, horn flies feed on blood and cause irritation by congregating on the belly, back, and sides of cattle. Irritation by either type of fly can reduce the time cattle spend feeding, and thus compromise gains or body condition. Both face and horn flies lay their eggs in fresh manure. They both can be controlled in part by face oilers, dust bags containing fly repellent, or ear tags impregnated with insecticide. Reduction of fresh manure piles eliminates fly breeding sites. Heel flies lay their eggs in the hair of cattle. After hatching, the larvae burrow through the hide and migrate to the back of the cow. The larvae, called cattle grubs, mature under the hide—usually during the winter. In spring, the larvae hatch and drop to the ground where they reside until molting into mature flies. The holes left by grubs cause a drastic reduction in the price of the hide after slaughter. Many compounds are available for controlling grubs, including ivermectin. However, it can only be used at certain stages in the grub's life cycle.

Anaplasmosis is caused by a parasite that infects the red blood cells of cattle. Most often seen in southern states, anaplasmosis causes lethargy and reduced body condition. Some animals show few signs of parasites, but act as reservoirs. Other animals may die. Cattle can be vaccinated against the symptoms of anaplasmosis, but identification and segregation of infected animals are better treatments. Infected animals can be treated with continual antibiotics to prevent the spread of anaplasmosis.

Coccidiosis is caused by tiny parasites that infest the lining of the intestine causing severe diarrhea. Cleanliness of surroundings, especially feeding areas, will greatly reduce the spread of *coccidia*. Several effective compounds are available for treatment.

Ringworm, a fungal infection of the skin, causes round, raised, bald

> **ANIMAL SCIENCE FACTS**
>
> Most steer shows require animals to have shipping fever vaccines as well as negative tests for tuberculosis.

areas. This infection is most common in winter when cattle are often confined in a dark environment. Treatment normally consists of the application of one of many effective solutions. Consult your veterinarian for specific formulas.

Mange mites infest the hides of cattle in much the same way as they infest hogs. Treatment consists of the use of one of several pour-on products or complete immersion. Ivermectin is effective in the treatment of cattle mange.

Stomach worms routinely infest beef cattle. Oral wormers, as well as ivermectin, are very effective on most species. Worm eggs reside on the bottom of grass plants. Overgrazing increases the chances of infestation. Pasture rotation will interrupt the life cycle of the worm.

Diseases affecting beef cattle will be divided into three categories: respiratory diseases, reproductive diseases, and gastrointestinal diseases.

Several respiratory diseases are lumped under a large umbrella called ***"shipping fever."*** Shipping fever consists of a complex of three separate diseases, hence the more accurate name "bovine respiratory disease complex (BRDC)."

The first segment of the complex is infectious bovine rhinotracheitis or IBR. IBR is caused by a herpes virus and results in reduced appetite, snotty noses, and difficulty in breathing. Treatment is difficult, but vaccination against IBR is fairly standard in the cattle industry.

The second segment of BRDC is BVD or bovine viral diarrhea. Symptoms are similar to IBR with the addition of high temperatures and severe diarrhea. Vaccination against BVD is a common practice.

The third segment of the respiratory complex is parainfluenza 3 or PI3. As the name suggests, parainfluenza is a flu-like viral disease often associated with IBR and BVD. Symptoms are interchangeable with those of IBR and BVD.

Vaccination for IBR, BVD, and PI3 can be accomplished with a single vaccine. Initial doses should be given two to four weeks apart, with annual boosters.

The first reproductive disease of national importance is brucellosis. Brucellosis is characterized by late-term abortions. Some states are nearly brucellosis free through vaccination programs, and testing and removal of infected animals. Brucellosis is a targeted disease for total elimination because it is infectious to humans. The human form of brucellosis is called "undulent fever."

Leptospirosis, called "lepto," is another easily prevented disease that can cause reproductive difficulties. Lepto can cause abortions at any time during pregnancy, and should be included in the vaccination program.

Trichomoniasis is technically a parasitic disease that causes early-term abortions and temporary sterility. Trichomoniasis is spread by infected bulls. Trichomoniasis can be controlled through the use of artificial insemination and uninfected bulls.

ANIMAL SCIENCE FACTS

American Angus Association
3201 Frederick Blvd.
St. Joseph, MO 64501
(816) 233-3101

American Brahman Breeders Association
1313 La Concha Lane
Houston, TX 77054
(713) 795-4444

American International Charolais Association
P.O. Box 20247
Kansas City, MO 64195
(816) 464-5977

American Chianina Association
P.O. Box 890
Platte City, MO 64079
(816) 431-2808

American Gelbvieh Association
10900 Dover Street
Westminster, CO 80021
(303) 465-2333

American Hereford Association
1501 Wyandotte
Kansas City, MO 64101
(816) 842-3757

North American Limousin Foundation
P.O. Box 4467
Englewood, CO 80155
(303) 220-1693

American Main-Anjou Association
760 Livestock Exchange Building
1600 Genesee Street, Suite 567
Kansas City, MO 64102
(816) 474-9555

American Polled Hereford Association
11020 NW Ambassador Drive
Kansas City, MO 64153-2034
(816) 891-8400

Red Angus Association of America
4201 N. Interstate 35
Denton, TX 76207-3496
(817) 387-3502

American Salers Association
5600 S. Quebec, #200
Englewood, CO 80111
(303) 770-9292

American Shorthorn Association
8288 Hascall Street
Omaha, NE 68124
(402) 393-7200

American Simmental Association
1 Simmental Way
Bozeman, MT 59715
(800) 548-0205

Texas Longhorn Breeders Association
P.O. Box 4430
Fort Worth, TX 76106
(817) 625-6241

Often, the first gastrointestinal disease to infect newborn calves is calfhood scours. Like BRDC, calfhood scours can be caused by any of several infectious agents. Bacterial causes include *E. coli*, *Salmonella*, and *Clostridium*. Viral agents that can cause scours include rotavirus and coronavirus. Scouring calves can dehydrate. This fluid loss can be fatal. Treatment with an electrolyte solution to replace lost fluids is generally recommended. Prevention starts with providing a clean place for cows to calve. Adequate colostrum within the first few hours of life is also critical. Vaccines are available, but should be used on an as-needed basis depending on the herd history of scours and veterinarian recommendations.

Johne's disease is one of the most difficult to control. The causing organism slowly thickens the intestine, preventing proper absorption of nutrients. Infected animals slowly lose condition while exhibiting a watery diarrhea. No treatment is available. Transmission occurs from the feces of infected animals. Although infection occurs shortly after birth, symptoms may not be present for several years. Prevention involves removing calves from their dams immediately after birth and raising them at a different site.

Tuberculosis (TB) is a wasting disease with symptoms similar to Johne's. Infection may also be present with no outward signs. Animals are infected through contaminated water. Herds should be periodically tested for TB with the removal of animals testing positively.

Other diseases of economic importance include pinkeye, an eye infection easily treated with antibiotics. Foot rot, a hoof ailment caused by wet, infected pastures, is also common. Treatment includes foot baths and topical treatments.

HOUSING

Housing requirements for beef cattle are not elaborate. Cows only need a sheltered place or windbreak. Many cattle are wintered in areas without a building. Thick trees or brush can provide adequate shelter. An ideal wintering location should be well-drained to keep mud to a minimum. In addition, newborn calves should have access to a portable shelter if born during the winter season.

Finishing cattle are fed in a variety of conditions. Small groups of cattle can be fed indoors on a manure pack. Many larger feedlots are entirely outdoors with only a windbreak for shelter and a high dry place on which cattle can lie.

MARKETING

Most cattle in the United States are marketed as calves or as finished cattle for slaughter. Furthermore, some cow/calf producers retain ownership of their calves and pay a feedlot operator to finish them.

Calves can be sold at the farm, feeder calf sales, or tele-auctions. Uniformity of size, good condition, and health are important. Feedlot operators want calves of similar frame size so they can be fed the same. They are then expected to finish at the same time. Condition is important because buyers want calves that are slightly thin. Thin calves that have not been fed much grain exhibit what is called ***compensatory gain.*** Compensatory gains are very profitable weight gains made at the start of feeding. Health is important for obvious reasons. Sick calves are not efficient.

Timing the marketing of finished cattle can make or break a finishing enterprise. Packers want cattle that are well ***marbled*** (intramuscular fat), but with little external fat. Genetics and feeding programs play the biggest roles in desirable fat condition. If cattle have too little fat, the price paid by buyers and packers is greatly reduced. If cattle are too fat, packers have to deal with the extra carcass fat. Extra fat costs the producer, since feed efficiency drops dramatically after cattle reach an ideal slaughter point.

Weight of finished cattle is also important. Live weights of finished cattle should fall between 1,100–1,350 pounds. Frame size (genetics) and feeding control finished weights. Heifers are sometimes discounted when sold on a live basis because the possibility of pregnancy could severely reduce the ***dressing percentage*** or carcass percent of live weight.

Cattle from small feed lots are often sold through local auctions. Larger western feedlots are visited directly by buyers. Bids are taken on a group of cattle. The cattle are sold to the highest bidder. Some feedlots sell cattle on a merit system, with high yielding, highly marbled cattle earning the most money.

As with hogs, cattle can be bought or sold on the cash market or through the futures market.

SUMMARY

The first beef breed imported to the United States was the Longhorn. Shorthorns, Herefords, and Angus, all from the British Isles, arrived in this country in the late 1700s and 1800s. These three breeds greatly enhanced the carcasses of commercial cattle. In the 1970s, several continental breeds were imported. These relatively new imports have become increasingly popular. Because of the availability of pasture and range grasses, beef producers often breed their cows to calve in the spring months. Because first calf heifers frequently take longer to re-breed, producers often schedule them to calve a month earlier than mature cows. A 60-day breeding season usually follows the calving period. Feeding programs differ according to sex, age, and purpose of cattle. Beef cattle require very little shelter. A windbreak to escape drafty conditions is all that is necessary. Commercial products can prevent and alleviate infestation of internal and external parasites. In addition, a herd health program can help ensure healthy cattle. Paying attention to local, cash, and futures markets, will assist producers in effectively marketing cattle.

Most registration numbers and addresses provided in this chapter were courtesy of The American Livestock Breeds Conservancy via the following publication:

Bixby, D. E., Christman, C. J. Ehrman, C. J., & Sponenberg, D. P. (1994) *Taking stock: The North American livestock census.* Blacksburg, VA: The McDonald & Woodward Publishing Company.

CHAPTER SELF-CHECK

___ terminal sire	1. offspring
___ commercial cattle	2. retained for herd
___ baldies	3. calving problems
___ polled	4. white face
___ performance testing	5. growth rate data
___ eared cattle	6. become pregnant
___ continental breed	7. moving from pasture to pasture over short periods
___ progeny	8. carcass percent of live weight
___ replacement heifers	9. profitable weight gain by thin cattle
___ feeder calf	10. bovine respiratory disease complex
___ dystocia	11. feeding cheap feed for first winter
___ lactating	12. Brahmans
___ settle	13. hornless
___ weaning	14. crossbred

___ backgrounding
___ intensive rotational grazing
___ finishing cattle
___ shipping fever
___ compensatory gain
___ marbled
___ dressing percentage

15. removal from the dam
16. giving milk
17. young cattle just weaned
18. intramuscular fat
19. feedlot animals fed high grain diets
20. from Europe
21. used in a crossbreeding system where all progeny are marketed

QUESTIONS AND PROBLEMS FOR DISCUSSION

1. Which breed of beef cattle was introduced to America by the Spanish in the 1500s?
2. Name the three beef breeds that originated in the British Isles and dominated the American beef cattle industry until the mid-1900s.
3. Crossbred ________________ (list breed) are referred to as eared cattle.
4. Importation of continental breeds mushroomed in this decade. List the decade. __________________________
5. Beef calves are usually born during these two seasons. List the seasons.

   ____
6. List three causes of dystocia.
7. The breeding season should last about _______ days.
8. Write an equation showing the days required in a 365-day beef production cycle.
9. Describe the winter diet of a mature pregnant cow.
10. A producer who uses a rotational grazing system can raise a cow/calf pair on as little as _______ acres.
11. List two parasitic flies that lay their eggs in manure.
12. True or False. Ringworm is a worm that habitually circles.
13. Name the three respiratory diseases that fall into the shipping fever category.
14. Describe the only housing requirement for beef cattle.
15. What is the current market price of finished steers?

ACTIVITIES

1. Have each class member select a beef breed for further research. Write to breed associations for information. Develop one-minute TV commercials that advertise each selected breed. Be creative. Videotape for later viewing.
2. Dressing percent in cattle averages 60 percent. Figure the dressed carcass weight of cattle weighing 1,000, 1,100, 1,200, and 1,300 lbs.
3. Develop a feeding program for a project heifer.
4. Make a chart of common beef parasites and diseases along with the respective symptoms.
5. Chart daily cash and future prices for cattle. Compare trends. Use *The Wall Street Journal* or an on-line computer network as a source.

LABORATORY ACTIVITY

FIGURING BEEF CATTLE FINANCES

Purpose

To compare profits/losses of cattle

Materials

pencils
paper
calculator (optional)
text as reference

Procedure

Use the listed data to answer the simulation questions in 1.

weight and prices received for three calves weaned on October 15

a. 450 lbs. (born June 1) @ .85/lb.
b. 550 lbs. (born April 1) @ .80/lb.
c. 650 lbs. (born Feb. 1) @ .75/lb.

Grass is available April 1. Extra feed cost for calving before April 1 is $10.00/month.

1. Answer the questions listed below.
   - A. In what month would calving be most profitable?
   - B. When should cows be bred in order to calve at the most profitable time?
   - C. If available, use local prices to do the same calculations.

Pretend you bought the feeder calves listed above and have finished them for slaughter. Use all available data to answer the questions listed below.

cost of feed = .07 lb.
slaughter weight = 1200 lbs.
selling price = .70 lb.
feed conversion = 7 lbs. of grain to 1 lb. of gain

daily gain of calves

a. 450 lbs. = 2.6
b. 550 lbs. = 2.8
c. 650 lbs. = 3.0
cost of facilities = .22/calf/day

2. Calculate total gain for each calf listed in 1. (Remember the slaughter weight is 1200 lbs.)

   a.
   b.
   c.

3. Calculate the feed cost for each calf.

   a.
   b.
   c.

4. Calculate the days on feed and facility cost for each calf.

	days on feed	facility cost
a.		
b.		
c.		

5. Calculate selling price for the calves.

   (Remember all weigh 1200 lbs. at slaughter.)

6. Add the total costs (initial purchase, feed, and facilities) for all three calves.

   a.

   b.

   c.

7. Which feeder calf would you want to finish?

Application of Laboratory Activity

Cattle producers need to estimate the profits and losses associated with their enterprise. In order to develop accurate estimates, they must have a firm grasp of the costs (initial purchase, feed, and facilities) as well as the price received at the time cattle are brought to market.

This laboratory activity gives you insight into the factors producers must take into account when determining profitability. Due to a number of factors, costs and market prices vary greatly across geographic areas. Because of this, a profitable enterprise in Oklahoma may be unprofitable in Pennsylvania.

Chapter 10

DAIRY CATTLE MANAGEMENT

From Moo to You

INTRODUCTION

Of all the species of livestock, lactating dairy cattle are perhaps the most difficult to manage properly. High-producing dairy cows are bred to give tremendous amounts of milk that, without pinpoint management, can overwhelm the physiology of the animal. Dairy producers constantly struggle with the complexities of breeding, feeding, and managing an animal pushed to the limit.

Figure 10-1.

OBJECTIVES

1. Categorize the physical and production characteristics of the six dairy breeds
2. Describe primary and secondary signs of estrus (heat) and identify the ideal time to artificially inseminate a cow
3. List the steps in the milking process
4. Develop a feeding program for a dairy calf
5. List two reasons feed intake is essential to a high producing dairy cow
6. Draw a lactation curve
7. List three practices that prevent mastitis
8. Discuss the housing requirements for dairy cattle
9. Identify three services offered through the Dairy Herd Improvement Association (DHIA)
10. Describe how most milk is marketed in the United States

TERMS

body condition
DHIA
displaced abomasum
dry cows
free-stall housing
grade
ketosis
mastitis
metritis
milk fever
placenta
somatic cell count
stanchion

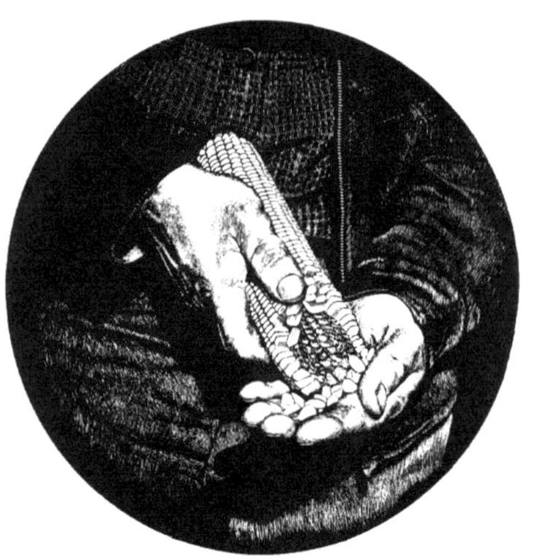

ANIMAL SCIENCE FACTS

What constitutes cows' milk? Mostly water. In fact, whole milk averages almost 88 percent water. Other components include:

Component	Percent
Fat	3.6%
Lactose (milk sugar)	4.7%
Protein	3.3%
Ash (calcium and other minerals)	0.8%

The above numbers may vary by breed. For example, Jersey milk averages over 5 percent fat. Lactose, protein, and ash are combined into a category called "solids-not-fat."

BREEDS

In the not-so-distant past, cattle were expected to provide meat, draft power, and milk for their caretakers. Over time, a few select breeds that excelled in milk production were established as "dairy breeds." As recently as the 1960s, significant populations of six different dairy breeds existed in the United States. However, since then, Holsteins have dominated the industry. Currently, Holsteins make up well over 90 percent of the dairy cattle in the United States. Holsteins are officially known as Holstein-Friesians and hail from the Netherlands and Northern Germany. Holsteins first arrived in the United States in the mid-1800s.

Since 1970, Holsteins have made tremendous genetic progress due to rigorous selection. Producers have chosen bulls proven to transmit genes for higher milk production. This selection process is one reason Holsteins produce significantly more milk per cow than the colored breeds.

Most Holsteins are black and white. However, like Angus cattle, this breed sometimes harbors a recessive red gene. Both horned and naturally polled Holsteins are included in the registry. Registrations of Holsteins in the United States totalled over 380,000 in 1990, easily making them the most popular dairy breed. See Figure 10-2.

Holsteins produce the most milk of all the dairy breeds. However, total solids percent (butterfat and protein percentages combined with minerals and lactose or milk sugar) are often lower than other breeds. Mature Holstein cows weigh 1,350 to 1,400 pounds.

On the other end of the size spectrum is the Jersey cow, weighing an average of just 1,000 pounds. Jerseys were developed on the island of

Figure 10-2. Holsteins are the most popular breed of dairy cow in the United States. Their black and white color pattern is unique to the breed. (Courtesy, Holstein Friesian Association of America)

Figure 10-3. Smallest of the dairy breeds, Jerseys are known for their high butterfat production. (Courtesy, American Jersey Cattle Club)

Jersey, located off the coast of France. Jerseys were first imported in the early 1800s. A Jersey's coat color ranges from light tan to almost black. Jersey registrations reached over 53,000 in 1990 ranking them a distant second in popularity. See Figure 10-3.

Jerseys' popularity stems from their ability to convert feed to milk efficiently. This efficient feed conversion is due to the lower body maintenance needs of the Jersey. Although the amount of milk produced by Jerseys is relatively low, the total solids' content is among the highest of all breeds.

The Guernsey was developed on the Island of Guernsey, which is also located off the coast of France. Guernseys were first imported to the United States in the early 1800s. Guernseys are a medium sized red and white breed (often appearing orange and white). Guernseys are larger than Jerseys. Mature cows weigh about 1,100 pounds. On average, Guernseys produce more milk than Jerseys, but "Golden Guernsey" milk is lower in total solids than Jersey milk. The deep yellow, or "golden" color of Guernsey milk is due to the presence of beta carotene. Guernseys ranked third in registrations in 1990 with 18,007. See Figure 10-4.

The fourth most popular dairy breed is the Brown Swiss (registrations totalled 12,473 in 1990). As the name implies, the breed originated in Switzerland. Brown Swiss first came to the United States in the mid-1800s. They are normally brown to grey in color and compare to Holsteins in size. Brown Swiss are known for their ability to produce milk in hot climates. Milk production is second only to Holsteins. Total solids content of the milk ranks in the middle of all breeds. See Figure 10-5.

Figure 10-4. Guernseys rank third in United States' popularity. They have a orange-red and white color pattern. (Courtesy, American Guernsey Cattle Club)

Figure 10-5. Brown Swiss are a large scaled breed. They are the fourth most popular dairy breed in the United States. (Courtesy, Brown Swiss Cattle Breeders Association)

Figure 10-6. Aryshires originated in Scotland. Their coat color is white and deep or cherry red. (Courtesy, Ayrshire Breeders Association)

Figure 10-7. Milking Shorthorns can be red, white, or roan. They originated from the same seedstock as the Shorthorn beef breed. (Courtesy, American Milking Shorthorn Society)

Ayrshires are a smaller breed in terms of registrations (9,539 in 1990), and size (mature weight averages about 1,200 pounds). Ayrshires are another red and white breed, but the reds are darker than those of the Guernsey. Ayrshires originated in the Ayr district of Scotland and were imported in the early 1800s. Milk production of Ayrshires ranks midway among dairy breeds. Total solids are relatively low. See Figure 10-6.

Another dairy breed in the United States is the Milking Shorthorn. Registrations totalled only 3,524 in 1990. Milking Shorthorns originated from the same base stock as beef Shorthorns and may be red, white, red and white, or roan. See Figure 10-7.

BREEDING

Unlike other species of livestock, crossbreeding is uncommon in dairy cattle. Crossbred dairy cattle appear to show negative heterosis, or, at best, no heterosis. In other words, there is no advantage in milk production or milk composition from crossbred dairy cows. Therefore, most dairy cows in the United States are purebreds or very high percentage ***grade*** (unregistered) animals.

Dairy producers were the first to adopt artificial insemination (AI) on a large scale. Most dairy calves are a result of AI, but some producers still keep a bull. Because of greater semen volume and multiple services during heat, a bull can help in settling problem breeders. However, dairy bulls are typically ill-tempered and dangerous. Since few bulls are needed by the dairy industry, most bull calves are sold for veal or as feeder calves at less than a week of age.

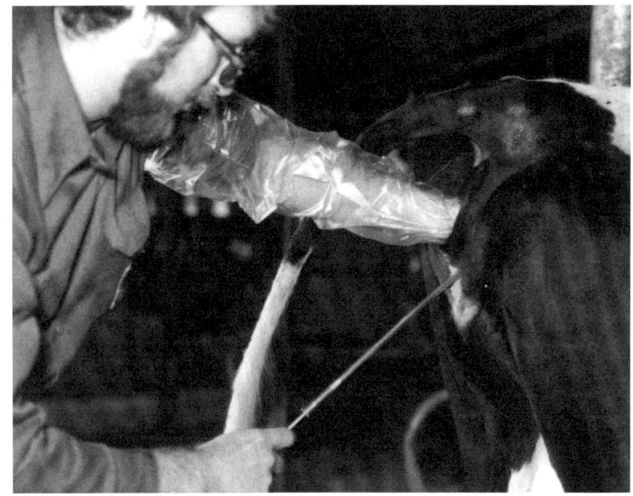

Figure 10-8. Artificial insemination should occur 12 hours after the first observation of standing heat. (Courtesy, American Breeders Service)

Dairy producers using artificial insemination release cows onto an exercise lot to watch for standing heat at least two times per day. Standing heat (standing still when another cow attempts to mount) is the primary sign that a cow is ready to conceive. However, several other signs may tip off an observant producer that a cow is near heat. Secondary signs of heat include nervous bawling, restlessness, attempts to mount other cows, clear mucous discharge from the vulva, and a sharp drop in milk production. Insemination should be timed 12 hours after the cow is first observed in heat. See Figure 10-8.

Figure 10-9. Inflations contained within the claw of a milking apparatus are placed over each teat. A vacuum draws the milk from the cow.

THE MILKING PROCESS

Cows are normally milked twice per day although some producers milk three times per day. Three times per day milking increases production by about 15 percent but requires more labor. At milking time, washing the cow's udder serves two purposes: disinfecting the udder and triggering the release of oxytocin, which initiates milk letdown. The udder is then dried with an individual disposable paper towel. One "inflation" of the "claw" of the milking machine is placed on each teat, or quarter (the udder is divided into four quarters). See Figure 10-9. A vacuum is applied to the inflation, which draws the milk from the udder. When milk flow stops, the claw is removed and each teat dipped in an iodine solution to prevent bacterial invasion of the udder. Total milking time is about seven minutes.

NUTRITION

Dairy calf nutrition starts at, or within, 24 hours of birth. This is necessary because calves are generally weaned immediately after receiving colostrum. Cows are then returned to the milking herd, and calves are raised entirely by humans.

Dairy calves must have colostrum as soon as possible after birth. Unlike humans who receive antibodies from the mother while in the uterus, the dairy calf is born without antibodies in the bloodstream to protect it from disease. The small intestine of the calf is very porous for the first 24 hours after birth and readily absorbs antibodies from the colostrum. Frozen colostrum is often substituted if the calf is removed before its first meal.

Six to eight pints of milk replacer should be fed to the calf daily for the first five to eight weeks of age. At one week of age, calves should have access to small amounts of high-quality grain calf starter. High-quality hay should be introduced at four weeks of age. Calves are usually weaned from the milk replacer when calf starter consumption reaches four pounds per day. The concentration of the milk replacer can be reduced for a week or two before it is replaced with clean water.

Calves are not born with a developed rumen capable of digesting forages. Newborn calves have a groove that transports milk from the esophagus, directly past the undeveloped rumen, and into the abomasum. The rumen slowly develops during the first 12 weeks of life. Calves fed hay containing saliva from mature cows will have access to essential bacteria to populate the developing rumen.

From 12 weeks to 1 year of age, heifers can be fed a grain mix containing a feed additive, such as monensin or lasalocid, and free choice high-quality hay or silage. The additives improve feed efficiency. These additives encourage the breakdown of methane gas bubbles in the rumen. Methane can then be absorbed by the rumen lining, allowing heifers to gather more energy from consumed feed. Heifers should gain about 1.5 to 1.8 pounds per day during this period (Jerseys and Guernseys a little less). Breeding size and puberty will be reached at 11 to 12 months of age.

After breeding, pregnant heifers should be fed free choice, high-quality forage. Several pounds of grain mix may be combined with the forage, to ensure proper development and provide needed trace minerals and vitamins. Heifers from large breeds should weigh 1,200 pounds at 24 months of age. This is approximately when they deliver their first calf. Smaller breeds should weigh 1,000 pounds. Care should be taken so that heifers do not become fat. Adipose tissue deposition in the udders of developing dairy heifers reduces lifetime milk production. See Figure 10-10 for examples of calf starter, and heifer diets.

Feeding lactating dairy cows is an art as well as a science. A brief guide for feeding follows. More information can be obtained from Cooperative Extension personnel or a publication called "Nutrient Requirements of Dairy Cattle" published by the National Research Council. Feeding programs have two segments: the needs of the cow and the nutrients provided by the feed.

The nutritional needs of the cow are dependent on body size (minor factor) and milk production (major factor). The amount of milk produced is dependent on the stage of lactation. See Figure 10-11. During early

Ingredient (lbs)	Calf Starter	Growing Heifers
Cracked, Shelled Corn	1040	
Ground Ear Corn		1818
Oats or Barley	400	
Soybean Meal	400	150
Molasses	100	
Dicalcium Phosphate	10	18
Ground Limestone	30	
Vitamin and Trace Mineral Salt	20	14
Total	2000	2000

1. Calves eating calf starter should be offered high quality alfalfa hay free choice.
2. Seven hundred pound growing heifers should be fed six pounds per head per day if good quality alfalfa hay is offered free choice. The grain mix will change depending on the forages used.

Figure 10-10. Common calf starter and heifer diets.

lactation (first 60–90 days after calving), milk production increases. Feed intake during early lactation may double from 25 to over 50 pounds. Rebreeding also takes place during early lactation. Peak lactation occurs about two to three months after calving. High milk production increases the nutrient needs to a point where the cow cannot physically consume enough feed to meet her demands. Therefore, cows often lose body weight during peak lactation. This loss amounts to 200 to 250 pounds from calving to peak lactation. Peak lactation for Holsteins averages about 120 pounds or 15 gallons of milk per day. Later lactation is a period of rebuilding ***body condition,*** or muscle and fat cover. Cows are "dried off" (milking is stopped) about 60 days before the next expected calving. Non-lactating cows are called ***dry cows.***

Nutrients provided by feed depend on the quality and amount of feed intake. Large quantities of milk require large amounts of high-quality feed. As a rule of thumb, lactating dairy cows can be expected to eat 3.5 percent of their body weight on a dry matter basis (see Animal Science Fact on page 177). Three and one-half percent of a 1,400 pound cow equals about 49 pounds of feed on a dry matter basis. High levels of feed intake are essential for lactating dairy cattle for two reasons. First, a pound of even the highest quality feed can only contain so many nutrients. More pounds of feed consumed means more nutrients available for milk production. Second, high feed intake is important for the maintenance of body condition. The metabolism of high producing dairy cows ranks milk production

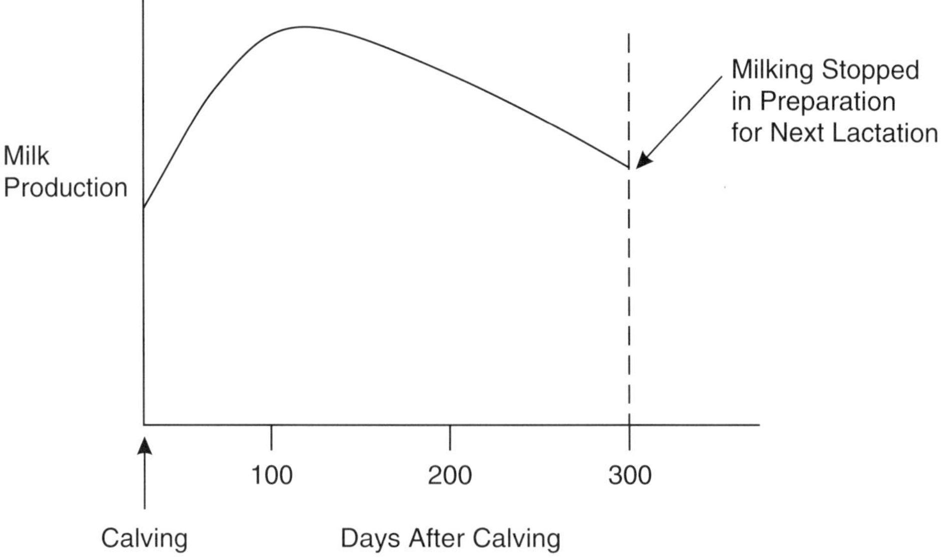

Figure 10-11. A lactation curve.

over maintenance of body condition. Good body condition is critical for successful rebreeding. If nutrients are limited by low intake, cows will still give large quantities of milk, but body condition and reproductive success will suffer. High feed intake ensures enough nutrients for milk and maintenance of body condition. Techniques used to increase feed intake include multiple feedings per day and feeding forages before grain.

Feeds used in lactating cow diets include forages. Forages may be dry hay, silage (fermented forages), or a mixture of hays and silages. Grain mixes may include any number of ingredients from the standard corn and oats to ingredients such as beet pulp (byproduct of the manufacture of beet sugar) and cottonseed. Grain mix ingredients are used based on the specific nutrients needed and the price of available feedstuffs.

Forage quality influences the cow's ability to produce maximum milk at peak lactation. The amount of nutrients per pound of feed is critical. In addition, cows must have at least 50 percent forage in the diet to maintain normal rumen function. During peak milk production, the amount of forage in the diet can be reduced to 40 percent for several weeks; however, a continued diet at that level will lead to ulcers in the rumen lining. Obviously, the amount of nutrients contained in forages is very important to total nutrient intake. In addition, nutrients (particularly protein and energy) accounted for in the forage portion of the diet are generally less expensive than nutrients supplemented in the grain portion of the diet.

Grain mixes supplement forages based on the needs and milk production of each cow. For instance, a dairy producer may have access to a silo full of high-quality alfalfa haylage. A cow producing 40 pounds of milk

ANIMAL SCIENCE FACTS

Dairy Facts

Heritability of milk production = 25 percent

Approximate average pounds of milk produced per cow per year considering all breeds = 12,000 pounds

Average weight of a mature Holstein cow = 1,350 to 1,400 pounds

Average weight of a mature Jersey cow = 1,000 pounds

Estrus cycle = 21 to 22 days

Gestation period = 283 days

Ideal calving interval = one year

per day in late lactation may only need 12 pounds of grain to supplement the haylage and meet her nutrient requirements. Another cow producing 110 pounds of milk per day at peak lactation may require 35 pounds of grain mix to supplement the same haylage. Part of the 35 pounds of grain mix may be a separate one fed only to high producing cows.

Lactating cows are fed in groups according to the amount of milk produced. High-producing groups of cows receive the most grain. This high producing group consists of cows in early and peak lactation. On the other hand, low producing groups (usually cows in later lactation) receive less grain. Computerized feeding systems on some farms offer individual cows grain, depending on their specific milk production.

Vitamin and mineral supplementation for lactating dairy cows also varies depending on the forages and grains used. For example, cows fed a diet high in corn silage require more supplemental calcium than cows fed a diet containing large amounts of young alfalfa hay (young alfalfa hay contains much higher levels of calcium than corn silage). Vitamins, minerals, and salt are normally added to the grain mix.

Other substances are often added to the grain mix of high producing dairy cows. For example, buffers, such as sodium bicarbonate, are added to the diet of cows eating finely chopped corn silage or haylage. Buffers help keep the rumen at the ideal pH of 7.0, which helps maintain a healthy bacterial population. Bypass, or rumen protected fat (chemically coated fat particles that resist rumen degradation) is another feed additive. Cows in peak lactation are normally deficient in energy. Fat is a very high energy feed ingredient, but the rumen is not equipped to process high

ANIMAL SCIENCE FACTS

Ayrshire Breeders Association
P.O. Box 1608
Brattleboro, VT 05302-1608
(802) 254-7460

Brown Swiss Association, USA
P.O. Box 1038
Beloit, WI 53511-1038
(608) 365-4474

American Guernsey Cattle Club
P.O. Box 666
Reynoldsburg, OH 43068-0666
(614) 864-2409

Holstein Association
1 Holstein Pl.
Brattleboro, VT 05301-0808
(802) 254-4551

American Jersey Cattle Club
6486 East Main Street
Reynoldsburg, OH 43068
(614) 861-3636

American Milking Shorthorn Society
P.O. Box 449
Beloit, WI 53511-0449
(608) 365-3332

levels of dietary fat. If too much rumen-available fat or oil is fed, it coats the forage particles, preventing the bacteria from digesting the forage. Rumen protected fat bypasses rumen fermentation. Absorption takes place in the small intestine.

Many successful dairy producers employ a professional consultant, feed company nutritionist, veterinarian, or extension person to balance dairy diets. Because stored forage quality changes from the top of a silo or hay mow to the bottom, forages should be tested for nutrient content at least every 60 days. Diets can then be updated based on the nutrient values obtained from these forage tests.

Dry cows are normally fed a diet almost entirely composed of forages. However, they are often fed a small amount of grain mix to provide vitamins, minerals and salt. Although body condition should have been recovered during late lactation, this time can be used to improve body condition of thin cows. Grain intake should be gradually increased two to three weeks before calving to reacclimate rumen bacteria to large amounts of grain.

PARASITES, DISEASES AND PREVENTION

Parasites and diseases of dairy cattle are much the same as those described in the beef cattle section. A few exceptions are described here.

ANIMAL SCIENCE FACTS

Many dairy feeds, particularly silages, contain a large amount of water. Corn silage may be 65 to 70 percent water, whereas hay contains normally only 10 percent water. In order to compare all feedstuffs on an equal basis, feed amounts are referred to on a "dry matter basis." That means, the weight of feed with all water removed. On the other hand, feed in as-is form is called "as fed basis." Conversion from as-fed to dry matter basis simply involves multiplying by the percent (in decimal form) dry matter. Diets are normally formulated on a dry matter basis, then converted to an as-fed before feeding.

Example

Corn silage = 70 percent water, 30 percent dry matter
50 lbs. corn silage (as fed) × .30 = 15 pounds dry matter basis

*Reverse the formula to convert from dry matter to as-fed basis.
15 lbs. dry matter / .30 = 50 lbs. corn silage as fed

Mastitis, an infection and inflammation of the mammary gland (udder), causes the greatest economic loss to the dairy industry. Mastitis can be acute or chronic. Acute mastitis (obvious or hot) is characterized by a hot, swollen udder and causes a noticeable drop in milk production. The milk may be bloody, clotted, or extremely watery. A cow with acute mastitis may be listless and refuse to eat. Acute mastitis can be treated with high dosages of antibiotics. However, with consumer concerns over antibiotics in milk, many producers are now treating acute mastitis by injecting doses of oxytocin at one hour intervals followed by stripping out the mastitic milk. This practice reduces the number of bacteria in the udder and allows the cow's natural defense system to fight the infection.

Subtle, or chronic mastitis is much less noticeable (subclinical). A few flaky appearing jets of milk at the beginning of milking may be the only sign. Chronic mastitis also causes a drop in milk production.

Prevention of either type of mastitis revolves around cleanliness. Controlling bacterial transfer through milking equipment, or while washing the udder in preparation for milking, is essential to controlling mastitis. Use of individual disposable paper towels to dry the udder after washing and before milking greatly reduces bacterial transfer. Dipping teats immediately after milking in an iodine-based solution kills bacteria spread during the milking process. Disinfecting all milking equipment between milkings keeps the bacterial load at an acceptable level. Also, cleanliness of the cows' resting area helps keep udders clean between milkings.

Other health problems in the dairy industry result from the stress of high levels of milk production and the feeding programs required to reach those levels.

Ketosis is a metabolic disorder of high producing dairy cows associated with a negative energy balance during early lactation. Usually ketosis is not caused by underfeeding, but a cow's sluggish appetite or being off feed due to a rough calving, uterine infection, stomach upset, or dystocia. Cows with ketosis become listless, stop eating, and their breath smells of acetone. Treatment of ketosis often involves intravenous glucose injections. Cows with ketosis experience a marked decrease in total milk production.

A ***displaced abomasum*** (DA) or "twisted stomach" occurs when the abomasum moves to an abnormal position in the body cavity. This condition is normally found in just-fresh cows fed too much concentrate or silage before calving. The ejection of a calf leaves a large space in the cow's body cavity. The space allows the stomach to swing from the right to the left side. Not more than 3 percent of a herd should have DAs in one year. A veterinarian should be consulted for treatment of DAs.

Milk fever is not really a fever, but results from an imbalance of calcium in recently fresh cows. The imbalance causes muscle paralysis and prevents affected cows from standing. The extremities of cows with milk fever will feel cold to the touch. Calcium and phosphorus supplements during the dry period will aid in prevention. Treatment consists of a transfusion of calcium salts. Recoveries after treatment can be miraculous — sometimes within minutes. Not more than 5 percent of a herd should experience milk fever within a year.

Cows normally expel the ***placenta*** (membrane in the uterus which protects and nourishes the unborn animal) within hours after calving. If the placenta is not shed, a condition known as retained placenta occurs. Retained placentas quickly become infected and begin to decay within the uterus of the cow. Manual removal of the placenta by a veterinarian was once commonplace. Research has now shown that the weight of the whole placenta hanging out of the vulva puts more pressure on the placental tissues to separate. Retained placentas are blamed on a variety of causes including heat stress at calving, twinning, and low vitamin E and selenium in the bloodstream. No more than 7 percent of a herd should experience retained placentas in one year.

Retained placentas are one of several ways for bacteria to invade the uterus after calving. The resulting uterine infection is called ***metritis.*** Cows with metritis will exhibit abnormal discharge from the vulva, go off feed, and stand with their backs arched. Treatment involves infusion of antibiotics directly into the uterine body by a veterinarian.

HOUSING

Housing for newborn and young dairy calves is often in individual stalls, either inside or, preferably outside. Because of improved ventilation, calves housed outdoors have a lower incidence of respiratory disease than those raised indoors. Calf hutches are popular for housing calves after weaning. See Figure 10-12. At eight weeks of age, heifers are normally grouped with other heifers of similar age. A separate "heifer growing" barn is standard equipment for many dairy operations. See Figure 10-13. Open fronted sheds are also popular heifer barns on many dairies.

Traditionally, dairy cows were housed in tie-stall or ***stanchion*** barns. In these facilities, cows were tied in individual stalls during milking and

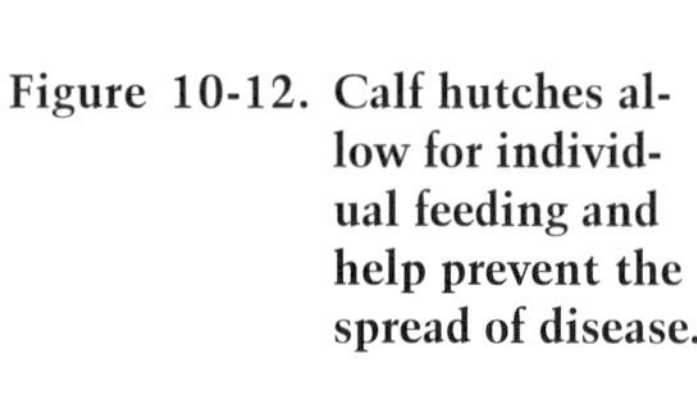

Figure 10-12. Calf hutches allow for individual feeding and help prevent the spread of disease.

Figure 10-13. This facility functions as a heifer growing barn.

remained there for most of the day. Cows may have been released into pasture at night during summer months, but were kept in their stalls during the winter.

Next came the advent of ***free-stall housing.*** See Figure 10-14. Free stalls, as the name implies, allow cows to enter and leave as they wish. A feed bunk is normally located in the center of the free-stall barn so that cows can eat at will. Cows housed in a free-stall arrangement are milked either in tie stalls, or in a milking parlor.

Milking in a tie-stall barn involves bending down to wash the cow's udder and bending again to attach and remove the milking machine. The person doing the milking has to work their way from one end of the barn to the other. This method involves many steps. The parlor system allows the cows to come to the milker instead of the other way around. See

Figure 10-14. Cows can come or go as they please in free-stall barns.

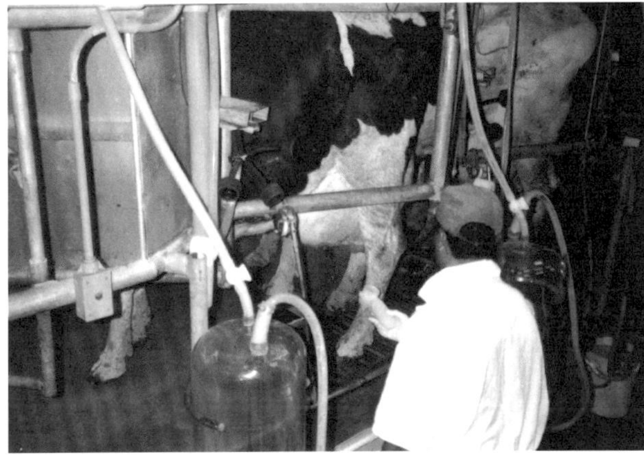

Figure 10-15. Parlor milking systems decrease milking time and increase labor efficiency. (Courtesy, Jasper S. Lee)

Figure 10-14. In a parlor system, 8 to 12 cows enter the parlor at a time. Positioning of the cows is such that the udders are at chest level for the milkers who stand in a pit behind the cows. All cows are washed and milked at the same time. When that group of cows is finished, they are released from the parlor, and the next group of cows enter. While parlors require a significantly higher investment, they double the number of cows a person can milk per hour.

RECORDS AND MILK MARKETING

Careful records are key to any dairy enterprise. Records of milk production from individual cows alert producers to the best genetics in the herd. Production records allow lower producing cows to be culled. Breeding records are important for keeping track of expected calving dates and keeping the average calving interval to an acceptable 14 months or less. With good records, feed costs and milk production can be monitored to determine profitability.

Dairy Herd Improvement Associations ***(DHIA)*** or similar record keeping organizations help producers keep track of important production records. Each month a DHIA tester weighs the milk produced and takes a milk sample from each cow. The samples are sent to a central laboratory where they are tested for protein, butterfat, and ***somatic cell count*** (an indication of mastitis and cleanliness). Somatic cells are white blood cells, which fight infection. Key information about breedings, calvings, and the feeding program are recorded. Each month, the producer receives a report detailing the production of each cow, where she is in the reproductive cycle, herd production averages, and other production information.

MILK MARKETING

Most milk is either marketed through farmer cooperatives or directly to processors. Producers in most dairy areas have a choice of two or more markets. Only those producers that pass a federal inspection for exceptional cleanliness and up-to-date facilities get to sell their milk as fluid milk used for drinking (Class I). Class I milk commands the highest price. Producers who do not pass the test must sell their milk as Class II. Class II milk is used for processed milk products, such as butter and cheese.

Milk price is set by the federal government through Federal Milk Marketing Orders. The basis for this control is to keep milk prices stable throughout the year and at an acceptable level for the consumer. Producers have traditionally been paid bonuses for high levels of butterfat. However, consumers now want low and nonfat milk and milk products. Milk buyers have responded by paying an incentive for high solids-not-fat (protein). Bonuses are also paid for low somatic cell counts, giving producers an incentive to produce milk under clean conditions.

SUMMARY

Breeds of dairy cattle in the United States are limited to only six: Holstein, Jersey, Guernsey, Brown Swiss, Ayrshire, and Milking Shorthorn. Holsteins are by far the most popular. Most dairy cows are bred artificially.

Crossbreeding is rarely practiced in the dairy industry. The milking process is designed to be easy on the milker, sanitary, and comfortable for the cow. Milking can be done two or three times per day. Dairy calves are fed milk or milk replacers after weaning, which occurs shortly after birth. Growing heifers should be fed limited amounts of grain and high-quality forages to ensure proper growth and development. Lactating cows are fed larger amounts of grain, but diets should always contain at least 40 percent forage. Forage quality and body condition are very important components in feeding the lactating cow. Cows are fed more grain in early and mid-lactation when milk production is highest and less in later lactation. Metabolic disorders are common in dairy cows because of the high demands of lactation. Mastitis is the most costly of all dairy cow ailments. Most cows are housed in either tie-stall barns or free-stall barns. Milking is normally done in tie-stalls or in a milking parlor. Production records, such as those provided by DHIA, are critical to profitability. Milk price is ultimately set by the federal government.

Most registration numbers and addresses given in this chapter were courtesy of The American Livestock Breeds Conservancy via the following publication:

Bixby, D. E., Christman, C. J., Ehrman, C. J., & Sponenberg, D. P. (1994). *Taking stock: The North American livestock census.* Blacksburg, VA: The McDonald & Woodward Publishing Company.

CHAPTER SELF-CHECK

___ placenta	1. indicates sanitation/mastitis
___ grade	2. tie-stall barns
___ body condition	3. twisted stomach
___ dry cows	4. uterine infection caused by retained placenta
___ mastitis	5. results from a calcium imbalance
___ ketosis	6. metabolic disorder
___ displaced abomasum	7. unregistered
___ milk fever	8. infection and inflammation of the udder
___ metritis	9. record keeping organization
___ stanchion	10. muscle and fat cover

___ free-stall housing
___ DHIA
___ somatic cell count

11. allows cows to freely enter and leave
12. not milking
13. membrane in the uterus which protects and nourishes the unborn animal

QUESTIONS AND PROBLEMS FOR DISCUSSION

1. Which dairy breed is the most popular in the United States?
2. Jerseys were first imported to the United States in the early _________ (date).
3. Name the primary sign of estrus (heat).
4. List three secondary signs of estrus (heat).
5. To achieve the best rate of conception, insemination of cows should occur ________ hours after the first observation of standing heat.
6. Describe the two main purposes for washing a cow's udder prior to milking.
7. What is the major factor considered when developing a dairy ration?
8. Lactating dairy cows can be expected to consume _____ percent of their body weight on a dry matter basis.
9. Cows are often fed in groups according to _______________.
10. Draw a lactation curve.
11. Which health problem in dairy cattle causes the greatest economic loss?
12. What causes milk fever?
13. The primary use for class I milk is __________.
14. Differentiate between free-stall and stanchion barns.
15. Most milk is marketed in the United States through _________.

ACTIVITIES

1. Four hundred pounds of blood must pass through the udder in order for one pound of milk to be produced. Calculate the daily amount of blood

that must pass through the udder for cows producing 50, 75, and 100 pounds of milk each day.

2. If a high producing cow gave 80 pounds of milk per day, figure how many one-half pint cartons of milk she could provide to your school cafeteria on a daily basis.
3. Research the milk pricing system used by the government. Report findings to the class.
4. Conduct telephone interviews of local dairy farmers. Inquire about their choice of breeds. Ask about their chosen breed's advantages and disadvantages.

LABORATORY ACTIVITY

MICROWAVE DRY MATTER CALCULATION

Purpose

To calculate the dry matter percent of various forages using a microwave oven

Materials

microwave oven
several silage, hay and grain samples (1 pound each)
paper plates
a metric balance capable of weighing samples to .1 gram
pencil and paper
calculator

Safety Precaution

Care should be taken that paper plates and forage samples do not catch fire or char when placed in the microwave. Microwaved samples may be hot. Care should be taken to avoid burns.

Procedure

1. Divide students into the same number of groups as there are available samples.
2. Have each group weigh out about 10 grams of their forage sample and record the weight. This weight is called the wet weight. Tear long hay samples into shorter pieces before weighing.

3. Place each sample on a paper plate and microwave on high, one at a time. Silage samples should be microwaved for four minutes, hay and grain samples for three minutes. (Times may vary depending on the power of the microwave. Remove immediately if sample starts to burn or char.)
4. Remove sample from the microwave, redistribute around the plate, and microwave on high again for half the original time. Repeat two times, reducing the time by half each time.
5. Remove sample from the plate and weigh each sample after the fourth heating.
6. Return to the plate and re-heat again for the same amount of time as the most previous heating.
7. Weigh sample again. If sample weight is the same as it was in #5, record this weight as the dry weight. If not, repeat heating and weighing until sample weighs the same two times in a row.

Analysis

1. Using the formula: (dry weight/wet weight) × 100, calculate the percent dry matter of each sample.
2. If a producer wanted to feed 20 pounds of dry matter per cow, how many pounds would need to be given to each cow on an as-fed basis? Do this calculation for each sample tested. Use the formula:

   Dry matter weight/(dry matter percent/100) = as fed weight.

Application of Laboratory Activity

Dairy rations contain many different types of feeds that vary widely in moisture content. In order to standardize feeds for ration formulations, producers and nutritionists calculate them on a dry matter basis. Finalized rations are converted back to an as-fed basis so they can be weighed for actual, on-farm feeding.

This laboratory activity demonstrates that various types of feeds contain different percentages of water. After doing the lab, you should be able to compare different feeds on an equal basis. This will give you a better understanding of the decisions producers make in formulating rations for their dairy herd.

Chapter 11

SHEEP MANAGEMENT

Baa Baa Black Sheep

INTRODUCTION

The history of the American sheep industry is clouded in conflict. For years, shepherds and cattle ranchers on the western range were at odds over whose animals would graze the available land. Both sheep and beef cattle convert forages into meat, but sheep take production one step further by providing an annual yield of wool. The feud has since cooled, but sheep still excel at converting forages into both food and fiber.

Figure 11-1. (Courtesy, Jasper S. Lee)

OBJECTIVES

1. Discuss the characteristics of the major sheep breeds
2. Explain the breeding systems used in the sheep industry
3. Develop a feeding program for a market lamb
4. List five management practices used in the sheep industry
5. Describe common sheep diseases
6. Make a chart of preventive health care measures used in the sheep industry
7. Discuss the housing requirements for sheep
8. Calculate the price for wool and a market lamb using current cash prices

TERMS

clip
creep feeding
crutched
docked
drench
fine wool breed
flushing
hothouse lambs
medium wool breeds
scur
seasonal breeders
stillborn
synthetic breed

ANIMAL SCIENCE FACTS

Research has shown that sheep and cattle can coexist. In intensively grazed pastures, an acre of land co-grazed by sheep and beef cattle produces more total weight gain than the same acre grazed solely by either species. Sheep and cattle eat different parts of forage plants, thus maximizing usage.

BREEDS

Nearly 50 different breeds of sheep comprise the United States' sheep industry. Of these, only nine have significant registrations. Breeds of sheep can be classified by place of origin, purpose (meat vs. wool production or sire vs. dam breeds), and type of wool.

According to 1990 registrations, the most popular breed in the United States is the Suffolk. Suffolks were developed in the late 1700s in a region of southern England called the Downs. Suffolks, along with other breeds originating in that region, are known as the Down breeds. The Down breeds were developed for meat production and are often used as sire breeds. Large framed and heavily muscled, Suffolks sire fast growing market lambs.

Suffolks and the other Down breeds are referred to as ***medium wool breeds.*** Medium wool breeds have average fleece quality. The head and legs of Suffolks are black and covered with hair instead of wool. Their ears are long and somewhat drooping. The only evidence of horns is an occasional ***scur,*** or small unattached horn. Suffolks were first brought to the United States in the late 1800s; however, most imports occurred during the 20th century. In 1990, there were 70,320 Suffolks registered in the United States. See Figure 11-2.

The second most popular breed in the United States is the Dorset. Dorsets are also a medium wool breed, but not a Down breed. They were selectively bred from native sheep in a small area located outside the Downs region. The average Dorset is smaller, but similarly muscled as

Figure 11-2. Suffolks are the most popular breed of sheep. (Courtesy, Jasper S. Lee)

Figure 11-3. Dorsets are white-faced sheep used as both a sire and dam breed. (Courtesy, Jasper S. Lee)

the Suffolk. Dorset ewes are popular in crossbreeding systems because of their ability to breed throughout the year. Most sheep are ***seasonal breeders*** (only able to conceive in the fall of the year). Because of their muscling and maternal abilities, Dorsets can be used as either a sire or dam breed. They are entirely white, free from wool around the eyes, and have short ears. Dorsets can be either horned, polled, or scurred. They were first imported in the late 1800s. See Figure 11-3.

Ranking third in popularity with 17,047 registrations in 1990, are Hampshires. A medium wool breed, they are also heavily muscled. Like Suffolks, Hampshires developed in the Down region. They sire fast gaining, heavily muscled market lambs and are one of the larger Down breeds.

Figure 11-4. Hampshires compete with Suffolks for use as market lamb sires. (Courtesy, American Hampshire Sheep Association)

They have a black face and legs, short ears, and are polled. The top of a Hampshire's head is covered with wool. Hampshires were also brought to the United States in the late 1800s. See Figure 11-4.

The Rambouillet, a ***fine wool breed,*** can trace its ancestry to the Spanish Merino breed (via France), which was known for superior wool quality. Since importation to the United States in 1840, Rambouillets have been selected for improvement of carcass qualities. They have been popular with western sheep producers as a dam breed. This is due to their ability to wean lambs with acceptable carcass merit and produce a high-quality, heavy fleece. Rambouillets are a large framed, white-faced breed that can be either polled or horned. There were 16,000 Rambouillets registered in 1990. See Figure 11-5.

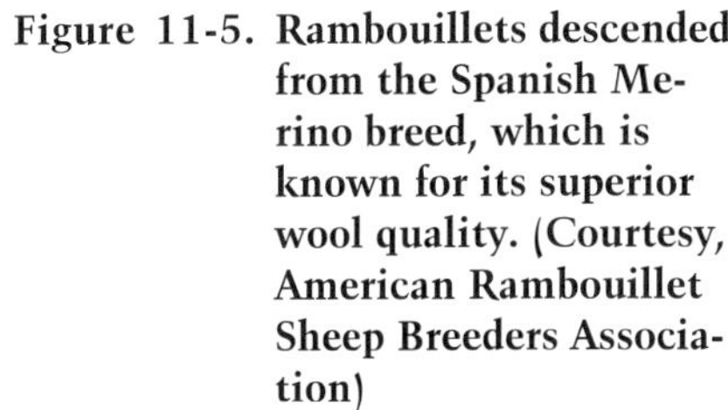

Figure 11-5. Rambouillets descended from the Spanish Merino breed, which is known for its superior wool quality. (Courtesy, American Rambouillet Sheep Breeders Association)

Polypays rank fifth in the United States with 11,429 registrations in 1990. Polypays are a new ***synthetic breed*** (planned crosses of established breeds), developed in the late 1960s and early 1970s. They accentuate the various strengths of four breeds, the Dorset, Finnsheep, Rambouillet, and Targhee. The resulting Polypay has excellent maternal abilities, high growth rate, good carcass qualities, and hardiness. They are often used as a dam breed because of their ability to conceive in all seasons. Polypays are white, medium-sized, have open eye channels (no wool around the eyes), and medium-sized ears. See Figure 11-6.

Figure 11-6. Polypays are utilized as a maternal breed. (Courtesy, Lane's End Farm, Jersey Shore, PA)

Another synthetic breed developed in the United States is the Columbia. Columbias were first produced in the early 1900s from a foundation of Lincoln and Rambouillet breeding stock. They are a white-faced breed with open eye channels, somewhat resembling the Rambouillet. Columbias are popular in western range operations, where they are often crossed with Suffolks or Hampshires. Registrations totalled 9,512 in 1990. See Figure 11-7.

One of the first breeds of sheep to be introduced to the United States is the Southdown. They are seventh in breed popularity with 5,800 registrations in 1990. Easily the smallest of the Down breeds, Southdowns

Figure 11-7. Columbia ewes populate many Western ranges. (Courtesy, Columbia Sheep Breeders of America)

Figure 11-8. Although small in size, Southdowns are known for their muscle volume. (Courtesy, American Southdown Breeders Association)

are extremely heavily muscled. Their face and legs are covered with brownish wool. Southdowns are popular on the East Coast for producing early maturing, heavily muscled, grain-fed lamb for the hotel, restaurant, and ethnic markets. See Figure 11-8.

Corriedales were developed in New Zealand in the late 1800s from Merinos, Lincolns, and Leicesters. Because of their heritage, their appearance is similar to that of Columbias. The idea behind such crosses was to reproduce the fine-wool qualities of the Merino and the carcass qualities of the Lincolns and Leicesters. Corriedales were first imported into the Western United States in the early 1900s. They have been used as a dam breed in range flocks. Registrations numbered 5,050 in 1990. See Figure 11-9.

Figure 11-9. Corriedales originated in New Zealand. (Courtesy, American Corriedale Association, Inc.)

Shropshires are a Down breed that can function in either a sire or dam role. They have brownish-black pigmented faces, ears, and feet. Shropshires are medium-sized sheep, ideal for eastern farm flocks. They were first imported to the United States in the late 1800s. See Figure 11-10.

Figure 11-10. Shropshires serve many functions in the sheep industry. (Courtesy, American Shropshire Registry)

BREEDING SYSTEMS

Sheep breeds and breeding systems vary greatly throughout the United States. The western plains and Rocky Mountain states are known collectively as the western range. Flocks in western range country often number in the thousands. Breeding systems rely on large, hardy, white-faced ewes with better than average wool production. Breeds such as the Rambouillet, Columbia, and Corriedale are popular as ewes in the western ranges. Wool production has traditionally been a major source of income for western sheep producers, hence the emphasis on ewes with a heritage of high-quality wool. The other source of income is from market lambs. Western range ewes are routinely bred to heavily muscled, growthy breeds, such as Hampshires or Suffolks. This breeding produces market lambs with improved growth rates and carcass qualities. Replacement ewes are either purchased or produced by mating a maternal, fine wool ram with a band (group) of ewes and retaining the offspring.

Ewes are shorn (wool removed) and lambed in the spring. Ewes and lambs are pastured during the summer. Sometimes this is done in high mountain meadows, often on public grazing lands. In late summer, the lambs are weaned and the rams are turned in with the ewes. The rams remain with the ewes for at least three estrus periods, or 48 days. Well grown, eight month old ewe lambs can be bred the first year. However, better reproductive performance results if ewe lambs mature to 1½ years of age. When summer grass conditions are adequate, weaned lambs are sold for slaughter. Lambs deemed too thin for slaughter are often sold to feedlots for finishing.

ANIMAL SCIENCE FACTS

Finnsheep have proven to be the most prolific of all sheep breeds. Instead of giving birth to twins or triplets, Finnsheep often deliver "litters" of four to six lambs. Synthetic breeds, such as the Polypay, have incorporated Finn genetics to capture some of this prolificacy.

Sheep flocks in the East and Midwest are much smaller than their western counterparts. They usually number fewer than 200. Their breeding programs are also different. Because of the proximity to populated areas and large quantities of grain, the majority of income is derived from the sale of market lambs. Ethnic groups concentrated in large cities comprise a major market for lamb. Therefore, Eastern and Midwestern producers have concentrated on meat breeds, such as the Suffolk, Hampshire, Dorset, and Southdown. Progressive producers primarily use maternal breeds (Dorset, Polypay, or crosses) that will conceive during any season. Ewes of these breeds produce a lamb crop every 8 months instead of every 12. As with Western producers, black-faced rams are used as market lamb sires. Lambs born "out of season" (in the fall or winter) can be marketed as ***"hothouse" lambs*** (raised indoors) at 9 to 16 weeks of age. Spring born lambs are normally weaned at 70 to 80 pounds and finished on grain.

ANIMAL SCIENCE FACTS

Lambing percentage indicates prolificacy. Lambing percentage equals the total number of lambs divided by the total number of ewes lambed, multiplied by 100. For example, if 100 ewes delivered 200 lambs, the lambing percent would be 200. If 100 ewes delivered 150 lambs, the lambing percent would be 150.

FEEDING PROGRAMS

As with the beef cow herd, feeding programs for the ewe flock are almost entirely forage-based. Sheep are pastured during the grazing season and fed stored forages the rest of the year. Feeding programs for western range ewes managed for an annual lamb crop are based on a yearly production schedule. Summer pastures for western range flocks are composed almost entirely of native grasses. Shepherds move the flock to a new area when the grass supply is exhausted. A grazing area may be used twice during the grazing season if sufficient time is allowed for regrowth. When the grazing season ends, sheep are moved to a winter range. The winter range is normally at a lower elevation or more sheltered location, where the grazing season may be longer. When pasture runs out, ewe flocks are wintered on a variety of stored forages, such as mixed legume hay or silage. Root crops, such as turnips, can be used as an early winter feed.

ANIMAL SCIENCE FACTS

Predators, such as coyotes and feral dogs, have always been a problem for sheep producers. Many western shepherds graze an Alpaca llama with each flock of sheep. Alpacas are extremely hostile toward canines and drive them away from their prospective meal.

After the lamb crop is weaned, the ewes are prepared for the breeding season by ***flushing.*** Flushing consists of increasing the energy content of the diet for 15 to 20 days before breeding. A ration of 1/2 to 1 pound per head per day of oats, corn, or a similar grain, in addition to pasture or stored forage is recommended. This practice increases the number of eggs

ANIMAL SCIENCE FACTS

Sheep Facts

Average body temperature = 102.3°F

Pounds of grain per pound of gain for lambs = 4:1

Gestation period = 148 days

Average ewes per ram (yearling or older) = 45

Dressing percent (carcass weight ÷ live weight) = 50%

ovulated and hence, the number of lambs born per ewe (lambing percentage). To calculate the lambing percentage, divide the total number of lambs by the total number of ewes lambed and multiply that number by 100.

Non-lactating, pregnant ewes in early gestation have lower nutritional requirements than at any other time during the production cycle. In some parts of the country, crop residue or wheat pasture is used as the sole feed. Free-choice legume or legume mixed hay is probably the most popular stored forage when pasture is unavailable.

Later in the 150-day gestation period, the rapidly growing fetuses require additional nutrients. Ewes also need to build body condition before the next lactation. About five weeks before lambing, ewes should be given supplemental grain to complement the available forage. For example, if ewes are being wintered on low protein forage, 1/2 pound per head per day of a high protein grain mix may be in order. Thin ewes receiving adequate protein from forage may simply require 1/2 to 1 pound of high energy grain per day. Decisions on the exact supplement should be based on the forage quality and body condition of the ewe.

Lactation places great nutrient demands on the ewe. Nursing lambs, wool production, body growth for young ewes, and the maintenance requirements of the ewe herself require the best forages available. The best quality legume hay should be reserved if lactating ewes are fed stored forages. Free choice hay and 1 to 2 pounds of high energy grain mix are necessary. See Figure 11-11 for an appropriate grain mix. Spring-born ewes and lambs should be released to high-quality pasture and the supplemental grain feeding eliminated, as soon as possible. Many managers separate ewes that have twins or triplets from those that have single lambs. The more productive groups of ewes can then be fed for maximum milk production. Ewes with single lambs can often go without supplemental grain.

Eastern farm flocks also utilize pasture when in season. Managing ewes for an eight-month production cycle means some ewes will lactate while being fed stored forage in the winter. Other dry, pregnant ewes will have access to lush spring and summer pastures. For these reasons, feeding ewes on an 8-month lambing cycle requires more nutritional management than feeding those on a 12-month cycle. The nutrition of ewes should be managed so that those with the highest demands (late gestation and lactation) receive the best pasture or stored forage, and grain. Dry, pregnant ewes in early gestation can be fed poorer-quality pasture or stored forage.

Ingredient (lbs)	Lactating Ewes	Creep	Finishing Lambs
Cracked, Shelled Corn	60	50	81
Oats or Barley	30	30	
Soybean Meal	10	15	16
Molasses		5	3
Total	100	100	100

1. Salt, vitamins, and minerals *manufactured specifically for sheep* should be offered free choice to all ewes. Vitamin/mineral premix *manufactured specifically for sheep* must be added to creep feeds and finishing diets in accordance to manufacturers recommendations.
2. Good quality legume or legume mix hay should be offered free choice for lactating ewes, creep fed lambs, and finishing lambs.

Figure 11-11. Rations for lactating ewes, creep fed, and finishing lambs.

Nutrition in the eastern flock starts with ***creep feeding*** (segregating lambs and feed) before weaning. Creep feed should be available free choice, in an area where only lambs have access. Creep feeding serves two purposes. First, it acclimates lambs to eating grain. Second, it increases weaning weights. Ewes nursing twins and triplets may not supply enough milk to fully satisfy each lamb. Creep feed can fill this milk shortage. Rolled oats, crimped corn, soybean meal, molasses, and other palatable feedstuffs are common creep feed ingredients. See Figure 11-11.

When lactating ewes have access to good pasture, lambs weaned at five or six months are often fat enough for slaughter. Weaned range lambs not ready for slaughter are normally sold to feedlots for two or three months of grain feeding. Eastern lambs that are not sold early as hothouse lambs are also fed a finishing diet for a two to three month period.

Finishing diets should contain at least 1 pound of high-quality legume forage per head per day. The protein and energy in such forages are normally less expensive than those in grains. Like finishing cattle, lambs require a small amount of forage to maintain normal rumen function. Grain mix ingredients are variable, but should be high in energy. A popular finishing diet consists of whole or cracked grains mixed with a fortified, pelleted supplement. See Figure 11-11. Lambs should be started on grain slowly. The initial feeding should be 1/4 pound per day. Feed allowance should be increased by 1/4 pound per day every three to five days until a maximum grain intake of 2 to 2 1/2 pounds per day is achieved.

Finishing lambs should gain at least 1/2 pound per day. Finished weight depends on genetics and the weight at which grain feeding begins. Most lambs require less than three months of grain feeding before slaughter.

Finished lambs do not require much exterior fat cover. Buyers want trim lambs in the 90 to 120 pound range. Fat cover should be sufficient to prevent excessive water weight loss (shrink) from the hanging carcass (at least .10 inch), but not enough to cause excessive trimming (less than .20 inch).

SHEEP MANAGEMENT

Sheep require several management practices not necessary in other species of livestock.

Tails are routinely ***docked*** (removed) in young lambs. Un-docked tails collect feces, which serve as an ideal place for maggots to grow. Tail docking should be completed within days of birth when lambs are easily handled. Tails may be cut with a hot pincher. This stops the wound from bleeding and helps prevent infection. Alternatively, tails can be banded with an elastrator. See Figure 11-12. This cuts the blood supply to the tail and the tail eventually falls off.

Castrated wether lambs entering a feedlot can be implanted with a growth promotant. Implants are injected just under the skin on the backside of the ear. Implants increase average daily gain and improve feed efficiency.

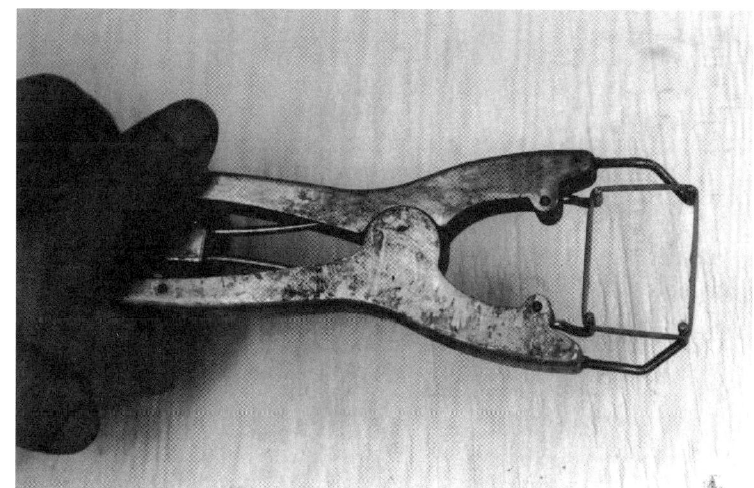

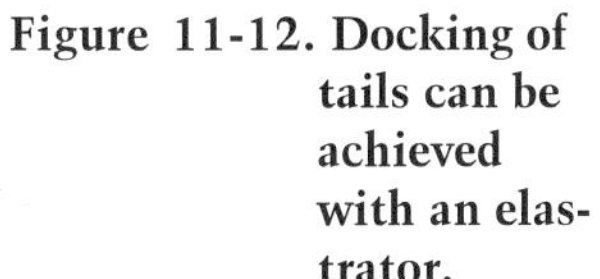

Figure 11-12. Docking of tails can be achieved with an elastrator.

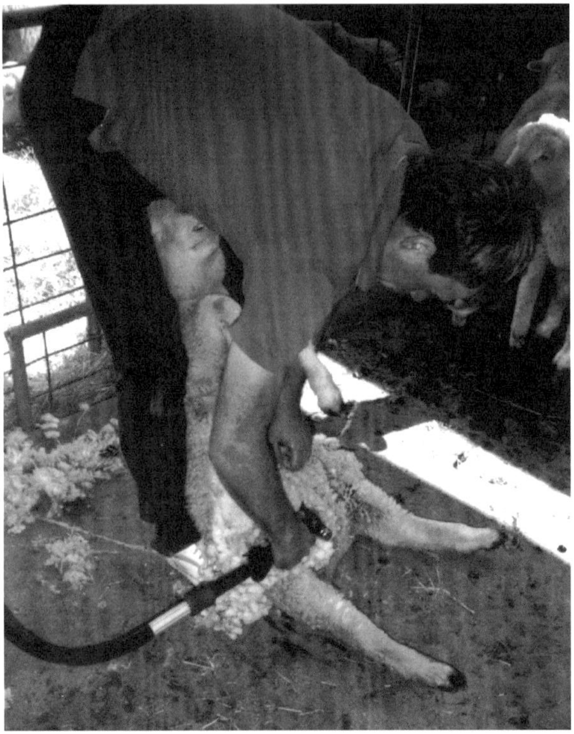

Figure 11-13. **Experienced shearers can remove a fleece in just a few minutes. (Courtesy, Jasper S. Lee)**

Some side effects, such as rectal prolapses, have been noticed because of using growth promotants. Obviously, care should be exercised.

Shearing is the removal of wool from the sheep. See Figure 11-13. Ewes are normally shorn once per year before lambing. If lambing is extremely early when inclement weather is still possible, shearing may be delayed until after lambing. After removal, the wool is called a fleece. The average full fleece for a mature ewe weighs about 8 to 10 pounds. However, fleeces from Down breeds average less than 8 pounds.

If shearing is delayed until after lambing, ewes should be ***crutched.*** Crutching is the process of shearing the dirty wool around the vulva and udder. Lambs from crutched ewes are born into a cleaner environment and have an easier time finding the udder. Ewes should also be crutched before breeding to aid in copulation.

If structurally correct, most adult sheep will show even wearing of the hooves. Those that do not should undergo corrective hoof trimming to avoid lameness. Both inside and outside hoof walls should be trimmed even with the fleshy center of the hoof. Rams' hooves should routinely be trimmed before the breeding season.

Ideally, each ewe should deliver and wean two lambs. Unfortunately, some ewes lamb only singles while others have triplets. Sometimes, one lamb in a set of twins is ***stillborn*** (born dead). In such instances, one lamb from a set of triplets may be placed with the ewe that has only one lamb. There are several methods of convincing the adoptive ewe to accept the new lamb. One method involves rubbing the lamb with placental fluids from the adoptive ewe. These fluids hide the odor of the lamb and make the ewe think it is her own. Another method involves skinning a ewe's stillborn lamb and tying the pelt around the new lamb. Again, the odor of her own lamb fools the ewe into accepting the new lamb.

PARASITES, DISEASES, AND PREVENTION

Sheep are subject to a variety of parasites and diseases. Those of economic importance are surprisingly few considering the adage, "a sick sheep is a dead sheep."

Parasite infestations in the form of worms are common, especially in sheep on permanent pastures. Worms steal feed nutrients. A heavy worm load causes sheep to lose body condition—even when on a high plane of nutrition. A light worm load can cause a decrease in gain and efficiency. Control of worms is a two-part process. The first is to rotate pastures to break the life cycle of worms. Often, the sheep ingest worm eggs or larvae while eating pasture grasses. The second is to routinely ***drench*** all sheep with an approved wormer. Drenching involves placing worming fluid in the back of the throat so the sheep is forced to swallow it. Lambs to be placed on finishing diets should be drenched at weaning.

A second parasite disease that often causes losses in feedlot lambs is *coccidiosis*. *Coccidia* also infest cattle, swine, and poultry. *Coccidia* are microscopic organisms that invade the intestinal lining, reduce nutrient uptake, and cause diarrhea. *Coccidia* reside in spores when outside the host animal. Spores are resistant to heat, as well as disinfectants, and are very difficult to eradicate. Prevention of coccidiosis involves reducing the stress on weaned lambs moved to the feedlot. Feed additives called coccidiostats may be fed in the grain mix during the finishing period to control coccidiosis.

Clostridia also reside in spores and cause a variety of diseases in sheep. *Cl. chauvoei* causes a disease known as blackleg. Blackleg is relatively uncommon, but results in the swelling of various parts of the body. These parts are often the site of a recent wound. *Cl. tetani*, the causative agent in tetanus, also enters through open wounds, such as those associated with castration or shearing. *Cl. Perfringens* has two types. Type C causes bloody scours or hemorrhagic enterotoxemia. Bloody scours most often affect young, rapidly growing nursing lambs. Mortality is high. Type D causes overeating disease or enterotoxemia. Overeating is so named because

ANIMAL SCIENCE FACTS

Columbia Sheep Breeders Association of America
P.O. Box 272E
Upper Sankusky, OH 43351

American Corriedale Association
P.O. Box 29C
Seneca, IL 61360

Continental Dorset Club
P.O. Box 506
Hudson, IA 50643

American Hampshire Sheep Association
P.O. Box 377
Whiteland, IN 46184

American Polypay Sheep Association
609. S. Central #6
Sidney, MT 59270

American Rambouillet Sheep Breeders Association
2709 Sherwood Way
San Angelo, TX 76901

American Shropshire Registry
P.O. Box 250
Hebron, IL 60034

American Southdown Breeders Association
HC 13 220
Fredonia, TX 76842

National Suffolk Sheep Association
P.O. Box 617
Columbia, MO 65205

it usually affects the biggest, fastest gaining finishing lambs. Affected animals usually die quickly.

Prevention of diseases caused by *Clostridia* involves vaccination. Vaccines for all Clostridial diseases are packaged separately or in combination. Ewes and lambs should be routinely vaccinated for *Cl. perfringens* types C and D. *Cl. tetani* vaccinations are also advised. Initial immunization requires two doses. Ewes require an annual booster shot.

Pneumonia is perhaps the most costly disease affecting feedlot lambs. Alternating weather conditions, poor ventilation, dampness, chilling, and poor nutrition contribute to pneumonia. Most pneumonia is caused by a bacterial invasion of the lungs. Symptoms include reduced feed intake, depression, and nasal discharge. Prevention involves reducing the factors that predispose lambs to pneumonia. Attention to comfort will prevent most pneumonia. Treatment with antibiotics is effective if the disease is detected early.

Foot rot is a bacterial infection of the hoof. Foot rot is most troublesome in wet pastures. Lameness is the most common sign, while a distinct rotting odor is also obvious. See Figure 11-14. If foot rot is diagnosed in a flock, all of the affected animals should be isolated in a dry environment and infected parts of the hoof should be trimmed. The entire flock should be forced to move through a foot bath containing a proscribed mixture of antibacterial ingredients. Trimming and footbaths should be continued weekly until signs are gone. To prevent reinfection, footbaths are recommended for several weeks after signs have disappeared. Treated sheep should not be restocked onto infected pastures for at least 25 days.

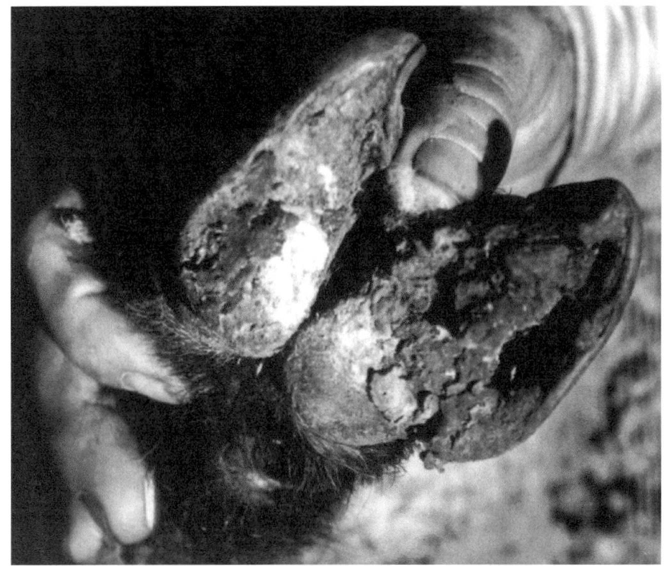

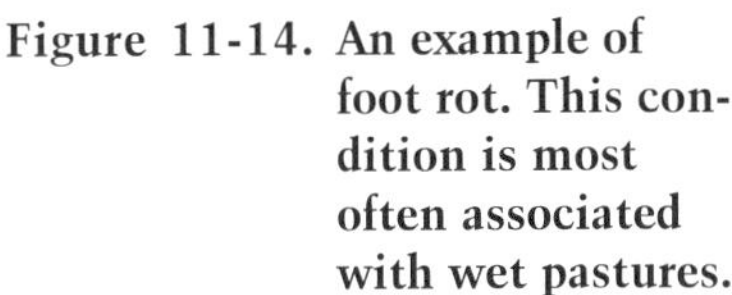

Figure 11-14. An example of foot rot. This condition is most often associated with wet pastures.

Sore mouth is a disease caused by a viral infection of the lips, tongue, and gums. The infection itself does not cause serious symptoms, but the resulting sores drastically decrease feed intake. Treatment is difficult. Sores normally heal in one to two weeks. Lambs will eat small amounts of a soft, palatable feed while sores heal. Vaccination is possible for flocks that have a history of sore mouth. However, it is not recommended for those that have never contracted the disease.

HOUSING

Mature sheep have no special housing requirements. Winter wool allows them to stay warm even in the most inclement weather if a suitable windbreak exists. Many producers provide an open fronted shed. Recently shorn sheep should be provided a warmer environment in cold, wet weather. This helps to prevent pneumonia.

When lambing season falls during the winter or early spring, ewes are often confined to a lambing shed. Most lambing sheds are enclosed, and some are heated. These sheds allow lambs to be born into a warm, low stress environment. Ewes are often penned by themselves during lambing for close observation. This also allows ewes and lambs to develop a firm bond. After several days inside, ewes and lambs can be returned to a paddock with access to an open fronted shelter. In especially inclement weather, lambs can be given access to a more confined shelter with openings too small for ewes.

Figure 11-15. This high tensile steel fence is more than sufficient for containing sheep.

Finishing facilities are normally a dry lot with access to a shed or cover. Some lambs are finished in confinement on slatted floors built over a deep pit. In such confined situations, mechanical or natural ventilation must be utilized to ensure acceptable air quality.

Sheep are more difficult to fence than cattle. Total fence height should be about 42 inches. Woven, barbed, or mesh wire and board fences are all used with success. The bottom strands of the fence are the most critical as they will be challenged most often. The first strand should be about 5 inches from the ground. The second strand should be about 13 inches from the ground. Electric fences should be charged by a powerful charger, because wool is a poor conductor of electricity. Sheep with long fleeces may push through poorly charged electric fences without getting shocked. High tensile strand fence, 5 to 7 strand (2 to 4 electrified), is the current vogue. See Figure 11-15.

MARKETING

Feeder lambs can be marketed through local sale barns, tele-auctions, organized feeder lamb sales, or directly to the feedlot. As mentioned previously, eastern producers market many feeder lambs as hothouse lambs. Most of these lambs move through sale barns or country buyers.

Finished lambs are sold through many of the same channels as feeder lambs. Lambs from large feedlots may be marketed directly to packing houses. Some lambs are sold on a weight and grade basis. In such arrangements, adjustments may be made to the price depending on carcass quality. Current prices for lambs can be found in local trade publications.

Wool is marketed through its own distinct channels. Depending on the part of the country and the amount to be sold, wool may be purchased by local or regional dealers, or wool cooperatives. Large amounts of high-quality wool may be sold directly to a mill. The wool's value is based on the fiber fineness, density, length, and cleanliness. The impurities found in wool are called grease.

Wool is graded according to the fineness of its fiber. Three different grading systems are used: blood system, count system, and micron system. The blood system computes a grade based on the percentage of Merino blood (ancestry) in a sheep. The blood system grades are fine, $^1/_2$ blood, $^3/_8$ blood, $^1/_4$ blood, low $^1/_4$ blood, common, and braid. The count system correlates grades based on the number of hanks (560 yards) spun from one pound of wool. Grades range from higher than 80 to coarser than 36. The USDA uses the count system. The micron system measures the average fiber diameter. Micron grades range from 17.00 to 40.21 plus. Wool is further divided by use. Fine wools (apparel wools) are used for clothing. Coarser wools (carpet wools) are made into rugs.

The market price of wool has traditionally been supported by the government. The government made up the difference between the actual market price and a predetermined "base" price. However, the wool price support program will cease in 1996.

SUMMARY

Of the nearly 50 sheep breeds, only nine have significant registrations. These nine breeds are divided among those having different uses in cross-breeding schemes. Breeding systems vary by geographical area. Western range flocks are managed for one lamb crop and one wool ***clip*** (fleeces from the flock) per year. Many eastern farm flocks use prolific ewes that breed year-round to produce a lamb crop every eight months. Ewes are fed a high forage diet. The highest nutrient requirements occur during late gestation and lactation, when some grain feeding may be necessary. Some just-weaned lambs are sold directly for slaughter, while others are fed grain for several months before slaughter. Tail docking, yearly shearing, and crutching are examples of management practices used only on sheep. Worms, coccidiosis, *Clostridia*, pneumonia, foot rot, and sore mouth are common health problems. Facilities for sheep normally consist of a wind break or open shed. Totally enclosed lambing facilities are often used. Lambs are marketed through channels similar to other livestock species.

Wool is sold by grade and the price of wool has traditionally been supported by the government.

Most registration numbers and addresses provided in this chapter were courtesy of The American Livestock Breeds Conserancy via the following publication:

Bixby, D.E., Christman, C.J., Ehrman, C.J., & Sponenberg, D.P. (1994). *Taking stock: The North American livestock census.* Blacksburg, VA: The McDonald & Woodward Publishing Company.

CHAPTER SELF-CHECK

___ medium wool breed	1. wool removed from vulva and udder
___ scur	2. planned crosses of established breeds
___ seasonal breeder	3. fleeces from the flock
___ fine wool breed	4. small unattached horn
___ synthetic breed	5. average fleece quality
___ hothouse lamb	6. dead at birth
___ flushing	7. raised indoors
___ creep feeding	8. superior wool quality
___ docked	9. liquid medicine placed in throat
___ crutched	10. conceives in fall of year
___ stillborn	11. lambs have access to feed, ewes do not
___ drench	12. tail removed
___ clip	13. increasing feed prior to breeding

QUESTIONS AND PROBLEMS FOR DISCUSSION

1. The Down breeds were developed in Southern _________.
2. Classify the following breeds as a medium or fine wool:
   Rambouillet— Hampshire— Suffolk—
3. List the two synthetic breeds discussed in this chapter.
4. Name three breeds often utilized as range ewes.

5. The major source of income for the Eastern and Midwest sheep producer comes from the sale of ________________.
6. Write the formula for calculating lambing percentage.
7. Explain when nutritional demands on ewes are the greatest.
8. Lambs consuming a finishing ration can be expected to gain ____ pound(s) each day.
9. Fat cover on slaughter lambs should range between _______ and _______ inch(es).
10. What are two benefits of crutching ewes?
11. Discuss three methods of pneumonia prevention.
12. What pasture condition aggravates foot rot?
13. True or False? Sheep with long fleeces may be able to push through an electric fence without getting shocked.
14. Name two primary factors used when computing wool price.
15. The government supported wool prices until ____. (List the year.)

ACTIVITIES

1. Interview local sheep producers. Ask them about the sheep industry in your area.
2. Conduct several class activities involving wool. Have a wool broker demonstrate grading. Research wool marketing in your area. Track the future prices for wool.
3. In cooperation with a food science class, prepare and sample several lamb dishes.
4. Write to breed associations. Inquire about registration procedures.

LABORATORY ACTIVITY

DOCKING LAMB TAILS

Purpose

To practice and compare the methods used to dock tails

Materials

elastrator and bands
burdizzo or emasculator and sharp knife
electric tail docker
young, live, unprocessed lambs (or sheep tails secured from a local producer after lamb processing)
disinfectant for open wounds

Safety Precaution

Use extreme caution when using the knife. Also, the bands cause a tourniquet effect. The electric docker will severely burn skin.

Procedure

After a tail docking lecture and demonstration from the instructor or a producer, each student should practice the three methods of docking: banding (elastrator), cutting (burdizzo or emasculator and sharp knife), and docking with electric tail docker. Remember to dock the tail just below the site where ligaments enter the tail.

Analysis

1. Which method of docking do you prefer? Explain your reasoning.
2. Which methods do you believe cause the least and the most stress on lambs? Why?
3. Name one consequence of leaving tails undocked.
4. Explain what could happen if open wounds are left untreated.
5. What other management practice could you perform using the same equipment?

Application of Laboratory Activity

The health of the flock directly affects the profitability of any sheep enterprise. One important aspect in protecting the health of the flock is tail docking. The removal of the tail prevents the buildup of fecal material and possible infestation by maggots. To insure the welfare of the flock, sheep producers are concerned with tail docking methods which cause animals a minimum of stress.

This laboratory activity is intended to give you hands-on experience in the removal of lambs' tails. Through conducting the docking using three different methods, you should come to an understanding of proper health management practices and stress levels associated with the different methods.

Chapter 12

HORSE MANAGEMENT

Hi Ho Silver — Away!

INTRODUCTION

Horses and humans have a long history together. As one of the first domesticated animals, horses have provided transportation, draft power, companionship, and even meat for their human caretakers. Until the advent of tractors and automobiles, nearly everyone used horses and knew how to care for them. Today, horses are used almost exclusively for recreational purposes including racing, riding, and exhibition. Although horses are not now as necessary to society, their popularity remains strong.

Figure 12-1.

OBJECTIVES

1. Categorize the characteristics of the major breeds of horses
2. Develop a feeding program for a pleasure horse
3. List four management practices used in the horse industry
4. Describe common horse diseases
5. Make a chart of preventive health care measures used in horses
6. Discuss the housing requirements for horses

TERMS

anthelmintic
distemper
draft
encephalomyelitis
equestrian
farrier
founder
hand mating
longeing
Palominos
sulky
tack

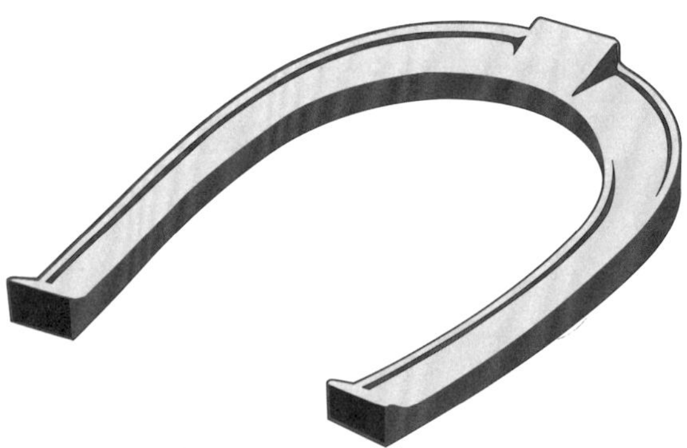

ANIMAL SCIENCE FACTS

Horse Facts

Average body temperature: 100.5°F

Average age of mares at breeding: two or, preferably, three years

Gestation period: 11 months

Time to run a quarter mile: 21 seconds

Number of pasture acres per horse: one to five

BREEDS

Individual breeds of horses are genetically the most closely related of the livestock species. Most breeds descend from, or are composites of, a handful of original breeds. To compound matters, some breeds are classified by color. Any horse of a certain color or color pattern can be registered as a "colored breed," though its genetics may be of an entirely different breed.

Horses are loosely divided by size into three main categories: light horses, heavy ***(draft*** or work) horses, and ponies. Light horses are further sorted by use. Some breeds are classified as riding horses. Riding horses may be used for pleasure riding, as cattle horses, or as mounts for equine sporting events. Other breeds are categorized as racehorses, either mounted or harnessed to pull a ***sulky*** or cart. Still other breeds are ordered as driving horses, distinctly developed for pulling carriages. A brief discussion of some popular breeds of light horses follows.

The Quarter Horse is the most popular breed in the United States. Registrations equalled 110,597 in 1990. Quarter horses are heavily muscled animals that excel in sprinting short distances. The name, Quarter Horse, originates from accomplishments in quarter mile races. The development of the Quarter Horse breed relied heavily on Thoroughbred bloodlines. However, the breed was entirely developed in the United States. The American Quarter Horse Association (AQHA) was first formed in 1940. Quarter Horse coat color often ranges in shades of brown or red. ***Palominos*** (yellow or golden color with light colored manes), blacks, and roans are also common. Quarter Horses are usually ridden as cattle horses or pleasure horses. See Figure 12-2.

Figure 12-2. Quarter Horses are the most popular breed in the United States. Their name is derived from their accomplishments in quarter mile races. (Courtesy, American Quarter Horse Association)

Figure 12-3. Due to their angular refinement, Thoroughbreds are without peers as race horses.

Thoroughbreds rank second in popularity with registration numbers of 44,000 in 1990. Known for their speed over long distances, Thoroughbreds were developed in England in the 1600s. The first Thoroughbred was imported to America in 1730. One of the oldest breeds of horses, Thoroughbreds have been used as foundation stock in the formation of many other breeds both in the United States and abroad. It is estimated that over 75 percent of all horses carry some Thoroughbred blood.

Thoroughbreds vary in color from many shades of brown, to black and even gray. They are refined, angular animals with long legs. Thoroughbreds are built for speed and used as racing horses. See Figure 12-3.

Figure 12-4. Arabians are generally recognized as the oldest breed of horse.

As its name implies, the Arabian breed was first developed in the desert region of Arabia. The natives of the area were warring nomads who bred horses to travel quickly over long stretches of open country. These horses were prized for their alert attitude and athletic endurance. Smaller than the Thoroughbred, Arabians are primarily used as a riding horse. Arabians were first imported to the United States in 1765. The Arabian Horse Registry was established in 1908. In 1990, 22,400 Arabians were registered in the United States. See Figure 12-4.

Paint horses, distinguished by their splotched coat color, classify as a colored breed. White areas on the Paints' hide are interspersed with large areas of another color. The colored areas are much larger in comparison to the color patterns of the Appaloosa (discussed later in chapter). Paints trace their ancestry to nondescript spotted horses found in America. To meet registration requirements, Paints' ancestry must be Paint, Quarter Horse, and Thoroughbred. Paints are commonly used as riding horses, cattle horses, or pleasure horses. The American Paint Horse Association was formed in 1965. There were 17,968 Paint Horses registered in 1990. See Figure 12-5.

Standardbreds were developed in the United States in the early 1800s through Thoroughbred crosses. They were originally used as driving horses in Colonial America. Today, Standardbreds often compete as harness racehorses. The Standardbred earned its name for the ability to move at two gaits, the trot and the pace, at the same or standard speed. Standardbreds come in the same color patterns as Thoroughbreds, but are smaller and more muscular. There were 16,556 Standardbreds registered in 1990. See Figure 12-6.

The ancestors of the modern Appaloosa first came to Mexico with Spanish explorers in the 1600s. Later, Native Americans in the Pacific Northwest acquired this spotted strain of horse. Appaloosas are mottled dark and white, have black and white striped hooves, and light pigment around the eyes. The Appaloosa Horse Club was formed in 1938 to preserve the remnants of these beautiful horses. Today, most Appaloosas are ridden for pleasure. There were 10,669 Appaloosas registered in 1990. See Figure 12-7.

ANIMAL SCIENCE FACTS

Horse Gaits

Walk = slow, four-beat gait

Trot or jog = rapid, two-beat gait (opposing front and rear hooves move in unison)

Pace = rapid, two-beat gait (front and rear hooves on the same side move in unison)

Canter or lope = slow, three-beat gait

Gallop or run = fast, four-beat gait

Stepping pace = similar to the pace, but hooves on the same side do not strike the ground at exactly the same time. Performed by five-gaited horses.

Running Walk = smooth, four-beat gait performed by five-gaited horses

Rack = quick, showy, four-beat gait performed by five-gaited horses

Fox Trot = Choppy trot

Figure 12-5. Famed in folklore, Paint Horses were commonly used by American Indians.

Figure 12-6. Standardbreds were developed primarily for speed at the trot and pace. They are commonly referred to as the American Trotting Horse. (Courtesy, U.S. Trotting Association)

Figure 12-7. Apaloosas were first brought to North America by Spanish explorers in the 1600s. (Courtesy, Apaloosa Horse Club)

The Tennessee Walking Horse or Tennessee Walker was developed in the Tennessee Valley in the 1800s. Based primarily on Thoroughbred blood from a variety of sources, the Tennessee Walker executes a gait called the running walk. Walkers carry this unique gait smoothly easing the physical stress on the rider. Currently, Tennessee Walkers are exhibited as riding horses and ridden for pleasure. The breed association was first formed in 1935. Registrations equalled 7,800 in 1990. See Figure 12-8.

Figure 12-8. First bred 150 years ago, the Tennessee Walking Horse is a rugged breed with outstanding stamina. (Courtesy, Tennessee Walking Horse Breeders and Exhibitors Association)

Morgan horses were the first truly American breed. They were developed in the late 1700s from a single stallion called Justin Morgan. Justin Morgan, named after his owner, is thought to have been of Thoroughbred and Arabian breeding. Morgans, known for their versatility, were an asset on the Eastern frontier. Their short, round appearance originally lent itself to work as a draft animal. However, Morgans were also used as riding horses and were even raced occasionally. Morgans have been useful in the development of subsequent American horse breeds, such as the Standardbred and Tennessee Walker. There were 3,600 Morgans registered in 1990. See Figure 12-9.

Pinto Horses are a true color breed. Despite size or breeding, almost any horse or pony with a spotted color pattern can be registered as a pinto. The pinto color pattern originated from horses of Spanish extraction. Most Pintos are currently used for pleasure or as cattle horses. There were 2,900 Pintos registered in 1990. See Figure 12-10.

American Saddlebreds were developed in the East in the mid-1800s. They are used as a smooth gaited saddle horse. Many American Saddlebreds are five-gaited horses (see animal science facts on horse gaits). The background of American Saddlebreds consists of mainly Thoroughbred as displayed by their large size. Today, American Saddlebreds are used primarily for pleasure riding or show. The American Saddlebred Horse Association was formed in 1891 and registered 2,680 horses in 1990. See Figure 12-11.

Figure 12-9. The first truly American breed, Morgan horses were developed from a single outstanding stallion named Justin Morgan.

Figure 12-10. The Pinto color pattern originated from horses of Spanish extraction. (Courtesy, National Pinto Horse Registry)

Figure 12-11. Although American Saddlebreds are extremely alert and curious, they are highly intelligent and people oriented. (Courtesy, American Saddlebred Horse Association)

Palomino horses descended from golden colored horses brought to the United States by early Spanish explorers. Palomino coloring does not breed true. For instance, the mating of two Palominos will result in only 50 percent Palomino offspring. However, mating Chestnut colored horses to albino horses results in 100 percent Palomino colored offspring. Only horses of Palomino color can be registered as a Palamino. Two breed associations, established in the 1930s and 1940s, exist. Palominos are currently used for pleasure riding and cattle purposes. There were 2,000 Palominos registered in 1990. See Figure 12-12.

Figure 12-12. Palominos are golden or yellow in color with light colored manes. (Courtesy, Palomino Horse Breeders of America, Inc.)

The preceding breeds are all recognized as light horse breeds, mainly used for transportation. The following are heavy or draft horse breeds developed for use as work animals. All draft breeds descended from the native Flemish horse, a low set, heavily muscled animal. Draft horses peaked in popularity before the advent of farm tractors. In those days, draft horses pulled plows and other agricultural equipment. Draft horses are still used today by the Amish, much the same as in the early 1900s.

The Belgian breed was developed in Belgium and was first introduced into the United States in 1886. Belgians are the most massive of the draft breeds. Mature stallions can weigh over a ton. Most Belgians are chestnut, roan, or bay in color. The Belgian Draft Horse Corporation of America was established in 1887 and registered 3,445 animals in 1990. See Figure 12-13.

Originating in France, the Percheron was first brought to the United States in the 1840s and 1850s. Percherons are smaller than Belgians and are predominantly black or gray in color. Registrations totalled 1,200 in 1990 making them the second most popular draft breed. See Figure 12-14.

Clydesdales, originating in Scotland, are perhaps the best known draft breed to the American public. Although they are well recognized, 1990 registrations totalled only 438. Clydesdales were imported into this country

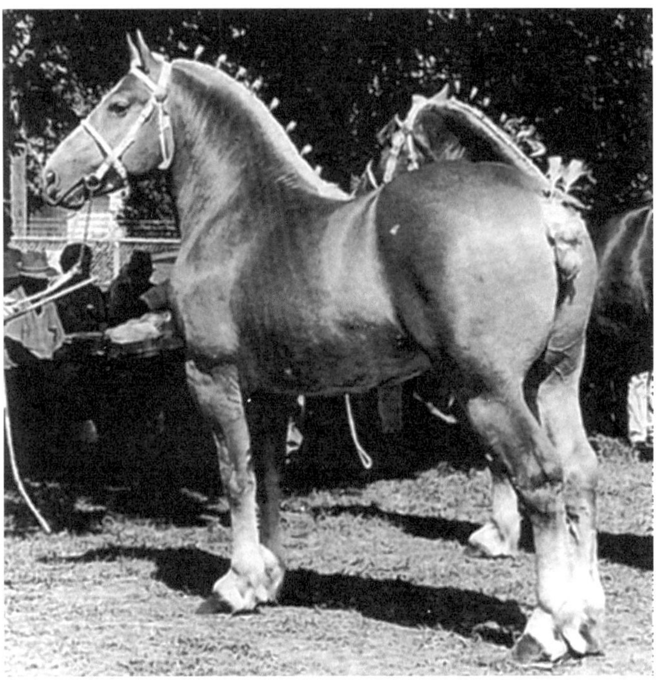

Figure 12-13. Belgians are by far the most numerous of all draft breeds in the United States. (Courtesy, Belgian Draft Horse Corporation of America)

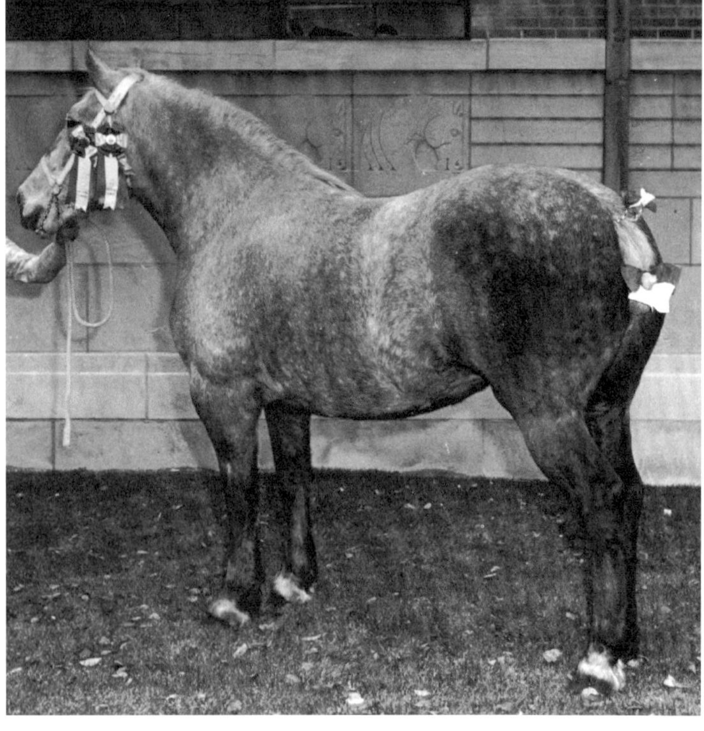

Figure 12-14. Percherons were the first draft breed to be imported. For their size, they are remarkably agile and mild-tempered.

in the 1870s. They are not as heavy as the Belgian and Percheron. Most are bay and brown with white markings, but blacks, grays, chestnuts, and roans are occasionally seen. See Figure 12-15.

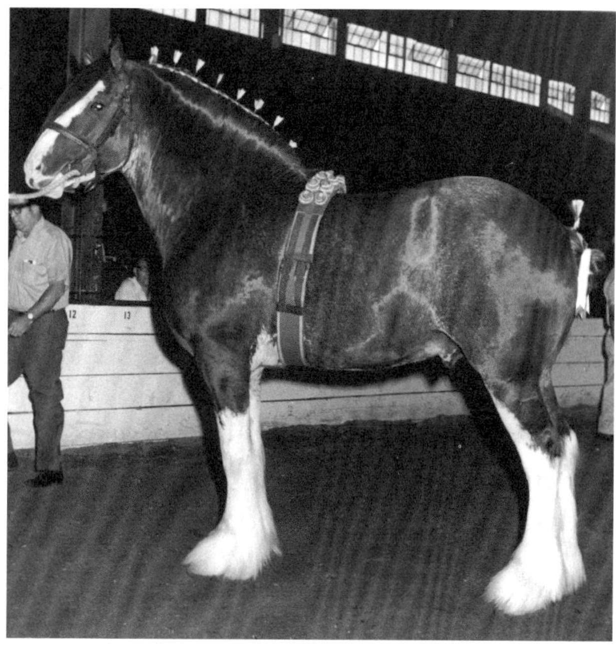

Figure 12-15. A Scotch breed, Clydesdales are noted for straight and high action.

According to recent registrations, several breeds of ponies show limited popularity. Ponies are less than 58 inches (14.2 hands) tall at the withers. There were 700 Shetland ponies and 597 Welsh ponies registered in 1990. Miniature Horses, which must meet stringent size requirements for registration (less than 30 inches at the withers), are exhibition animals popular in petting zoos and parks. There were 7,000 Miniature Horses registered in 1990. Ponies and Miniature Horses differ in their body proportions.

BREEDING

Large scale horse breeding is a highly specialized field. There are few producers in relation to other species of livestock. Farms and ranches in the business of breeding horses must have a genetically sound breeding herd and a planned marketing strategy for selling progeny.

Some herds may be in the business of producing registered Thoroughbred racehorses to train or sell to investors. Other breeders may breed a specific sire to a band of broodmares. That sire may be selected for his proven abilities in ***equestrian*** (horse and rider) events or harness races. Progeny from this type of breeding operation are not necessarily registered or purebred horses, but are simply bred to perform a desired task.

The biology of horse breeding is similar to other species of livestock. Horses are seasonal breeders like sheep. An increase in day length triggers the onset of estrus in mares. If left to natural means, heat periods begin in March or April and cease in September or October. Mares normally cycle every 21 to 23 days during the breeding season, but some mares are very irregular. Many producers use artificial lighting to trick mares into coming into heat at other times of the year. Heat detection for mares was described in chapter five.

The mating of a receptive mare is normally done naturally, either in a pasture setting or, more often, through ***hand mating.*** During hand mating, a handler prepares the mare for breeding by washing her vulva and wrapping the tail to keep it out of the way. The mare's rear legs are

ANIMAL SCIENCE FACTS

American Saddlebred Horse Association
4093 Iron Works Road
Lexington, KY 40511-8401
(606) 259-2742

Appaloosa Horse Club
P.O. Box 8403
Moscow, ID 83843
(208) 882-5578

Arabian Horse Registry of America
12000 Zuni Street
Westminster, CO 80234
(303) 450-4748

Belgian Draft Horse Corporation of America
P.O. Box 335
Wasbash, IN 46992
(219) 563-3205

American Morgan Horse Association
P.O. Box 960
Shelburne, VT 05482
(802) 985-4944

American Paint Horse Association
P.O. Box 961023
Fort Worth, TX 76161
(817) 439-3400

Palomino Horse Association
P.O. Box 324
Jefferson City, MO 65102
(314) 635-5511

Palomino Horse Breeders of America
15253 East Skelly Drive
Tulsa, OK 74116-2620
(918) 438-1234

Percheron Horse Association of America
P.O. Box 141
Fredricktown, OH 43019
(614) 694-3602

National Pinto Horse Registry
P.O. Box 486
Oxford, NY 13830
(607) 334-4964

Pinto Horse Association of America
1900 Samuels Avenue
Fort Worth, TX 76102-1141
(817) 336-7842

American Quarter Horse Association
P.O. Box 200
Amarillo, TX 79168
(806) 376-4811

Tennessee Walking Horse Breeders and Exhibitors Association
P.O. Box 286
Lewisburg, TN 37091
(615) 359-1574

The Jockey Club (Thoroughbred)
821 Corporate Drive
Lexington, KY 40503
(800) 444-8521

often hobbled during breeding so the mare does not move or kick and risk injury to the stallion. The mare is tied to a stall and the stallion is brought to the mare. The handler normally assists in copulation.

Artificial insemination (AI) technology is available for horse breeders. However, many breed associations will not allow the registration of progeny produced by artificial insemination. Other associations will allow registration, but semen may not be frozen or transported from where the stallion was collected. AI greatly reduces the chance of injury to both the stallion and mare during mating. It also allows more mares to be mated to a single, high quality stallion.

Gestation period for horses is about 11 months. Most foals are dropped in April, May, and June. Mares exhibit a "foal heat" one to two weeks after foaling. Most producers wait until the second heat, 25 to 30 days after foaling, to rebreed mares. This allows the mare time to recover from the stress of foaling.

NUTRITION

Survey 100 people on the street about what horses eat and the most likely answer will be "oats." The second most popular answer will probably be "hay." These two are excellent horse feeds, but others are routinely fed to horses as well.

As mentioned in chapter four, horses are post-gastric fermenters. The first part of the digestive tract is similar to that of a human or pig. However, the cecum of the horse is developed for the digestion of forages. Thus, forages are a preferred feedstuff for horses.

Good quality pasture is preferred as a forage for horses. Bluegrass, orchardgrass, bromegrass, fescue, and reed canarygrass function as good pasture species. Legumes, such as alfalfa, trefoil or clover, may be alone used as pasture or mixed with grasses. When stored forage must be fed, hay made from any of the above pasture species makes excellent feed if cut at a young stage of maturity. Young timothy hay is preferred by many equestrians, but timothy makes poor pasture.

Grains often added to horse diets include oats, barley, corn, wheat, and milo. Protein supplements include linseed meal, cottonseed meal, canola meal, and, of course, soybean meal. See Figure 12-16. Bran is often used as a source of fiber to maintain digestive tract function.

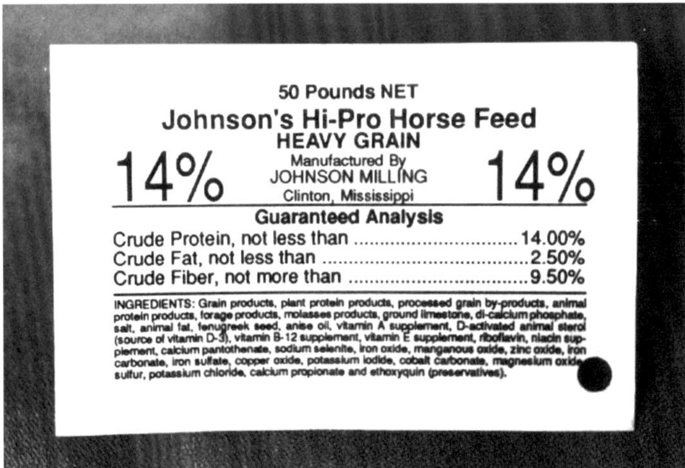

Figure 12-16. Grain-based horse feeds are often used to supplement a forage diet. (Courtesy, Jasper S. Lee)

Nutrition for breeding mares, foals, and yearling horses is similar to other species of livestock. Mares in late gestation and lactation have the highest nutrient demands and require top-notch nutrition. Both quantity and quality of feed are critical. Bran is often used in mare diets around foaling time to prevent constipation.

Foals should be creep fed beginning at two weeks of age. The nutrient density of mares' milk decreases drastically one to two months after birth.

After weaning, foals must continue to grow and develop at a rapid pace. A 400-pound, weaned foal should consume about 8 pounds of hay and 6 pounds of grain per day. Growth requirements indicate the need for high levels of protein and good quality hay or pasture. Rations should be gradually increased as the foal grows through its first winter.

Developing yearlings can survive on good quality pasture their first summer, but many producers supplement pasture with a small amount of grain. Yearlings entering their second winter will require about 8 pounds of grain and 12 pounds of hay per day.

Feeding pleasure horses presents some challenges. The nutrient needs of mature pleasure horses depend on the animal's size and level of activity. Work is divided into three categories. Light work is less than three hours of riding per day. Horses worked lightly need only $^{1}/_{2}$ pound of grain and $1^{1}/_{2}$ pounds of hay per day per 100 pounds of body weight. Medium work is three to five hours per day. Medium-worked horses require 1 pound of grain and 1 pound of hay per 100 pounds of body weight. Heavy work is more than five hours per day. Horses worked heavily require $1^{1}/_{2}$ pounds of grain and 1 pound of hay per 100 pounds of body weight. The amount and proportion of grain versus hay always increase with increased demand for energy.

Salt, vitamins, and minerals should be included in the grain mix or provided free choice for horses of all ages.

Changes in horses' diets should be made slowly. A gradual adaptation from one grain mix or forage to another over seven to ten days reduces the risk of an upset digestive system. Horses should initially be released onto lush spring pasture for short periods. The consequences of overconsumption of lush pasture will be discussed later in this chapter.

MANAGEMENT

Identification of horses was traditionally difficult. If a horse was stolen, the only way an owner could identify it was through some distinguishing physical marking. Some horses were hot branded (hot iron applied to the hide leaving a permanent scar) at weaning as a means of permanent identification. More recently, freeze branding (applying a cold iron to the hide, which causes hair to grow in white), and lip tattooing have become more popular identification methods.

ANIMAL SCIENCE FACTS

A Horse of a Different Color

Horse colors, such as black, white and gray, are self explanatory, however some other colors are a bit more vague.

Albino = White
Appaloosa = Mottled over rump or loin
Bay = Brown with black points (legs, mane, and tail)
Blue Roan = Dark and white hair intermixed
Buckskin = Tan with black points and dorsal stripe
Chestnut = Deeper reddish brown than sorrel
Sorrel = Reddish brown
Paint = Distinct patches of dark and white
Palomino = Golden
Red Dun = Reddish tint with dark points
Red or Strawberry Roan = Red and white hair intermixed

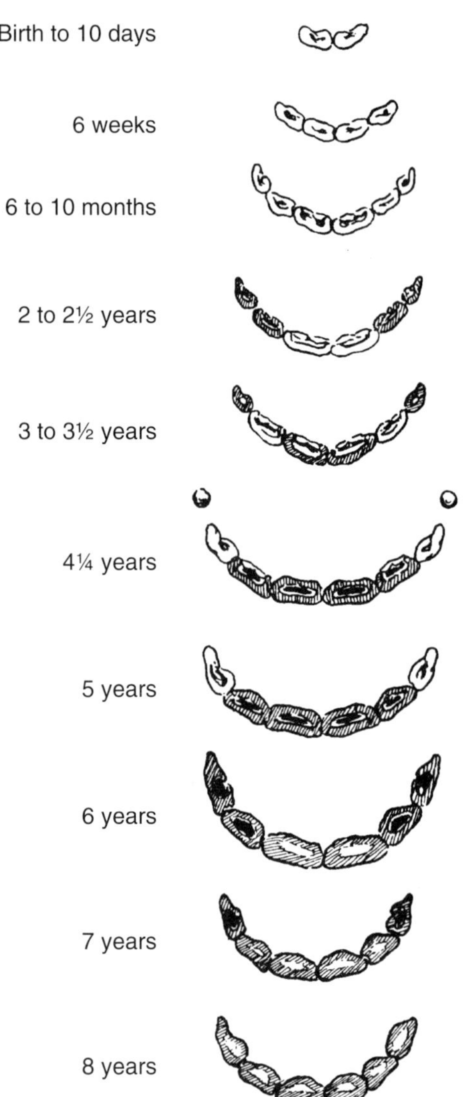

Figure 12-17. The age of a horse is determined by examining its teeth. This figure relates the age of a horse to teeth development.

The old saying, "Never look a gift horse in the mouth" is sound advice because, the age of a horse can be learned by looking at the teeth. A gift horse may be very old and of little value. However, since it is a gift, it is best to accept it "as is." If you are buying a horse, looking in the mouth is very important. Unscrupulous horse traders have made a good living passing off old horses to novices (those new to business). Temporary teeth appear in foals at a very young age. The entire complement of eight temporary teeth is in place by ten months of age. These temporary teeth wear until about two and one half years of age when the front set of permanent teeth replace the front set of temporary teeth. From two and a half until four years of age, all eight temporary teeth are replaced by eight permanent teeth. At four or five years of age, canine teeth appear behind the teeth already present. From that time on, these permanent teeth show varying degrees of wear from which an experienced equestrian can determine the age of any horse. Also, the angle of the teeth, when viewed from the side, changes with age. Teeth of young horses are nearly perpendicular, while older horses' teeth angle forward. Upper and lower teeth should meet evenly. An overbite or underbite is considered an undesirable conformation trait. See Figure 12-17.

Horses on pasture normally get enough exercise. However, horses confined in stalls must have some exercise every day or two in order to remain healthy. If horses are not ridden or hitched, they may be exercised using a ***longeing*** rope. Longeing a horse consists of attaching one end of the rope to the horse's halter. The other end is held by the caretaker in the center of a circle. The horse is then allowed to exercise around the perimeter of the circle while the caretaker retains control. See Figure 12-18.

Foot and hoof health and maintenance keep horses sound and functional. Horses must have healthy feet and hooves to provide locomotion for humans. Hoof trimming should be done every six to eight weeks by an experienced ***farrier*** (hoof caretaker). The need for trimming depends on the amount of hoof wear and the presence or absence of shoes. A good farrier can correct damaged or uneven hooves. Shoes should be applied to horses residing on rocky ground or horses asked to travel on roads. Shoes protect the hoof from excessive wear and deterioration. Hooves should be inspected and cleaned daily to avoid infections and spot possible causes of hoof damage. See Figure 12-19.

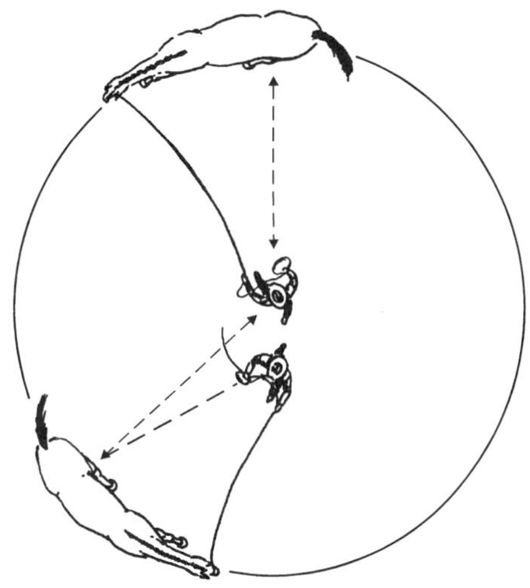

Figure 12-18. Longeing is one way of providing less active horses with exercise.

Figure 12-19. Farriers play an important role in proper hoof care and maintenance.

PARASITES, DISEASES, AND PREVENTION

Parasites in horses are similar to those of other livestock species. Flies cause considerable discomfort to horses. The elimination of breeding areas and good sanitation help keep fly numbers at manageable levels.

Mange mites can cause serious itching when infestation occurs. Mites can be prevented by avoiding contact with infested horses. Mange can be treated with a variety of insecticide powders and sprays.

Ringworm infests horses as well as cattle. It is most prevalent in winter when horses are confined to stalls. Horses infested with the ringworm fungus show round circular, hairless patches. Individual patches should be treated with an antifungal creme or iodine.

Figure 12-20. Ticks rob horses of nutrients and may transmit diseases. Fortunately, they can be treated with a variety of insecticides.

Ticks rob horses of nutrients by tapping the horse's blood supply. They can also harbor and transmit other horse diseases. The blood-filled bodies of ticks are easily seen protruding from the horse's hide. See Figure 12-20. Ticks can be controlled by many topical insecticides.

Many types of worms infest horses and rob them of body condition and energy. Horses should be wormed with an approved ***anthelmintic*** (worming medicine) every two months to keep worms under control. Worms can build up a resistance to an individual wormer, so anthelmintics containing different active ingredients should be used in rotation. Pastures should also be rotated to break the life cycle of worms.

Large worm loads in the digestive tract can result in colic, or abdominal pain. Worms cause colic by essentially plugging up the digestive system. Other causes of colic include a rapid change of diet, too much grain at a single feeding, or too much water after a period of heavy work. Colic can even be caused by improper tooth wear, which prevents the horse from chewing properly. A horse with colic may act nervous, kick at its side, or attempt to roll over. Colic can be prevented by frequent wormings, tooth inspection, and close attention to the horse's diet. Horses with colic should be treated by a veterinarian.

Encephalomyelitis (en-sef-a-lo-my-el-itis), also called sleeping sickness, is a horse disease carried by mosquitos. The disease infects the central nervous system causing the horse to act sluggish and sleepy. Death is common. Encephalomyelitis is easily prevented through an annual vaccination of all horses.

Equine infectious anemia (EIA) is a serious viral disease transported to horses by insects. Infected horses act depressed, lose weight, and may die. Those that live act as reservoirs for the virus, allowing it to be spread to other horses. The Coggins Test can detect equine infectious anemia before symptoms appear. Horses that test positive are quarantined, permanently identified as a carrier and often destroyed. Horses sold across state lines must have a recent negative Coggins Test. Yearly testing of all horses is recommended.

Equine influenza is a viral disease that attacks the respiratory system causing loss of appetite, depression, coughing, and nasal discharge. It is usually associated with an elevated temperature. Equine influenza is common anywhere horses from different farms are gathered, such as races and shows. Prevention consists of two initial vaccinations administered one to three months apart followed by an annual booster.

Founder, or laminitis, is an inflammation of the hoof caused by overconsumption of water, grain, or lush forage. Besides tender feet, a high temperature is usually associated with this ailment. Founder is prevented by restricting feed consumption to the amount required for maintenance, plus, growth or work. Water should be given a little at a time, especially in hot weather.

Potomac horse fever is a new disease causing high temperatures and severe diarrhea. Death losses can be significant. In areas where the disease is prevalent, a two-dose vaccination given three weeks apart is recommended. Treatment with tetracycline antibiotics is effective if a definite diagnosis has been made.

Rabies infections in horses are common, but can easily be prevented. Bites from infected animals, such as raccoons, skunks, or dogs, lead to violent behavior or, later, paralysis. Two initial vaccinations administered one month apart and an annual booster protect against this viral disease.

Distemper or strangles most often affects young horses and is caused by *Streptococcus* bacteria. The bacteria invade the lymph glands under the jaw. The glands enlarge and may break, producing pus. The disease is often accompanied by a cough and high fever. Distemper can be prevented through immunization every six to twelve months.

Tetanus is caused by the bacteria *Clostridium tetani,* which invades the body through open wounds. Initial symptoms include stiffness and progress to severe muscle spasms. Treatment is difficult, but prevention can be accomplished through two initial vaccinations spaced three weeks apart and accompanied by an annual booster.

Some horse vaccines are sold in combination, reducing the number of shots required.

HOUSING

Horse housing generally comes in two types: open shed and box stall. Open sheds are often used for horses on pasture or an exercise lot. The open face of the shed should be to the south or east so that the shed can be used as a windbreak. Open sheds should be constructed on high ground so that water from rain or snow drains away. The shed should be at least 8 to 9 feet high. About 150 square feet of floor space should be provided per horse. A feed storage area may be built above or as an addition to the shed. See Figure 12-21.

Box stalls are constructed inside a barn. Generally 12 feet square in size, the walls of box stalls are made of horizontal planking to about 5 feet high. Above the horizontal planking are vertical bars making the total stall height 7 feet. Box stalls normally have clay floors. Stalls should always be clean and dry with adequate bedding.

Proper ventilation of box stalls should be given serious consideration. A constant supply of fresh air keeps humidity at acceptable levels, reducing the incidence of respiratory disease. Adequate ventilation also reduces possible disease spread between horses.

In many horse barns, box stalls line both sides of an interior alleyway. In the center of the barn, one or two box stalls may be replaced by a ***tack*** (harnessing, riding, and grooming equipment) and feed room. Hay is often stored on the second floor.

Horses are most often fenced with board, welded pipe, woven wire, or pole fences. Vinyl covered high tensile wire fences, which simulate painted boards, are gaining popularity because of reduced maintenance.

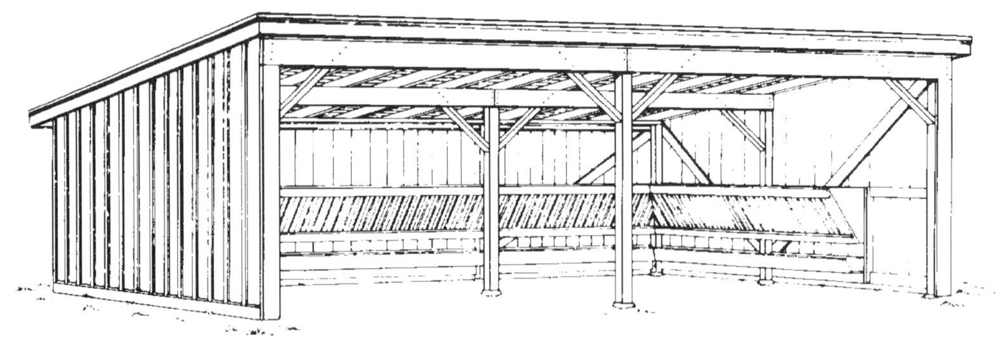

Figure 12-21. Open sheds are often used for horses on a pasture or an exercise lot. Simple in construction, they are effective as a windbreak.

SUMMARY

Most breeds of light horses are descendants from only a handful of original breeds. All draft horses descended from the Flemish horse. Horses are often bred to perform a certain task regardless of any purebred pedigree. Mares are seasonal breeders, normally conceiving only in the summer. Nutrition for horses relies on pasture (in season), good quality grass or legume mixed hay and grain. The amount of feed and proportions of hay and grain required depend on the size of the horse and its activity level. Young growing horses and lactating mares have high nutrient requirements. Methods of horse identification, age determination, exercising, and hoof care are all management practices to be mastered by equestrians. Parasites infesting horses are similar to those of other livestock species. Vaccines are available for most horse diseases. Many horses are housed in open sheds or in box stalls.

Most registration numbers and addresses provided in this chapter were courtesy of The American Livestock Breeds Conservancy via the following publication:

Bixby, D. E., Christman, C. J., Ehrman, C. J., & Sponenberg, D. P. (1994). *Taking stock: The North American livestock census.* Blacksbrug, VA: The McDonald & Woodward Publishing Company.

CHAPTER SELF-CHECK

____ draft	1. infection caused by *Streptococcus*
____ sulky	2. work
____ Palomino	3. cares for hooves
____ hand mating	4. riding and grooming equipment
____ equestrian	5. cart
____ longeing	6. golden or yellow with light colored mane
____ farrier	7. worming medicine
____ anthelmintic	8. laminitis (inflamed hoof)
____ encephalomyelitis	9. natural service, human assisting
____ founder	10. horse and rider
____ distemper	11. sleeping sickness carried by mosquitoes
____ tack	12. horse exercises by circling handler

QUESTIONS AND PROBLEMS FOR DISCUSSION

1. Explain the origin of the Quarter Horse name.
2. Over 75 percent of all horses carry this breed's blood. Name the breed.
3. To meet registration requirements, Paint horses must trace their ancestry to the following three breeds: _______________, _______________, and _______________.
4. Which breed is credited with being the first to be developed in the United States?
5. Are mares naturally receptive to stallions during the winter?
6. When in season, mares normally cycle every ____ to ___ days.
7. Explain post-gastric fermentation.
8. Creep feeding in foals should begin at ___ to ___ weeks of age.
9. How many pounds of grain should a 400 pound, weaned foal consume?
10. List two methods of identifying horses.
11. Teeth angle sharply forward in young/old horses. Select one.
12. Founder is caused by _________.
13. Tetanus invades the body via _______________.
14. Write a common box stall dimension.
15. Define tack.

ACTIVITIES

1. Horses are measured in hands. One hand equals 4 inches (approximately the same size as a clenched fist measured vertically from little finger to thumb). Compare your fist measurement with actual inches. Is it close to four? Measurements of horses are taken at the base area of the neck called the withers. Ponies are generally considered to be 13 hands 2 inches or less. (54 inches). Practice converting the following measurements into hands: 69", 88", 63", 44". Placing one fist on top of another measure an object in hands. Convert to actual inches. Compare results.
2. Inquire about local horse shows held in your locale. Visit one, if possible.
3. Report on the various rodeo associations.
4. Write to breed associations. Inquire about registration procedures.

LABORATORY ACTIVITY

ESTIMATING ANNUAL HORSE EXPENSES

Purpose

To investigate the yearly costs of keeping a 1,000 pound pleasure horse

Materials

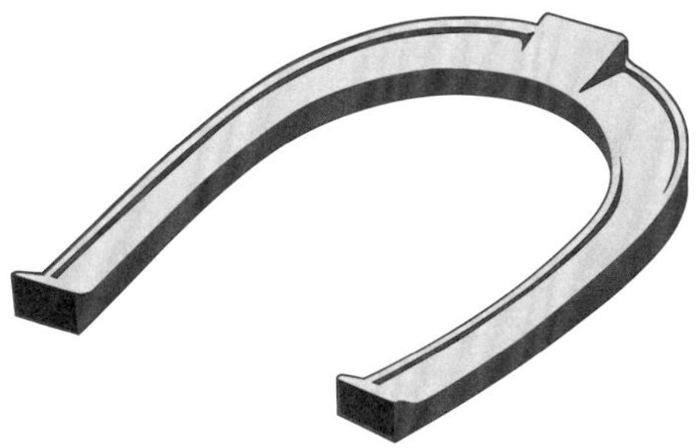

pencil
paper
calculator
farm paper or reference containing current hay and grain prices

Procedure

1. One or two days before the assigned laboratory period, divide students into five groups. Assign each group one of the following topics: grain, hay, housing, veterinary, foot care.
2. Grain group instructions:
   a. Using information from the nutrition section in chapter 12, calculate the amount of grain required per year for a 1,000 pound horse under a medium work load.
   b. Using the following grain mix, calculate the cost per 100 pounds. Use a farm paper for current corn and oats prices.

Shelled Corn	30
Oats	70
Total	100

   c. Multiply the amount of grain required per year (in pounds) by the cost of the grain mix per pound (cost per 100 pounds/100) to calculate the yearly grain cost.
   d. Assume salt, minerals, and vitamins cost $30 per year.
   e. Add salt, mineral, and vitamin cost to total yearly grain cost for a total yearly non-forage feed cost.
3. Hay group instructions
   a. Using information from the nutrition section of chapter 12, calculate the amount of hay (in tons, 1 ton = 2,000 pounds) required per year for a 1,000 pound horse under a medium work load. Assume no pasture.

b. Using the farm paper, look up the price per ton of excellent quality grass hay or good quality grass and legume mixed hay.

c. Multiply the number of tons needed by the price per ton to calculate the total yearly forage costs.

4. Housing group instructions

   a. Contact one or two local horse boarding establishments or a horse breeder. Ask for a price on boarding a horse for a year *not including feed costs.* Obtaining an accurate cost may require some detective work.

5. Veterinary group instructions

   a. Contact a local horse veterinarian and ask for an annual cost of checkups and vaccinations for a single horse.

6. Foot care group instructions

   a. Contact a farrier or horse breeder in your area for an estimate on yearly foot care costs for one horse, include hoof trimming and shoeing.

Analysis

1. Each group should bring their findings to class the day of the scheduled laboratory.
2. In succession, each group should report on the steps they used to calculate or investigate their topic.
3. Calculate a total yearly cost for a 1,000 pound pleasure horse by adding the costs obtained by each group.
4. Discuss any possible income that could be obtained from the horse to help offset your calculated expenses. What additional expenses would these income activities incur?

Application of Laboratory Activity

First time owners sometimes do not realize the expenses incurred to feed and care for a horse. Often, they only realize these costs after the horse has been purchased. Such expenses as feed, housing, and medical and foot care must be taken into account before the purchase. Exercises such as this laboratory activity minimize surprises and maximize pleasure from owning a horse.

This laboratory activity details all the major costs associated with owning a horse. Also, it demonstrates the scope of allied horse industries, and how such businesses as feed suppliers, boarding operations, veterinarians, and farriers impact the condition of a horse and the costs of ownership.

UNIT III

Evaluation

Chapter 13

JUDGING CONTESTS

Eenie Meenie Minie Mo

INTRODUCTION

Judging contests serve many purposes. Along with teaching students to evaluate animals or cuts of meat, they teach many important life skills. The decisions made during the judging contest require a contestant to draw from all of the pertinent information available and use logical problem solving skills. This decision making process must be clarified, organized, and defended when giving oral or written ***reasons***, an explanation of the logic used for one's placing order. Finally, the oral delivery of reasons improves public speaking skills, while written reasons help one's writing and organizational skills. Few students may evaluate livestock or meat after their scholastic days, but the skills and self-confidence gained in judging contests will remain with them for life.

Figure 13-1.

OBJECTIVES

1. Explain the format of a judging contest
2. Correctly complete a placing card
3. Take notes for use in oral reasons
4. Give a set of oral reasons
5. Score an incorrectly placed class given an official placing and cuts

TERMS

cull
cuts
grants
keep–cull
reasons

ANIMAL SCIENCE FACTS

In the 1920s, vocational agriculture students were invited to participate in livestock judging contests at the American Royal Rodeo, held annually in Kansas City, Missouri. This event was the precursor to the current National FFA Livestock Judging Contest.

CONTEST FORMAT

Judging contests are normally based on classes (groups) of four. Four exhibits (animals, carcasses, or cuts of meat) are presented in a group. Each exhibit is randomly identified by a number. Depending on the size of the contest and the amount of space available, half the classes are often displayed at once. For example, if the contest contains six classes, three may be displayed at one time. After all three classes have been evaluated, the contestants take a break while the other three classes are prepared for evaluation. If enough space is available, all classes can be presented at once. See Figure 13-2.

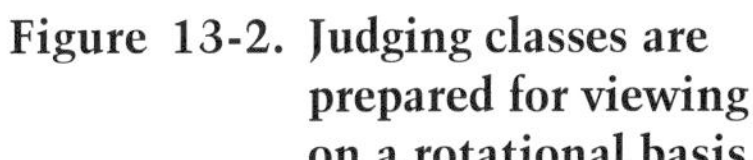

Figure 13-2. Judging classes are prepared for viewing on a rotational basis.

Upon arrival at a contest, judging team members are divided into groups, usually designated by color or number, and given a placing card for each class to be judged. If possible, team members are assigned to different groups. Each group will have a contest assistant acting as a group leader. When the contest begins, the group leader moves the group to the first class. Contestants' backs should be turned to the class upon arrival. The group leader or a loud speaker will announce when judging can begin. See Figure 13-3.

The amount of time allowed for placing each class varies. Twelve to 15 minutes is standard. During the allotted time, contestants evaluate the class presented, and rank the exhibits from best to poorest. For instance, if a contestant thinks number four is the best, number two the second best, number three the third best, and number one the poorest, the

Figure 13-3. Contestants begin the contest with their backs to the first class.

contestant's placing would be 4–2–3–1. If the contestants will give reasons on the class, notes are taken to help contestants remember what the animals, carcasses, or cuts of meat look like.

At the end of the allotted evaluation period, time is called and contestants are instructed to turn their backs to the class. Each contestant then fills out a placing card and gives it to the group leader. The contestant's ranking of the class will be compared to an official placing, and a score calculated. A perfect score (correct placing) is worth 50 points. After cards are collected, the group moves to the next class.

Contestants are not allowed to talk or use reference materials during the contest. Infractions of these two rules will result in expulsion from the contest or a zero score on the class. A zero score is also given to contestants who improperly complete the placing card. Contestants routinely forget to record a placing or fill in their contestant number. Contest scorers have no way of knowing the card's owner or an intended placing.

Placing cards come in several varieties. Manually scored cards contain a space for contestant number, class name, and four blanks for the contestants' placing. See Figure 13-4. Placing cards scored by computer contain the same blanks for contestant number and name of the class. However, instead of writing the placing, all possible placings of the class are listed on the card. The contestant must find the desired placing and simply mark that placing with an X. Similar cards may require the contestant to circle the desired placing. See Figure 13-4.

Another type of class found in many contests is known as a ***"keep–cull"*** class. In a keep–cull class, eight animals are presented instead of four. The objective of this of class is to simulate decisions livestock producers make in selecting replacement breeding stock. The contestant must select four animals to "keep" as replacements and four animals to ***"cull"*** or eliminate or sell. Contestants mark the four numbers of the animals to keep (or cull, depending on the contest—some contests require the four cull animals to be selected) on their placing cards. Again, a correct placing would result in 50 points.

While the contest is in progress, experienced judges rank each class and take notes on the important characteristics of the class and its placing. At the end of the contest, these official placings and reasons will be shared with the participants.

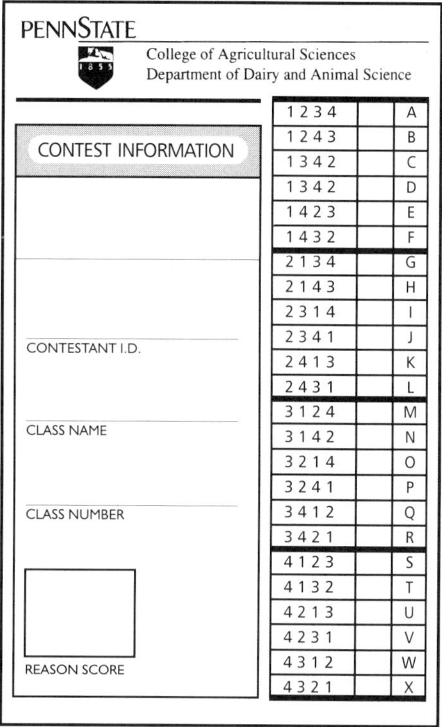

PENNSTATE

College of Agricultural Sciences
Department of Dairy and Animal Science

CONTEST INFORMATION

CONTESTANT I.D.

CLASS NAME

CLASS NUMBER

REASON SCORE

Placing		Code
1 2 3 4		A
1 2 4 3		B
1 3 4 2		C
1 3 4 2		D
1 4 2 3		E
1 4 3 2		F
2 1 3 4		G
2 1 4 3		H
2 3 1 4		I
2 3 4 1		J
2 4 1 3		K
2 4 3 1		L
3 1 2 4		M
3 1 4 2		N
3 2 1 4		O
3 2 4 1		P
3 4 1 2		Q
3 4 2 1		R
4 1 2 3		S
4 1 3 2		T
4 2 1 3		U
4 2 3 1		V
4 3 1 2		W
4 3 2 1		X

Figure 13-4. Sample judging card. Courtesy of the Department of Dairy and Animal Science, College of Agricultural Sciences, The Pennsylvania State University.

REASONS AND QUESTIONS

Many contests, especially for experienced participants, require contestants to defend their placings through oral (livestock and dairy contests) or written (meat contests) reasons. Oral reasons are given to an official judge in a one-to-one setting. Notes taken during the class may be used as an aid in preparation for a set of reasons. The presentation of reasons must be made from memory.

There are four components to a good set of reasons: accuracy, organization, terminology, and delivery.

Accuracy is the first and most important component of a good set of reasons. Tell the truth. Do not try to invent reasons just to impress the judge. If there was no difference between two animals concerning a certain characteristic, it should not be discussed in the reasons.

Organization is the second component in a good set of reasons. A set of reasons is organized into three pairs. The top pair compares the first and second placing, the middle pair compares the second and third placing, and the bottom pair compares the third and fourth placing. The order of reasons always follows the sequence: top pair, middle pair, bottom pair.

Within each pair, there is a logical order of discussion. In this example pair, 1 placed above 2.

1. The important reasons why 1 placed over 2. Major reasons are mentioned first in an opening statement, followed by smaller details that support the opening statement.
2. Any characteristics about 2 that were better than 1 (called ***grants***).
3. Faults of 2 (optional).

The basic organization of an entire set of reasons follows:

1. Introduction:
   "I placed this class of vermites (an imaginary animal) 4–3–2–1." Note, that numbers are used as identification.
2. Opening statement, detailed reasons, grants, and faults for the top pair.
   a. Opening statement: "In my top pair, 4 placed over 3 because 4 was bigger and had more desirable fur."
   b. Detailed reasons: "4 was taller and had longer legs. In addition, 4 had longer, more luxurious fur, particularly between its ears and on its front legs."
   c. Grants: "I grant that 3 had a longer, bushier tail."
   d. Faults: "However, she had bad breath."
3. Transition to middle pair: "Moving to my middle pair, I placed 3 over 2."
4. Opening statement, detailed reasons, grants, and faults for middle pair.
5. Transition to bottom pair.
6. Opening statement, detailed reasons, and grants for bottom pair.
7. Reasons why 1 was last. "Nonetheless, 1 was the skinniest, smelliest, most uncoordinated vermite in the class."
8. Conclusion: "Thank you."

Obviously, the terminology used for an actual set of reasons would be appropriate for the species in question.

Correct terminology is the third component of a good set of reasons. Use of incorrect terminology quickly identifies the contestant as a novice.

The final component in a set of reasons is the delivery. The contestant should enter the reasons room in a confident manner, remain 8 to 12 feet from the judge, stand with feet squarely placed, and hands held behind the back. Hats and notes should be left outside the reasons room. Reasons should be delivered in a slow, smooth, confident, and authoritative manner. Voice should be conversational in tone, loud enough to be understood, and show good voice inflection. Good eye contact with the judge improves the impression of confidence. Practicing reasons in front of a mirror can help in gaining confidence and improving eye contact. See Figure 13-5.

Figure 13-5. Correct reasons stance.

A well delivered and organized set of reasons should last less than two minutes.

Taking notes serves to remind the participant of important physical characteristics, which influenced class placing. Logical arrangement of notes eases the organization process in preparing a set of reasons. Figure 13-6 shows a sample note taking format, and Figure 13-7 shows sample notes for the class of vermites. Judges rely less and less on their notes as they gain more experience. After much practice, classes of livestock seem to imprint themselves in the "mind's eye," where they can be "seen" and accurately described during a set of reasons.

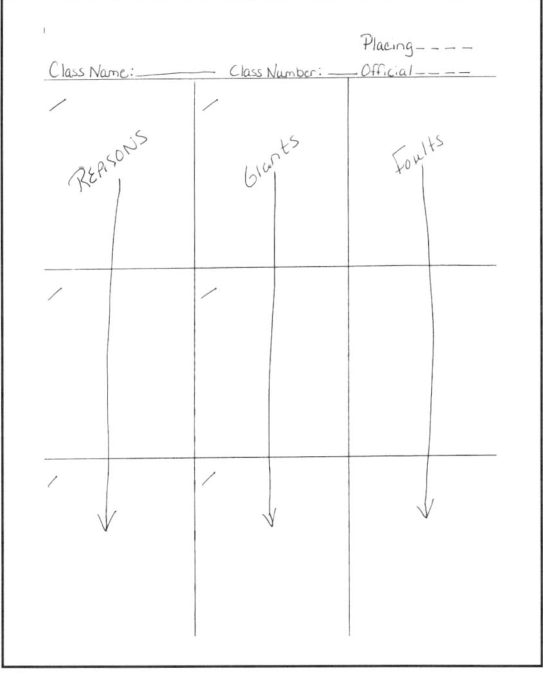

Figure 13-6. Organized note taking format.

Placing 4 3 2 1

Class Name: Vermites Class Number: 1 Official _ _ _ _

4/3 Bigger - long legs - taller Better fur - longer - ↑quality	3/4 Long, Bushy Tail	3/ Bad Breath
3/2 Movement - quicker - takes longer steps Sharper Teeth	2/3 Pointier ears	2/ Limps
2/1 Longer eyelashes - more refined Toes - Longer - more - Better Pedicure	1/2 Longer tongue	1/ Skinny Smelly Uncoordinated

Figure 13-7. Sample notes for the animals.

Younger participants in a judging contest may be asked questions about each of the designated reasons classes. An example question follows: "Between 1 and 4, which vermite had longer whiskers?" Questions are answered without the use of notes.

CUTS AND SCORING

Between each pair in the placing, officials assign a ***"cut."*** Each cut, a number, indicates the difficulty of that placing. For instance, if the top pair was very close, the cut would be small (one or two). A large cut would be assigned when one animal of the pair easily places over another animal. Six or seven points would be a large cut. Placings of intermediate ease receive intermediate cuts. Total cuts cannot exceed 15.

Cuts are used to score incorrectly placed classes. To illustrate, the official placing of a class was 4–3–2–1, with cuts of 2–5–4. The contestant placed the class 3–4–2–1. The contestant switched the top pair and would lose the top pair cut (2). Thus, the contestant's score would be 50 – 2 = 48. A middle pair switch (4–2–3–1 placing) would score 45 (50 – 5 = 45), while a bottom pair switch (4–3–1–2 placing) would score 46 (50 – 4 = 46). A correct placing would receive 50 points.

Other incorrect placings are more difficult to score. Every possible placing must be analyzed and the proper cuts deducted. In all situations, the process seen in Figure 13-8 must be followed.

ANIMAL SCIENCE FACTS

Livestock judging remains the most popular FFA skills contest. More FFA members compete in livestock judging at all levels of competition, local through national, than any other contest. Information courtesy of the National FFA Organization via Carol Duval.

Questions:	If Yes,	If No,
1. Was 1 placed over 2?	Subtract 0	Subtract a
2. Was 1 placed over 3?	Subtract 0	Subtract a + b
3. Was 1 placed over 4?	Subtract 0	Subtract a + b + c
4. Was 2 placed over 3?	Subtract 0	Subtract b
5. Was 2 placed over 4?	Subtract 0	Subtract b + c
6. Was 3 placed over 4?	Subtract 0	Subtract c
Use this formula to compute scores after a contest. Fortunately, contest officials have computer programs and precalculated scoring grids to help score contests.		

Figure 13-8. Official Placing 1–2–3–4 cuts a,b,c,

SUMMARY

Judging contests assist students in developing critical thinking skills. Along with placing classes, students are asked to justify the placing with reasons. The four key components of a good set of reasons include: accuracy, organization, correct terminology, and delivery. Accuracy is the most important component. Keep-cull classes require students to select animals to keep as herd replacements. Classes are scored with a system of cuts that more harshly penalizes easy pair switches.

CHAPTER SELF-CHECK

___ reasons	1. eliminate or sell
___ keep–cull	2. indicates difficulty of placing
___ cull	3. simulation of replacement selection
___ grants	4. reasons why second place in pair may be better than first
___ cut	5. explanation of placing logic

QUESTIONS AND PROBLEMS FOR DISCUSSION

1. Judging classes usually contain _____ (number) animals, carcasses, or cuts to place.
2. Animals, carcasses, or cuts are designated by a _____ within each class.
3. A perfect score for a correct placing earns the contestant _____ points in a judging contest.
4. True or False? Judging contests normally begin with participants facing the classes.
5. Contestants in judging contests can earn a zero score for three basic infractions. Name them.
6. Most judging classes last for _____ to _____ minutes.

7. A __________ class simulates producer selection of herd replacements.
8. May judging contest participants take notes during the classes?
9. Explain the four components to a good set of reasons.
10. Define cull.
11. What is a grant?
12. Describe the stance competitors should take when delivering oral reasons.
13. If a contestant switched the top pair in class with cuts of 6–2–1, what score would the contestant earn?
14. What FFA skills contest remains the most popular?
15. Describe the origin of the National FFA Livestock Judging Contest.

ACTIVITIES

1. Organize a judging contest for class. Formulate classes of four items (boots, candy bars, books). Have selected students be "official" judges for classes. They should place the classes and assign cuts. Designate the officials to hear and score reasons. Have other students act as a scoring committee. Compute a class winner. Award prizes.
2. Sponsor a judging contest for the school faculty during FFA week. Award a significant prize to the winner.
3. Contact state FFA and 4-H offices for information on judging contests.

LABORATORY ACTIVITY

DELIVERY REASONS

Purpose

To practice delivering a set of reasons

Materials

paper and pencils for notes
one shoe from each student in class

Procedure

1. Arrange student shoes in classes of four.
2. Using correct format, take notes for giving reasons on all classes of shoes. Allow 12 minutes per class.
3. Remove classes from sight.
4. Have students write reasons for each class.
5. Randomly select students to orally deliver reasons.

Analysis

Score students on notes, as well as written and oral reasons. Consider format, organization, and delivery in evaluation.

Application to Animal Science

Consumers judge goods constantly. Being able to critically select products and services is essential for success in everyday life.

Chapter 14

PERFORMANCE DATA

Crunching Numbers

INTRODUCTION

Livestock and most livestock products are sold by the pound. However, individual animals within a species differ in their genetic ability to produce pounds of product. Performance data are used by livestock producers to identify animals that carry desired genes for growth, leanness, milk production, or other economically important traits. Animals with the best performance data can then be retained as sires and dams for the next generation. Rapid genetic improvement on any trait relies heavily on the use of performance data.

Figure 14-1.

OBJECTIVES

1. Explain the significance of performance data to livestock breeders
2. Compare EPDs, EBVs, indexes, and actual data
3. Discuss why selection may vary depending on environmental conditions
4. Use performance data in judging classes and on-farm selection

TERMS

average daily gain (ADG)
backfat (BF)
birth weight EPD (BW EPD)
days to 230 pounds (DAYS)
estimated breeding value (EBV)
expected progeny difference (EPD)
FEPDs or flock EPDs
general purpose index (GPI)
individual data
litter weight at 21 days (LW21)
maternal line index (MLI)
maternal milk (MM)
multiple trait indexes
number born alive (NBA)
percent difficult births in heifers (%DBH)
predicted transmitting ability (PTA)
predicted transmitting ability for type (PTAT)
production type index (PTI)
PTA$ for cheese yield (CY$)
PTA$ for milk and fat (MF$)
PTA for milk, fat, and protein (MFP$)
ratio
scenario
sow productivity index (SPI)
terminal sire indexes (TSI)
type-production index (TPI)
weaning weight EPD (WW EPD)
yearling weight EPD (YW EPD)

ANIMAL SCIENCE FACTS

It used to be necessary to slaughter an animal to collect data relating to carcass traits, such as backfat and loineye. Obviously it was impossible to use a carcass as breeding stock even if it turned out to be exceptionally lean or heavily muscled. Until the advent of ultrasound technology, this type of data had to be collected from close relatives and progeny, and assumptions made about the animal in question. Real-time ultrasound allows seedstock producers to literally see and measure the amount of fat and muscle in a live animal. Many prospective sires and dams can be ultrasonically scanned and only the best retained as parents. A real-time ultrasound probe emits sound waves through the skin, fat, muscle, and bone. Sound waves reflect any change in tissue type. The reflected sound waves are collected and processed by a sophisticated machine, and the resulting cross-sectional picture is displayed on a screen. Actual measurements can be taken from this realistic picture. EPDs for the ribeye area in beef cattle have evolved because of the use of ultrasound. See Figure 14-9.

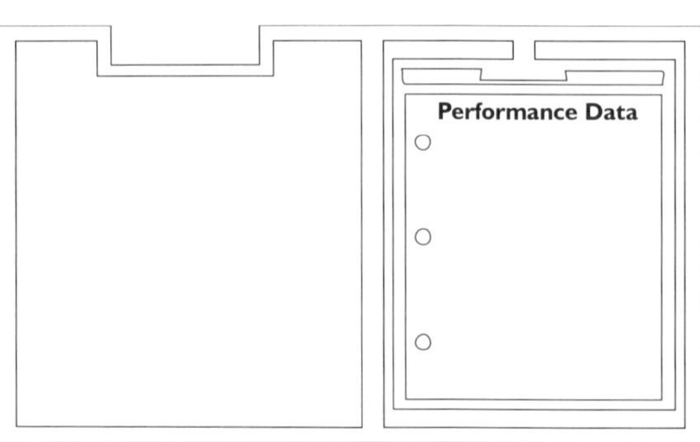

TYPES OF PERFORMANCE DATA

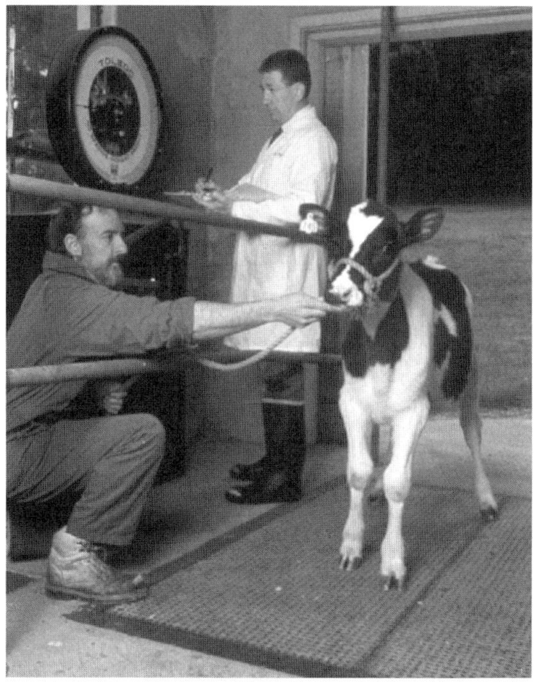

Figure 14-2. Accurate scales are critical in data collection. (Courtesy, Agricultural Research Service, USDA)

Individual data form the bedrock upon which all other performance data categories are based. See Figure 14-2. ***Individual data*** reflects the performance of an individual animal for a specific trait. For example, a bull's individual performance for ***average daily gain (ADG)*** may be 4.2 pounds per day.

Calculation of individual data is based on accurate records. Three pieces of information are needed to calculate the above bull's average daily gain. First, a beginning, or on-test weight at the start of the feeding period is needed. Second, an ending, or off-test weight at the conclusion of the feeding period is needed. The third piece is the length of the feeding period in days. Average daily gain is calculated using the formula:

(off-test weight – on-test weight) ÷ days on feed

Other individual data require only one measurement. Birth weight, number born, number weaned, backfat, loineye size, weaning weight, and milk production are all traits that require only one measurement. Other types of performance data rely on individual performance data from more than one animal.

Individual data from a contemporary group of animals allows comparison within the group by a ***ratio***. Ratios compare animals using a percentage approach for a single trait. Using the above bull example, if the average of the bull's contemporary group was 4.2 pounds of gain per day, the previously mentioned bull's ratio would be 100. Most ratios (where higher numbers are more desirable) are calculated using the formula:

$$\frac{\textbf{selected animal's measurement}}{\textbf{group average}} \times \textbf{100}$$

Another bull from the same contemporary group with an average daily gain of 4.0 pounds per day would have a ratio of 95 (5 percent below average). A bull that gained 4.5 pounds per day would ratio 107 (7 percent above average).

Indexes can be formulated by geneticists to select for several traits simultaneously. These indexes are called ***multiple trait indexes***. For example, terminal sire boars may be selected for average daily gain, backfat, and loineye size. A terminal index would compare all three measurements to the contemporary average and weight each trait according to its economic importance. For instance, if a 1 percent change in backfat is worth $.10, and a 1 percent change in loineye size is worth $.05, backfat would be weighted twice as heavily as loineye size. Like single trait ratios, multiple trait indexes are based on 100. Maternal indexes can be calculated emphasizing traits such as weaning weight and milk production with a lesser emphasis on carcass traits. Ratios and indexes more accurately show genetic value than individual data because of the comparison to contemporaries.

An ***estimated breeding value (EBV)*** takes the index concept one step further. EBVs include the heritability of a trait. Also, EBVs present an estimate of the animal's genetic worth as a parent in comparison to other animals in a group. EBVs use all available records of relatives within the same herd or flock. EBVs are presented in index form with 100 as an average. An EBV of 102 means that the animal's progeny should perform 2 percent above the average for that flock or herd for a specific trait. Estimated breeding values are more useful than either indexes or individual

ANIMAL SCIENCE FACTS

Ultrasound technology cannot accurately predict the amount of intramuscular fat (marbling). To obtain EPDs for marbling, producers still must rely on progeny or relative slaughter data.

data because records from more individuals are incorporated into the calculation. See Figure 14-3.

The most recent and useful performance data for meat animal species is the ***expected progeny difference (EPD)***. Like EBVs, EPDs also include all available records of relatives and incorporate heritability. However, through the use of the BLUP (best linear unbiased predictor) computer program, EPDs can also include records from relatives outside the herd. Even records of dead ancestors can be used.

Estimated Breeding Values		Acc.
Birth	95	54%
Weaning	101	42%
Maternal	97	64%
Yearling	99	38%

Figure 14-3. Beef pedigree showing EBVs. (Courtesy, American Maine Anjou Association)

EPDs are presented in practical units such as pounds, days, or inches. For example, an Angus bull with an EPD for a weaning weight of +10 pounds should sire calves with weaning weights 10 pounds heavier than those sired by an Angus bull with a WW EPD of 0. In contrast, an Angus bull with a –10 EPD for weaning weight should sire calves with weaning weights 10 pounds lighter than those sired by an Angus bull with a WW EPD of 0. If presented with individual data, indexes or EBVs, EPDs should be given more emphasis in selection. EPDs can be used to compare animals of the same breed in different herds. See Figure 14-4. The breed average EPD for a given trait is not necessarily zero. For example, the average weaning weight EPD for Angus bulls is 22.2 pounds.

Sold to: White Oak Mills
419 W. High St.
Elizabethtown, PA 17022

Performance Record

Traits	EPD	Indexes	EPD
Pigs Born Alive	0.08	Maternal Line	106.5
21 Day Weight	-0.01	Terminal Sire	113.6
Days to 230	0.90	SPI	100.8
Backfat Thickness	-0.06		

Reg. No. Date Bred

Figure 14-4. Swine pedigree showing EPDs. (Courtesy, White Oak Mills)

Dairy geneticists have improved on the EPD concept. In the dairy industry, EPDs are known as ***predicted transmitting ability (PTA)***. Using current market prices for milk, percent butterfat, and percent protein, PTAs are translated into a dollar value that predicts the added income to be gleaned from an individual sire's daughters. For example, a bull with a ***PTA for milk, fat, and protein (MFP$)*** of $100 would sire daughters whose milk (because of increased milk production, fat and protein content) would be worth $100 more than daughters of a bull with an MFP$ of $0. Again, these dollar values (or the raw PTAs) can only be used to compare sires within the same breed. PTAs can also be negative. See Figure 14-5.

ACTIVE A.I. PTI ORDER — JERSEY BULL EVALUATION LIST FOR JANUARY 1995

NAME OF BULL	NUMBER	NAAB CODE	NO HRDS	NO DTRS	% RIP	% REL	MILK	% FAT	FAT	% PROT	PROT
MASON BOOMER SOONER BERRETTA	651835	7J254	232	467	41	96	+1481	-.22	+35	+.07	+66
COMFORT ROYAL ALF–ET	651068	1J382	21	32	22	71	+1188	+.09	+70	+.08	+57
DUNCAN DUKE OF GLENWOOD	649231	29J2910	34	67	9	81	+1642	-.04	+70	-.10	+45
COMFORT PAL ADONIS–ET	651974	14J216	22	41	34	75	+1177	+.11	+71	+.03	+49
OSBS MISTER T	652229	14J218	23	33	24	71	+1409	-.17	+40	-.02	+50
AVON ROAD TRADER–ET	652247	7J252	97	150	23	91	+1254	+.10	+74	-.01	+45
HL DUNCAN TRUSTEE	651346	8J295	50	65	34	82	+1364	-.15	+41	-.02	+47
HIGHLAND DUNCAN LESTER	645454	29J2875	746	2865	72	99	+1174	-.11	+38	-.02	+41
PURPLE SOONER KYLE–ET	652914	122J4227	13	33	30	71	+1567	-.25	+35	-.09	+44
CLOVER FARMS PROTEIN	652401	7J258	40	54	52	79	+1691	-.32	+29	-.08	+51

(continued)

NAME OF BULL	NUMBER	CY $$	PROT $$	REL SCS	PTA SCS	REL PL	PL	% US	PCT ILE	PTI
MASON BOOMER SOONER BERRETTA	651835	+243	+198	91	3.47	74	+2.9	100	98	+397
COMFORT ROYAL ALF–ET	651068	+233	+189	58	3.34	47	+1.1	100	96	+340
DUNCAN DUKE OF GLENWOOD	649231	+191	+197	65	3.25	58	+2.0	100	97	+324
COMFORT PAL ADONIS–ET	651974	+206	+178	61	3.31	45	+1.5	100	91	+316
OSBS MISTER T	652229	+191	+174	56	3.47	43	+1.7	100	87	+301
AVON ROAD TRADER–ET	652247	+194	+178	81	3.45	63	+1.3	100	83	+289
HL DUNCAN TRUSTEE	651346	+182	+167	66	3.19	50	+2.0	100	93	+284
HIGHLAND DUNCAN LESTER	645454	+160	+146	98	3.52	84	+2.3	100	75	+281
PURPLE SOONER KYLE–ET	652914	+167	+171	58	3.37	44	+1.7	100	88	+272
CLOVER FARMS PROTEIN	652401	+188	+184	63	3.47	43	+1.2	100	90	+269

Figure 14-5. Many types of performance data are available in a dairy sire listing.

The numbers generated by a BLUP are only as accurate as the number of records incorporated into the program. If few records are available, the accuracy (reported as reliability in dairy) is low. As more records are added to the animal's performance profile, accuracy of the BLUP increases. Increasing accuracy means the estimate of a trait by performance data is a true genetic estimate, and not a result of environmental conditions. EBVs, EPDs and PTAs are usually presented with some decimal measure of the accuracy of the number. The closer the decimal accuracy number is to 1, the more accurate the number. Compare the following two bulls.

	Weaning Weight EPD	Accuracy (Acc.)
Bull A	+10	.15
Bull B	+10	.89

Although the weaning weight EPDs are the same, the accuracy of Bull B is much higher. More emphasis should be placed on the weaning weight EPD of Bull B. Chances are that Bull B is an older sire with many progeny records upon which to base the EPD. Bull A is most likely a younger sire whose EPD is based on a relative's records. As more records are added to Bull A's performance profile, the accuracy will increase. A good chance exists that Bull A's EPD will move up or down dramatically. Bull B's EPD would be much less likely to change as more records are added.

USES OF DATA IN DIFFERENT PRODUCTION SITUATIONS

Performance data is used to predict the genetic differences among animals. Producers and judges identify the genetic traits that are important in a given production situation. Students must learn to analyze production situations, identify the combination of performance data that best fits that situation, then select livestock based on that decision.

Judging contests that include performance data for a class normally include a ***scenario***. Scenarios tell the judges the production situation in which the animals will be placed. The scenario influences how the judge analyzes the performance data and phenotypic traits. A properly constructed scenario should contain the following information:

1. Use of animals in a pure or crossbreeding program.
2. Availability of resources and environment for the herd or flock.
3. Marketing strategy for the progeny.

The scenario gives clues to which data should be emphasized in selection. Beef cattle perhaps offer the best illustration of interpreting scenarios. Other livestock species require similar thought processes. Following are two beef cattle scenarios accompanied by an analysis. See Figures 14-6 and 14-7.

Figure 14-6. Range cattle require performance data to fit their low maintenance lifestyle.

Figure 14-7. Cattle raised on Midwestern or Eastern farms require different performance data to fit their environment.

Scenario 1: Limousin Bulls

Bulls will be bred to registered mature cows. Feed and labor resources are adequate and typical of those on a small, Northeastern farm. Some heifer calves will be retained as replacements. All bull calves will be retained, performance tested, and sold as commercial bulls to producers selling slaughter cattle on a grade and yield program.

Analysis

1. *Bulls will be bred to registered mature cows.*

   Any time bulls will be bred to heifers, birthweight data should be a concern. Heifers are prone to dystocia if calves are too big.

Birthweight EPD predicts the size of the calves sired by a bull. Since bulls are bred to mature cows, higher birthweight EPDs could possibly be tolerated.

2. *Feed and labor resources are adequate and typical of those on a small, Northeastern farm.*

   Eastern and Midwest producers often have access to higher quality forages and more available labor at calving time than producers in range states. Therefore, larger framed, later maturing, higher birthweight EPD bulls can be tolerated than if feed and labor resources were typical of those on the Western range.

3. *Some heifer calves will be retained as replacements.*

   Since some daughters of these bulls will be retained as replacements, milk production and weaning weight EPDs should be considered. Birthweight EPDs should not be excessively high in comparison with other bulls.

4. *All bull calves will be retained, performance tested, and sold as commercial bulls to producers selling slaughter cattle on a grade and yield program.*

   This is perhaps the most important sentence in this scenario. Bulls used to sire calves sold under a grade and yield program must be fast growing (high weaning and yearling weight EPDs) and heavily muscled (high ribeye EPDs, if available).

The overall evaluation of this scenario should lead the judge to this conclusion: The ideal bull for this scenario should be fast growing, heavily muscled and have adequate milk production EPDs. Birthweight EPDs should not be excessively high in comparison with other bulls in the class.

Scenario 2. Angus Heifers

These heifers will be naturally mated to Simmental bulls. Female offspring will be retained as commercial females to be bred to Maine-Anjou bulls. All male offspring will be castrated and sold as feeder steers. These heifers and their progeny will be raised on range conditions where feed and labor resources are limited.

Analysis

1. *These heifers will be naturally mated to Simmental bulls.*

   As a breed, Simmentals are not known for their calving ease, but are heavily muscled and fast growing. Therefore, birthweight EPDs should be emphasized over growth EPDs for the Angus heifers.

2. *Female offspring will be retained as commercial females to be bred to Maine-Anjou bulls.*

   The Angus × Simmental female progeny will be used as commercial cows, so emphasis should be on maternal EPDs.

3. *All male offspring will be castrated and sold as feeder steers.*

   A significant proportion of income will come from the sale of feeder steers. The heavier the steer calves at weaning, the bigger the producer's paycheck; so weaning weight EPDs are important.

4. *These heifers and their progeny will be raised on range conditions where feed and labor resources may be limited.*

   Since feed and labor may be limited, low maintenance, easy keeping heifers should be emphasized. Limited feed resources suggest that heifers with excessively high milk production EPDs should be avoided. The nutrient demands of high milk production combined with limited feed can lead to thin heifers that will not re-breed. Again, low birthweight EPDs are emphasized because of the lack labor at calving. Heifers must be able to calve unassisted.

Evaluation of this scenario points toward the selection of Angus heifers that have low birthweight EPDs, moderate milk production EPDs, and high weaning weight EPDs. Yearling weight EPDs should receive the lowest consideration.

DATA FOR INDIVIDUAL SPECIES

The performance data presented to contestants in judging contests varies by species. Following is a discussion of the types of data likely to be encountered for each species.

SWINE

Individual data encountered may include birth date, ***number born alive (NBA)*** in the litter, number weaned in the litter, ***litter weight at 21 days (LW21)***, average daily gain (ADG), ***days to 230 pounds (DAYS)***, and ***backfat (BF)***. For the first four types of individual data, higher numbers are desirable. Days to 230 and backfat are two traits for which lower numbers are more desirable.

Contemporary ratios may be calculated for any of the above traits. Multiple trait indexes are also used in swine selection. ***Sow productivity index (SPI)*** combines NBA and LW21 into a gauge of maternal performance. ***Maternal line index (MLI)*** emphasizes maternal records with minor emphasis on growth and carcass records. ***General purpose index (GPI)*** gives equal weight to maternal, growth and carcass traits. ***Terminal sire indexes (TSI)*** use days to 230 and backfat. See Figure 14-8.

Pietrain x Danish Duroc
Terminal Sires
(All Data Adjusted to 260 Pounds)

Test Date	Tag#	B-date	Fat	Loin	Days to 260	Contemporary Index
2/7	Y1837	8/22	.41	7.5	173	134
2/7	Y1867	8/23	.42	7.8	175	135
2/7	Y1894	8/17	.45	7.2	168	130
2/1	Y1896	8/15	.49	8.3	173	132
2/1	Y1704	8/16	.48	7.4	160	138
2/1	Y1738	8/8	.44	7.4	165	140
1/17	Y1711	7/26	.31	6.6	188	137
1/17	Y1705	7/26	.40	7.0	164	151
1/17	Y1723	7/26	.42	6.6	160	149
1/12	Y1659	7/14	.47	7.8	171	141
1/5	Y1625	7/1	.43	8.1	174	142
1/5	Y1614	7/5	.56	8.5	156	143
1/5	Y1698	7/5	.37	6.9	175	137
12/27	Y1447	7/9	.53	8.1	165	136
12/27	Y1584	7/5	.41	7.1	179	130
12/27	Y1672	7/5	.45	6.9	169	132
12/22	Y1557	6/15	.40	7.7	169	147
12/22	Y1572	6/15	.45	6.9	165	136
12/22	Y1519	6/15	.40	6.5	169	136
12/22	Y1547	6/21	.45	7.5	168	139
12/15	Y1487	6/12	.47	7.4	165	138
12/15	Y1488	6/9	.48	7.4	170	132
12/15	Y1437	6/14	.42	7.5	180	131
12/15	Y1469	6/27	.49	7.3	165	134
12/8	Y1456	5/29	.49	7.7	176	135
12/8	Y1482	5/26	.38	7.3	168	155
12/8	Y1479	6/6	.45	6.9	179	130

Figure 14-8. Listing of terminal sire data and indexes for crossbred boars.

EPDs for swine include NBA and LW21 for which higher EPDs are more desirable. Lower numbers are preferable for days to 230 and backfat.

BEEF

Actual data encountered in beef performance classes include birth date, birth weight, weaning weight, yearling weight, frame score or hip height, and the scrotal circumference for bulls.

Contemporary ratios of all the above traits may also be encountered.

EPDs for beef cattle include ***birth weight EPD (BW EPD)*** (lower numbers mean lower birth weights), ***weaning weight EPD (WW EPD)***, ***yearling weight EPD (YW EPD)***, and ***maternal milk (MM EPD)*** (difference in pounds of calf weaned due to milk production). Recently, EPDs for the ribeye area and a marbling score (intramuscular fat) have become available for certain beef breeds. See Figure 14-9.

PEDIGREE RELATIONSHIP	CARCASS EPD			
	CARCASS WT.	MARBLING	RIBEYE AREA	FAT THICKNESS
	ACC.	ACC.	ACC.	ACC.
Individual	I +3 .29	I -.06 .29	I +.02 .29	I +.01 .29
Progeny				
Sire Progeny	+1 .94	-.12 .94	-.02 .94	+.01 .94
Dam Progeny				
Mat. G Sire Progeny	+12 .75	-.01 .75	+.12 .74	+.00 .73

Figure 14-9. EPDs for marbling and ribeye area set an example of producers selecting livestock based on consumer needs. (Courtesy, American Angus Association)

SHEEP

Individual data for sheep is much the same as other species. Birth date, birth weight, type of birth (Single = S, Twin = TW, Triplet = TR, Quad = Q), type of rearing, weaning weight (adjusted to 30, 60, 90 or

120 days), and yearling weight are all data that could be included in a judging class. Data for classes of wool breeds may include grease fleece weight (weight of uncleaned fleece), clean fleece weight, staple length, or grade.

Indexes can be calculated for the above data except the birth date.

EPDs are available for sheep, but too little data is available for most breeds to compare EPDs with animals outside the flock. Therefore, EPDs for most breeds of sheep must be calculated within a flock and are called ***FEPDs or flock EPDs***. Sheep FEPDs are available for number of lambs born, pounds of lamb weaned, weaning weight (adjusted to a constant weight), grease fleece weight, clean fleece weight, staple length, and fiber diameter.

DAIRY

Individual dairy production data are available for milking records and type. Pounds of milk, protein, and fat per lactation or lifetime are common individual production records. Average percent protein and fat in the milk are also frequently used. Type scores include a variety of visual evaluations including udder quality, feet and legs, dairy character, and body depth. Calving ease is reported as ***percent difficult births in heifers (%DBH)***. This number is an estimate of the number of calving problems expected per 100 heifers bred to a specific bull.

Type-production index (TPI) combines production with type into a general purpose index for Holsteins. Other breeds use a ***production type index (PTI)*** that accomplishes the same purpose. Both indexes use 100 as average.

Common PTAs available for dairy cattle include pounds of fat, percent fat, pounds of protein, percent protein, and pounds of milk. PTA$ data are available for milk, fat, and protein MFP$. Portions of MFP$ can be isolated depending on production emphasis. ***PTA$ for cheese yield (CY$)*** predicts the value of milk protein content. ***PTA$ for milk and fat (MF$)*** predict the dollar difference due to milk production and butterfat content. ***Predicted transmitting ability for type (PTAT)*** indicates the number of type points that will be gained or lost compared with the breed average for a specific sire's daughters. PTAT is the only performance data that is subjective in nature. See Figure 14-5 for dairy performance data.

SUMMARY

Several types of performance data are available to aid in selection of breeding livestock. Actual data is the easiest to gather, but the least accurate. Ratios and multiple trait indexes compare animals within a contemporary group. EBVs, EPDs, and PTAs offer a more accurate assessment of an animal's genetic potential because they include individual data from the animal and all related family members. The accuracy or reliability of EPDs, and PTAs increases with the number of records available. When using performance data for the selection of livestock, the judge must consider the environment and intended use of the animals in a crossbreeding scheme. This information is presented to judging contest contestants in the form of a scenario for each performance class. Critical thinking skills must be used to analyze scenarios and select the animals with the correct balance and fit of performance data. Data for certain species may include a variety of information including individual data, indexes, EBVs, EPDs, PTAs, or PTA$.

CHAPTER SELF-CHECK

___ individual data	1.	dollar value of progeny cheese
___ average daily gain (ADG)	2.	emphasizes growth/carcass traits
___ ratio	3.	dollar value of progeny milk and contents
___ multiple trait indexes	4.	combines NBA and LW21
___ estimated breeding value (EBV)	5.	performance of single animal
___ expected progeny difference (EPD)	6.	expected progeny weaning weight
___ predicted transmitting ability (PTA)	7.	measures rate of gain
___ PTA for milk, fat, and protein (MFP$)	8.	measures milk production in beef
___ scenario	9.	compares animals outside herd
___ number born alive (NBA)	10.	measures lifetime growth rate in pigs
___ litter weight at 21 days (LW21)	11.	production situation
___ days to 230 pounds (DAYS)	12.	change in progeny type score

___ backfat (BF)	13. dollar value of progeny milk and fat
___ sow productivity index (SPI)	14. allows comparison within a group, average equals 100
___ maternal line index (MLI)	15. emphasizes maternal traits
___ general purpose index (GPI)	16. compares all animals within a herd or flock
___ terminal sire indexes (TSI)	17. subcutaneous fat
___ birth weight EPD (BW EPD)	18. dairy equivalent of EPD
___ weaning weight EPD (WW EPD)	19. weight of litter of pigs
___ yearling weight EPD (YW EPD)	20. selects for several traits
___ maternal milk (MM)	21. pigs born per litter
___ FEPDs or flock EPDs	22. general purpose Holstein index
___ percent difficult births in heifers (%DBH)	23. evenly weights maternal, growth and carcass traits
___ type-production index (TPI)	24. expected progeny birth weight
___ production type index (PTI)	25. general purpose index for other dairy breeds
___ PTA$ for cheese yield (CY$)	26. sheep EPD
___ PTA$ for milk and fat (MF$)	27. expected progeny yearling weight
___ predicted transmitting ability for type (PTAT)	28. indicator of calving ease in dairy

QUESTIONS AND PROBLEMS FOR DISCUSSION

1. List an example of individual data that requires three pieces of information.
2. Give five types of individual data that use only one measurement.
3. A ratio compares animals within the same ________ group.
4. The average ratio score is _____.
5. ______________ presents an estimate of the animal's genetic worth as a parent in comparison to others in the same herd.
6. ______________ presents an estimate of the animal's genetic worth as a parent when compared with animals inside and outside the herd.

7. True or False? EPDs are presented as a number with 100 being the average score.
8. The _______ industry uses predicted transmitting ability to forecast the added income to be gained from an individual's daughter.
9. Explain why EPDs and PTAs become more accurate as more individual records are used in computation.
10. Given individual data, ratios, EBVs, and EPDs, which should be given the most consideration in selection? Why?
11. How can a scenario help a judging contestant consider performance data?
12. List four types of performance data commonly used in the swine industry.
13. List four types of performance data commonly used in the beef industry.
14. List four types of individual data commonly used in the sheep industry.
15. List three types of performance data commonly used in the dairy industry.

ACTIVITIES

1. Collect sire catalogs. Compare sires using information learned in this chapter.
2. Write to a breed association listed in the text. Request performance data from sires within the breed. Develop production scenarios to fit listed sires.
3. Calculate performance data from records kept in Supervised Agricultural Experience (SAE) programs.

LABORATORY ACTIVITY

ANALYSIS OF PERFORMANCE DATA

Purpose

To analyze performance data for livestock selection

Materials

paper and pencils for notes

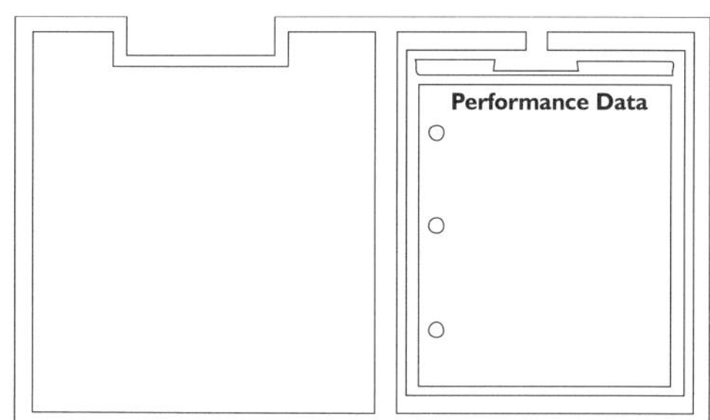

Procedure

Use the following data set and scenario to answer questions 1 through 5.

Yorkshire Gilts

#	B-Date	NBA	$NBA_{(EPD)}$	Acc.	$LW21_{(EPD)}$	Acc.	BF	Dam's SPI	DAYS
1.	3/28	10	+.2	.17	+1.0	.15	.80	101.0	165
2.	3/31	11	+.3	.17	+2.4	.15	.82	103.5	160
3.	3/26	8	0	.17	−1.1	.15	.75	98.5	170
4.	3/31	11	+.3	.17	+2.4	.15	.70	103.5	157

Scenario:

These gilts will be mated to Landrace boars to produce F-1 or crossbred breeding gilts for total confinement swine operations. The prolific progeny will be mated to exceptionally lean, heavily muscled terminal sires for the production of terminal market hogs.

Analysis

1. According to the scenario, are maternal or growth and carcass traits more important to these Yorkshire gilts? Why?
2. Using what you know about the various kinds of performance data, which four performance categories would you weight most heavily? Why?
3. Which is the better indicator, NBA or NBA EPD, to predict the gilt's genetic ability to farrow large litters of pigs? Why?
4. What does the LW21 EPD column tell you about the expected 21-day litter weight of gilt #3 versus gilt #4?
5. Rank these Yorkshire gilts according to the scenario and defend your ranking.

Application of Laboratory Activity

Livestock producers continually use performance data when making breeding decisions. Valid performance data can expose an animal's true genetic worth to producers. When combined with phenotype, performance data helps to isolate the genetic best from the rest.

Chapter 15

MEATS JUDGING

Cut the fat

INTRODUCTION

Market livestock are visually evaluated for the contents of the final carcass, namely meat. A basic knowledge of meats evaluation helps livestock judges better select animals that, when slaughtered, will produce carcasses containing large amounts of high quality, lean meat. Breeding livestock are evaluated on their ability to produce progeny that will eventually yield higher quality carcasses. Therefore, muscle content and quality are important in breeding stock selection. Clearly, a knowledge of carcasses and meats is key to the evaluation of all species of slaughter livestock.

Figure 15-1.

OBJECTIVES

1. Sketch beef, pork, and lamb carcasses and label the primal cuts
2. Differentiate between yield and quality grades
3. Place classes of beef, pork, and lamb carcasses
4. Compare ideal cooking requirements and conditions of roasts vs. steaks

TERMS

break joint
buttons
cutability
Institutional Meat Purchasing Specifications (IMPS)
kidney, pelvic, and heart fat (KPH)
lean-to-fat ratio
preliminary yield grade (PYG)
primal cuts
quality grades
spool joint
subprimal cuts
yield grades

ANIMAL SCIENCE FACTS

Some FFA meats judging contestants do not give reasons. Instead, they answer a series of questions based on observations made of a class. Accurate notes are essential to either reasons or questions classes.

BASIS FOR MEAT EVALUATION

Meat is judged using three parameters: amount of muscle present in the carcass or cut (more is better), amount of exterior fat cover (less is better), and quality of the lean meat in terms of color, texture, and marbling (intramuscular fat). The first two parameters reflect economics. A higher ***lean-to-fat ratio*** (more lean to less exterior fat) translates into more product (meat) to sell from each carcass. ***Cutability*** refers to lean-to-fat ratio. A carcass or cut with a high lean-to-fat ratio is said to have high cutability. The third parameter, quality, attracts consumers. Consumers want bright colored, fine textured, marbled meat. Marbling, desirable to consumers, indicates flavor and juiciness of meat.

A dilemma arises because, physiologically, many animals must deposit an excessive amount of exterior fat before depositing an appreciable amount of marbling. Meat cutters trim exterior fat before sale, but this results in waste and higher prices for consumers. Ideally, meat should be highly marbled, but with a small amount of exterior fat.

CUTS OF MEAT

Whole carcasses are often judged in meats contests. However, during processing, carcasses are broken into large sections called ***primal cuts***. See Figures 15-2 through 15-4 and the color section for primal cut diagrams. Different primal cuts vary widely in their retail value. Cuts from the back are the most valuable because they tend to be the most tender, juicy, and flavorful. Cuts from the hindquarters rank second in value. Shoulder, neck, and other cuts are less valuable. When whole carcasses are evaluated, more emphasis should be placed on the primal cuts of the highest value.

Primal cuts themselves may also be judged in a meats judging contest. In high-priced cuts (such as pork loins, and beef or lamb ribs), muscle quality must be emphasized, but not necessarily at the expense of cutability. Primals of intermediate price, such as beef rounds, pork legs/hams or lamb legs should be judged on cutability, with muscle quality taking secondary importance. Lower priced primals, such as pork shoulders or beef chucks, should be judged with a heavy emphasis on cutability.

Primal cuts are often further broken into ***subprimal cuts*** before being packaged, boxed, and shipped to restaurants, supermarkets, or other institutions, such as schools and hospitals. Upon arrival at their destination, subprimals are further reduced into retail cuts before cooking and consumption. By looking at a retail cut of meat, contest participants may be asked to identify the species, primal cut, subprimal cut, and the best way to cook that certain piece of meat.

Institutional Meat Purchasing Specifications (IMPS) have been adopted by the meat industry to standardize various primal and subprimal cuts. Standardization arose from increased sales of boxed meat. For instance, a restaurant that specializes in prime rib does not need a whole beef carcass, so only beef ribs, which come packaged one or two to a box, are purchased. IMPS were adopted to ensure consistency of boxed meat. IMPS cuts are numbered in a series. The 100 series is reserved for fresh beef cuts; 200 fresh lamb and mutton; 300 fresh veal and calf; 400 fresh pork; 500 cured, smoked, or cooked pork; 600 cured, dried or smoked beef; 700 edible by-products; 800 sausage by-products; 1000 portion-cut meat products. Numbers within the series identify specific cuts.

FFA meat judges are asked to identify several IMPS cuts. Current examples include the 107 beef rib, 112 ribeye roll, 120 boneless brisket, and 170 bottom round. The cuts used in contests change over time.

Individual retail cuts are also fair game in a judging contest. A class of four retail cuts may be judged using the same three criteria—muscling, trimness, and quality—used to judge carcasses or primal cuts. Refer to Figures 15-2,15-3, and 15-4 and the color section for a breakdown of beef, pork, and lamb carcasses into retail cuts.

COOKING DIFFERENT CUTS

Judging contestants are sometimes asked the best way to cook a certain piece of meat. Generally, a steak or thinly-sliced pieces of meat should be broiled, panbroiled, or panfried. These three methods all use high heat, short duration, dry cooking methods. Steaks are naturally juicy and tender. Short duration, high heat cooking seals in the juiciness and retains the inherent tenderness.

Roasts or large, tender, pieces of meat should usually be roasted. Roasting involves a moist, long duration, lower heat cooking method. During the roasting process, the roast constantly self-bastes, retaining the meat's original moisture content.

Less tender pieces of meat can be braised. Braising involves seasoning and browning of the meat, followed by simmering in a covered, liquid-filled container. This helps break down the connective tissue, to make the meat more tender.

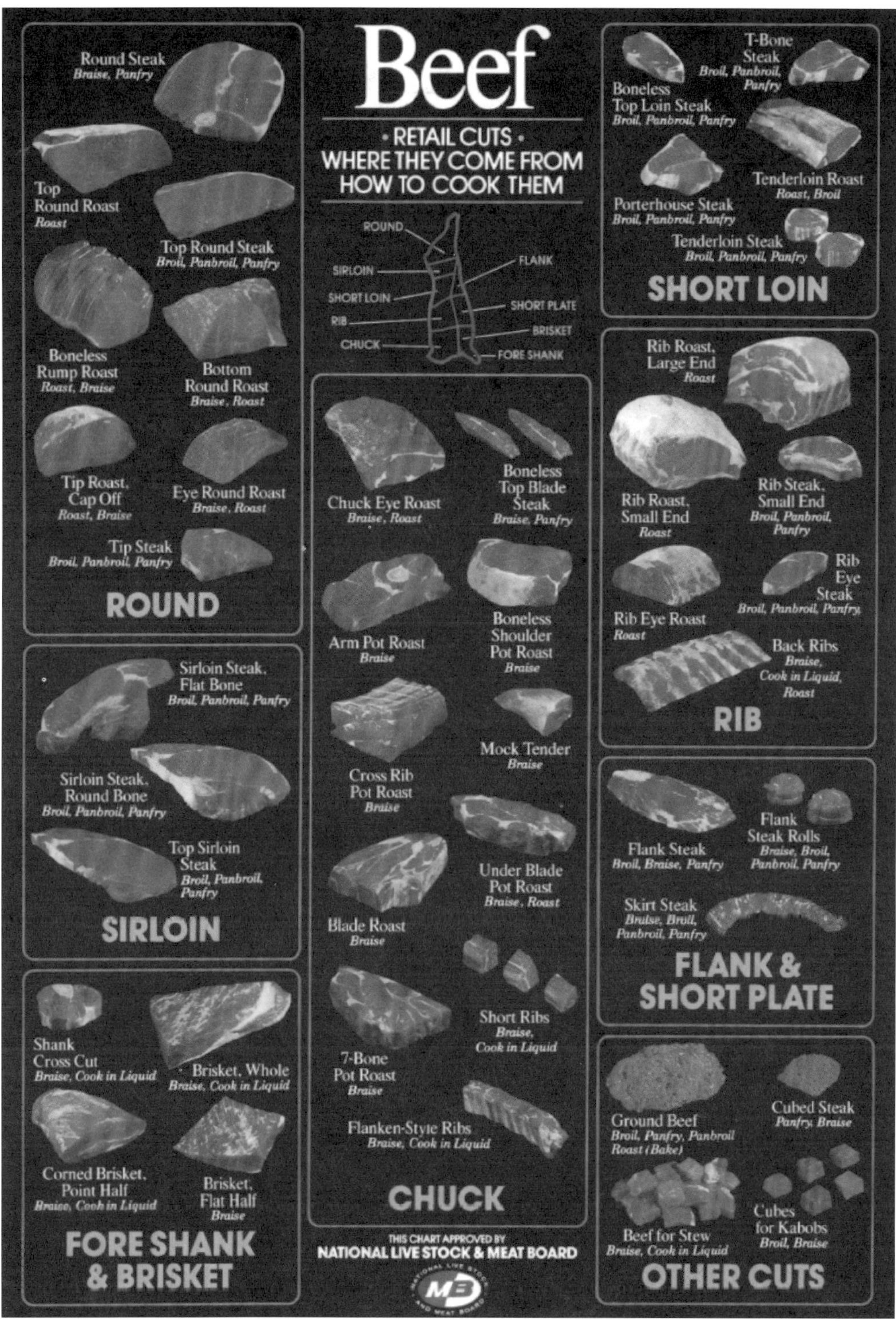

Figure 15-2. Beef primal and retail cuts—also located in color section. (Courtesy, National Live Stock and Meat Board)

Figure 15-3. Pork primal and retail cuts—also located in color section. (Courtesy, National Live Stock and Meat Board)

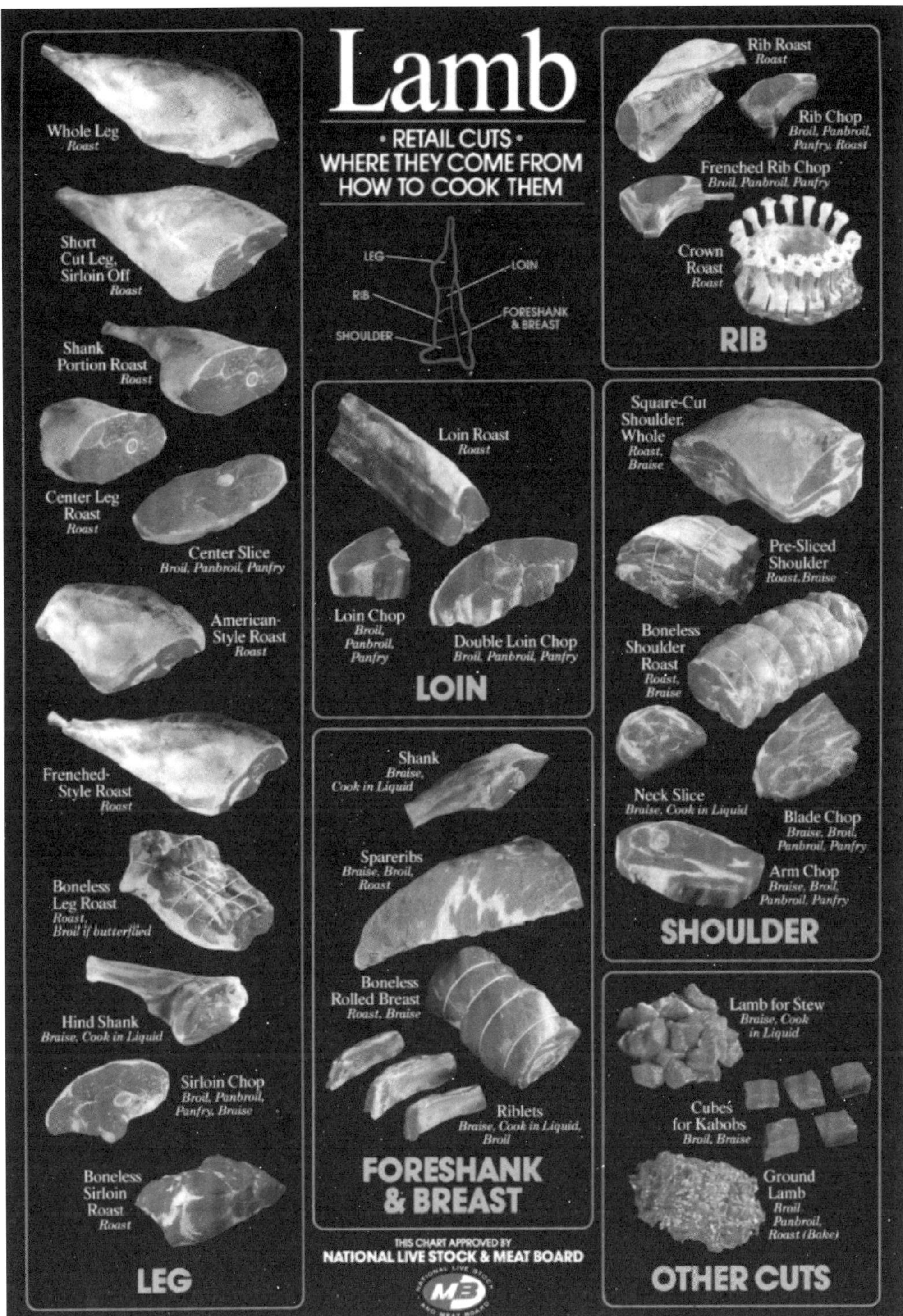

Figure 15-4. Lamb primal and retail cuts—also located in color section. (Courtesy, National Live Stock and Meat Board)

BEEF JUDGING

Beef carcasses are judged on a combination of cutability and quality. The beef industry has made the evaluation of beef carcasses very objective. Through ***yield grades*** (estimate of cutability) and ***quality grades*** (estimate of marbling and age), the value of an entire beef carcass can be predicted. Contestants in a meats judging contest are asked to yield and quality grade a rail (15 carcasses) of cattle for a score. The basis for both grades relies primarily on an appraisal of the ribeye muscle cut between the 12th and 13th rib.

YIELD GRADING

Yield grades are based on a score of one to five with one being an extremely lean, heavily muscled carcass, and five being an extremely fat, lightly muscled carcass. Most slaughter cattle have yield grades between two and three.

Step I

The amount of fat covering the ribeye gives a good approximation of the amount of fat on the entire carcass. Therefore, a measurement of the exterior fat cover, 3/4 the distance over the ribeye at the 12th rib, is used to calculate a ***preliminary yield grade (PYG)***. See Figure 15-5. If the 3/4 measurement is drastically more or less than the fat cover over the lower rib, the PYG may be adjusted up or down.

Fat Thickness	PYG
0	2.0
0.2	2.5
0.4	3.0
0.6	3.5
0.8	4.0
1.0	4.5

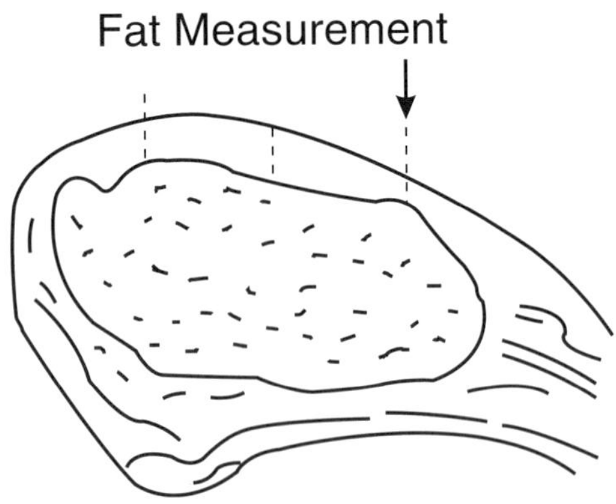

Figure 15-5. Fat thickness is measured 3/4 the distance over the loin muscle and converted to PYG.

Step 2

The PYG is adjusted for the size of the ribeye compared with carcass weight. If the ribeye is bigger than expected for its carcass weight, the PYG is adjusted down (remember, a lower numerical yield grade is better). If the ribeye is smaller than expected, the PYG is adjusted up. See Figure 15-6 for standard ribeye areas.

PYG is adjusted by .3 for every square inch of ribeye area above or below the expected target (or .1 for every .3 square inch of ribeye area). For example, a 600-pound carcass with a 12.0 square inch ribeye would result in a subtraction of .3 from the PYG. A 600-pound carcass with a 10.0 square inch ribeye would result in the addition of .3 to the PYG. Remember, more muscle means subtract from the PYG.

Carcass Wt.	Expected Ribeye (square inches)
500	9.8
550	10.4
600	11.0
650	11.6
700	12.2
750	12.8

Figure 15-6. Expected ribeye areas for various carcass weights.

Step 3

The PYG is adjusted based on the amount of internal fat present in the carcass. Internal fat is more often called ***kidney, pelvic, and heart fat (KPH)***. Three and a half percent of carcass weight is considered average. If the carcass contains more than 3.5 percent, the PYG is adjusted up. If the carcass contains less than 3.5 percent, the PYG is adjusted down. See Figure 15-7 for adjustments.

KPH%	Adjustment to PYG
1.0	–.5
1.5	–.4
2.0	–.3
2.5	–.2
3.0	–.1
3.5	0
4.0	+.1
4.5	+.2
etc.	

Figure 15-7. KPH adjustments.

Example yield grade calculation:

Carcass weight	700	pounds	
Adjusted fat thickness	.40	inches	
Ribeye area	11.6	square inches	
KPH%	2.5	percent	
PYG (from table)			3.0
Ribeye adjustment			
expected ribeye	12.2		
actual ribeye	11.6		
Adjustment (12.2 – 11.6) × .3			+.2
KPH adjustment (from table)			–.2
Final Yield Grade			3.0

Note–When calculating ribeye adjustments, round all numbers to the nearest .1.

Obviously, calculating yield grades takes practice. In a contest, participants must estimate fat, ribeye and KPH measurements, then calculate a yield grade based on their own estimates. Carcass weights will be provided.

QUALITY GRADING

Beef quality grades are based on two factors: the amount of marbling and the physiological age of the carcass.

Marbling is a subjective measurement based upon the amount of intramuscular fat visible on the cut surface of the ribeye. The amount of marbling depends on feeding programs and genetics. In general, carcasses with higher degrees of marbling are more valuable. Marbling is divided into nine categories listed from the most marbling to the least: abundant, moderately abundant, slightly abundant, moderate, modest, small, slight, traces, practically devoid. See Figure 15-8 in the color section.

Each category further subdivides into percentages from 0 to 100 percent in 10 percent increments. For example, small marbling could be designated small10, small20, small30, etc. Small30 is a higher marbling score than small10.

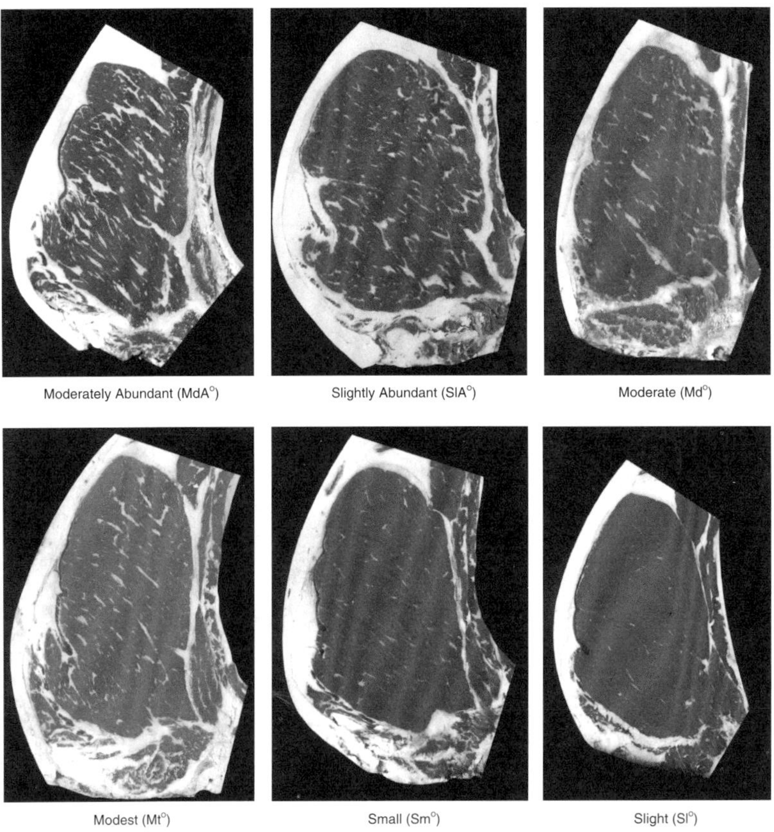

Figure 15-8. Least amount of marbling for each category *located in color section

The second factor in quality grading is physiological maturity. As cattle age, the tenderness of the muscle decreases. Quality graders attempt to estimate the age of a carcass by looking for clues hidden in the bones.

Buttons are pieces of cartilage along the split surface of the carcass over the rib. Buttons become harder and more calcified as cattle age. Along with buttons, the appearance of the ribs inside the carcass changes with age. As cattle get older, ribs that were once round and red, become flat and white. Also, as cattle age, the vertebrae in the loin region fuse. See Figure 15-9 for locations of bones used in aging carcasses.

Bone clues are used to give each carcass a maturity score. Maturity scores range from A (youngest) to E (oldest). Each score is further divided into thirds. The youngest third of the A maturity would be classified A–, the middle third of the A maturity would be A0, and the oldest third, A+. All other maturity levels are divided similarly. See Figure 15-10 for

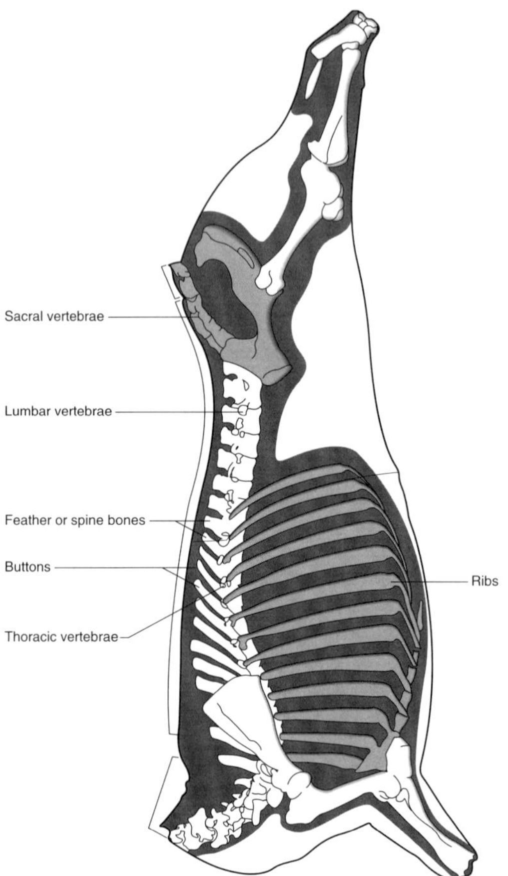

Figure 15-9. Bones used in aging beef carcasses.

USDA Maturity Group	Sacral Vertebrae	Lumbar Vertebrae	Thoracic[1] Vertebrae	Ribs	Age[2]
A-	Distinct separation	No ossification	No ossification	Slight tendency toward flatness	9 mos.
A+/B-	Completely fused	Nearly completely ossified	Cartilages show some evidence of ossification (5% ossified)	Slightly wide and slightly flat	30 mos.
B+/C-	Completely fused	Completely ossified	Cartilages are partially ossified (20-30% ossified)		42 mos.
C+/D-	Completely fused	Completely ossified	Outlines of cartilages are plainly visible (65-75% ossified)	Moderately wide and flat	72 mos.
D+/E-	Completely fused	Completely ossified	Outlines of cartilages are barely visible (95% ossified)	Wide and flat	96 mos.

[1] Descriptions refer to the uppermost three thoracic vertebrae in the forequarter (in the region of the 10th, 11th, and 12th ribs.

[2] Approximate chronological age equivalent

Figure 15-10. Comparison of bone clues with maturity scores

RELATIONSHIP BETWEEN MARBLING, MATURITY, AND CARCASS QUALITY GRADE*

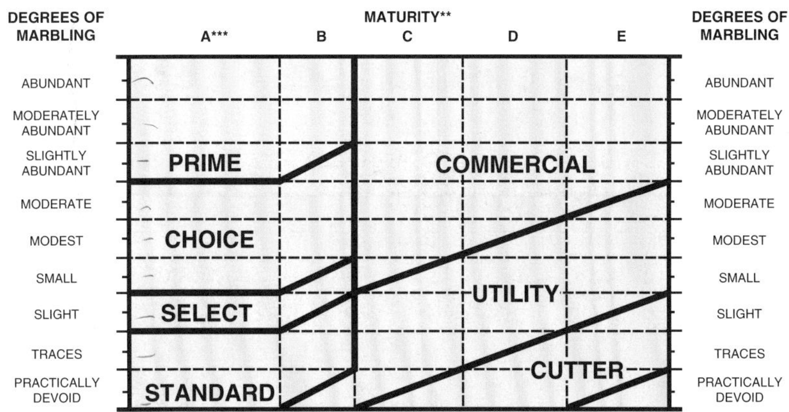

*Assumes that firmness of lean is comparably developed with the degree of marbling and that the carcass is not a "dark cutter."

**Maturity increases from left to right (A through E).

***The A maturity portion of the Figure is the only portion applicable to bullock carcasses.

Figure 15-11. USDA quality grades combine marbling score with maturity score.

comparison of bone clues with maturity scores. Color and texture of lean are used as tiebreakers if a carcass is on the borderline between two maturities. Carcasses with bright, finely textured lean receive the younger score. Carcasses with dark, coarsely textured lean receive the older score.

Marbling score combined with maturity score determine the final quality gradc. Thc USDA quality gradcs (from most to lcast dcsirablc) arc: Prime, Choice, Select, Standard, Commercial, Cutter, and Canner. See Figure 15-11. The Prime, Choice, Select, and Standard grades can only be applied to cattle with A or B maturity scores.

Many quality grades are further broken down into halves or thirds of a grade. Compare the following carcasses and refer to Figure 15-11. Line up marbling score on the left column with maturity score along the top. Trace both lines into the chart until they intersect. Use Figure 15-11 to determine quality grade from the several examples listed below.

Marbling score	Maturity score	Quality Grade
Moderate10	A0	Choice +
Modest10	A0	Choice 0
Small10	A0	Choice –
Small10	B0	Select +
Small10	B+	Select –

Successful meat graders memorize Figure 15-11 and can accurately quality grade cattle without visual aids.

JUDGING BEEF CARCASSES

Once yield and quality grading are mastered, judging beef carcasses becomes very easy. Simply yield and quality grade the carcasses, then rank them on a combination of yield and quality. The carcass with the best combination of yield and quality grade would be placed first, the second best combination, second, and so on. See Figure 15-12 for parts of a beef carcass.

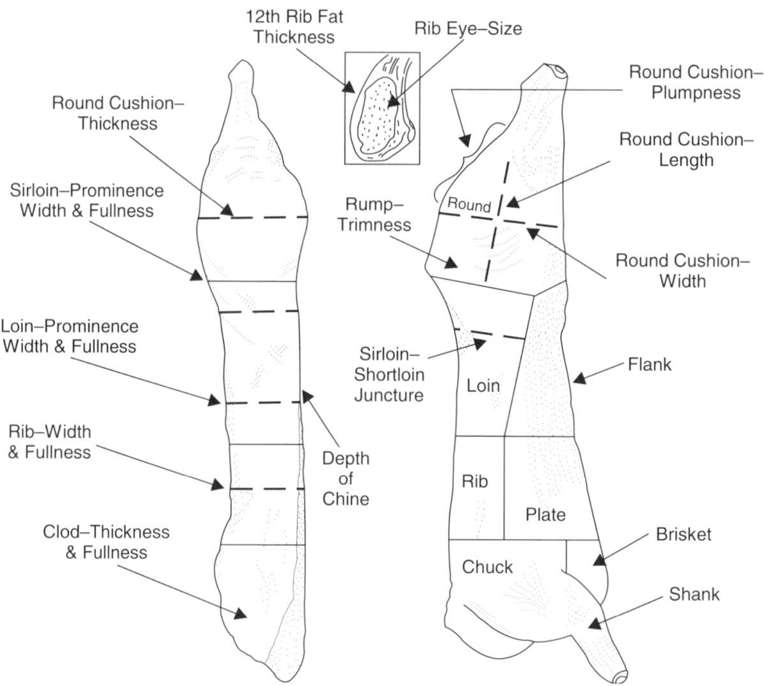

Figure 15-12. Parts of a beef carcass.

BEEF TERMINOLOGY AND REASONS

Terminology for beef judging relies on a certain comfort with the descriptive names of carcass parts. For instance, if a certain carcass is fatter over the sirloin, the person delivering the reasons has to know the sirloin location. Study and memorize the parts presented in Figures 15-12.

Beef reasons differ little from the generic reasons used in Chapter 13. Opening statements for each pair should reveal the most important reasons for the placing. Broad opening statements should be followed by specific examples that reinforce the opening statement. The three key areas that must be covered in each pair are differences in muscling, trimness, and quality. Comparative terms should be used within the discussion of each pair. Grants should be given when necessary.

CARCASSES

Use Figure 15-12 to assist in the following section. The following are places to look for muscling differences on beef carcasses and some descriptive terms to use in a set of reasons:

chuck—heavier muscled, meatier, thicker

ribeye—larger, fuller, more oval-shaped

sirloin—wider

rump—wider, fuller

round—deeper, wider, or thicker cushion

The following are places to look for trimness differences on beef carcasses, and some descriptive terms to use in a set of reasons:

KPH fat—lower percent KPH, less kidney, pelvic, and heart fat

brisket, chuck, lower rib, flank, loin edge, sirloin-shortloin junction, sirloin, rump, round—trimmer

less fat over clod, less cod or udder fat

The following are places to look for quality differences on beef carcasses, and some descriptive terms to use in a set of reasons:

roundness of ribs—more youthful bone

buttons—showed less ossification

color of fat—whiter, flakier

ribeye—higher degree of marbling, more evenly distributed marbling, brighter cherry red color, more youthful color, finer-textured lean

JUDGING PRIMAL CUTS

Chucks, ribs, loins, or rounds can be judged using the principles learned from yield and quality grading. Chucks and rounds are judged with a heavy emphasis on yield or cutability. Classes of chucks and rounds should be placed almost entirely on cutability with the heaviest muscled, leanest cuts placed at the top of the class. Loins and ribs however, are judged with an emphasis on quality. Marbling, color, and texture of lean are more important than yield on these more expensive primal cuts.

CHUCKS

Use Figure 15-13 to assist with the following text. The following are places to look for muscling differences on beef chucks and some descriptive terms to use in a set of reasons:

blade end or face—bigger blade eye, wider or deeper blade face

clod—fuller

neck—fuller, thicker

arm end or face—wider, deeper

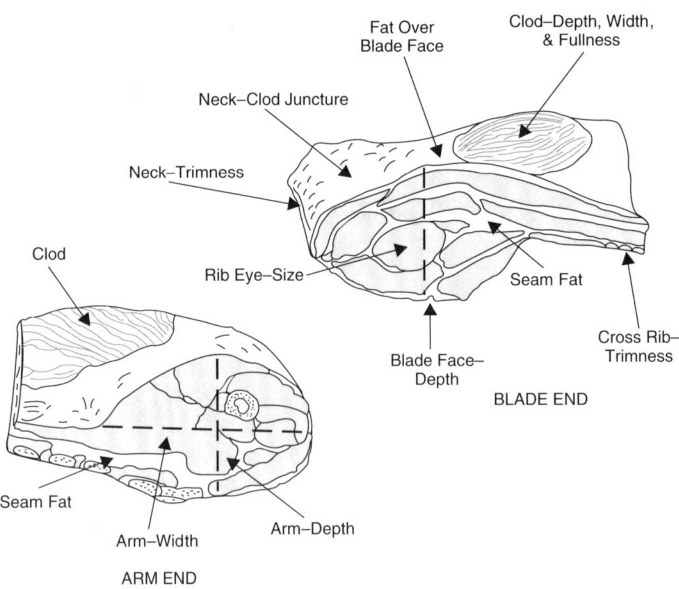

Figure 15-13. Parts of a beef chuck.

The following are places to look for trimness differences on beef chucks and some descriptive terms to use in a set of reasons:

blade end or face—less fat over blade end, less seam fat in the blade face
neck-clod junction—trimmer
arm end or face—less fat over arm end, less seam fat in the arm face

The following are places to look for quality differences on beef chucks and some descriptive terms to use in a set of reasons:

blade end or face—brighter colored lean, higher degree of marbling
arm end or face—brighter colored lean
fat color—whiter, flakier

RIBS

Use Figure 15-14 to assist with the following text. The following are places to look for muscling differences on beef ribs and some descriptive terms to use in a set of reasons:

ribeye—larger, fuller, more oval-shaped, longer, wider
chine—deeper

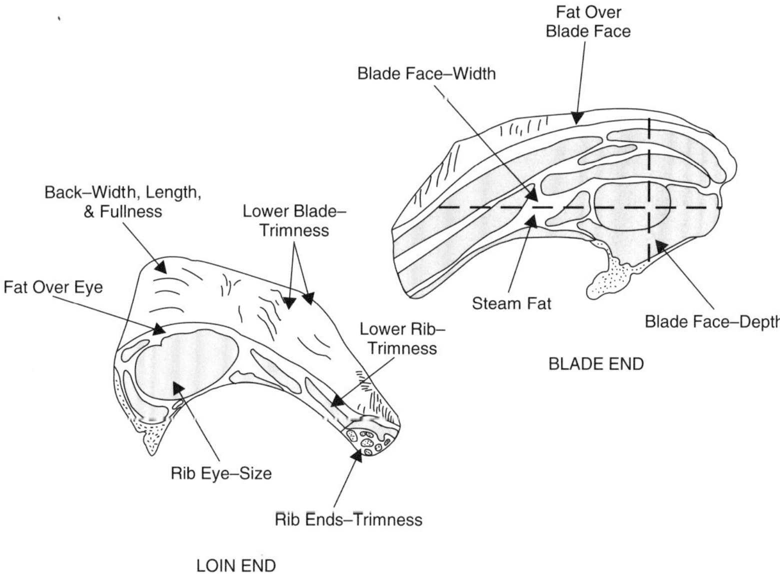

Figure 15-14. Parts of a beef rib.

blade end or face—wider, deeper, more exposed lean, bigger blade eye

The following are places to look for trimness differences on beef ribs and some descriptive terms to use in a set of reasons:

loin end, ribeye, lower rib, back, lower blade—trimmer
blade end—trimmer, less seam fat

The following are places to look for quality differences on beef ribs and some descriptive terms to use in a set of reasons:

ribeye—same as for beef carcasses

LOINS

Use Figure 15-15 to assist with the following section. The following are places to look for muscling differences on beef loins and some descriptive terms to use in a set of reasons:

loineye—wider, deeper, larger
back—wider, longer
chine—deeper

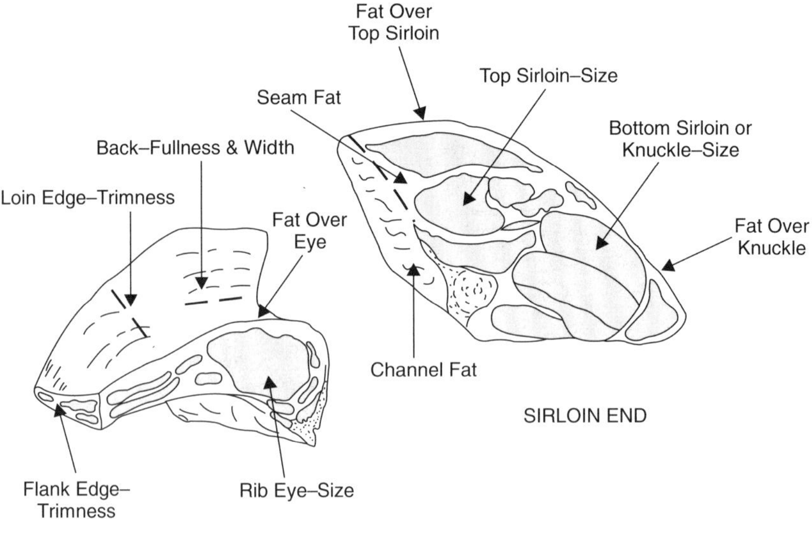

Figure 15-15. Parts of a beef loin.

sirloin—fuller

sirloin end—more exposed lean in top or bottom sirloin

The following are places to look for trimness differences on beef loins and some descriptive terms to use in a set of reasons:

loineye, loin edge, flank edge, sirloin-shortloin junction—trimmer

sirloin end—trimmer, less seam fat

pelvic area—less pelvic or channel fat

The following are places to look for quality differences on beef loins and some descriptive terms to use in a set of reasons:

loineye—same as for beef carcasses

sirloin end—same as for beef carcasses

ROUNDS

Use Figure 15-16 to assist with the following section. The following are places to look for muscling differences on beef rounds and some descriptive terms to use in a set of reasons:

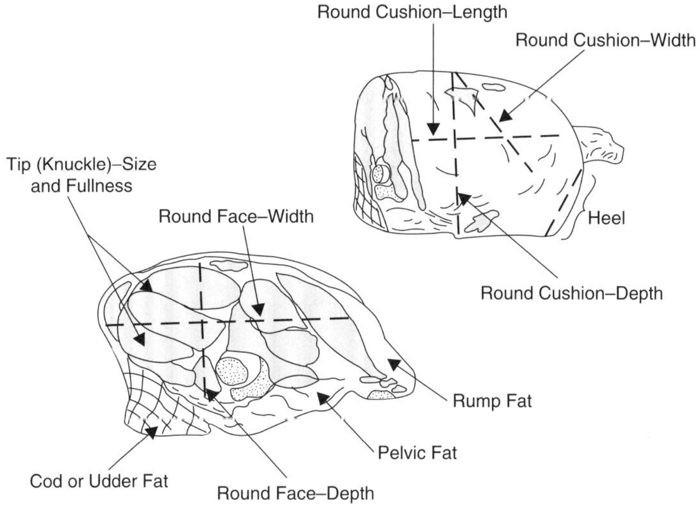

Figure 15-16. Parts of a beef round.

sirloin end or face—more lean area exposed, deeper, wider
cushion—deeper, wider, longer
heel—meatier, fuller, plumper
shank—shorter (indicates more muscle)

The following are places to look for trimness differences on beef rounds and some descriptive terms to use in a set of reasons:

sirloin end or face, rump, knuckle, cushion, heel—trimmer
sirloin end—less seam fat
cod or udder, pelvic fat—less

The following are places to look for quality differences on beef rounds and some descriptive terms to use in a set of reasons:

sirloin end or face—marbling, firmness, texture, color (same as beef carcasses

GENERAL INTRODUCTORY BEEF TERMS

higher percentage boneless, closely trimmed retail cuts
higher cutability
higher percentage trimmed steaks and roasts
more total pounds of trimmed steaks and roasts
higher percentage of higher priced cuts
higher yielding
have a more desirable yield grade
higher, more desirable quality grade

SAMPLE BEEF REASONS

Notes for all meats judging should be recorded for the three main criteria (muscling, trimness, and quality) for each pair. In a given pair, only the criteria that were important to placing the class should be discussed.

Also, meats reasons may be written rather than give orally, so good penmanship and grammar are essential.

Beef Carcasses Placing: 1–2–3–4

FIRST 1/2 In this class of beef carcasses, 1 placed over 2 as 1 more ideally combined cutability and quality. 1 was heavier muscled as demonstrated by a thicker, fuller cushioned round, which carried into a deeper heel. In addition, 1 was a higher quality carcass as shown by a more youthful, cherry red, finer textured lean. 1 had an additional advantage in trimness particularly over the ribeye, lower rib, loin edge, and chuck. Thus, 1 would yield a higher percentage of closely trimmed, consumer acceptable cuts. I realize 2 had a clear advantage in the degree of marbling and was trimmer internally, displaying less kidney and pelvic fat.

SECOND 2/3 2 easily placed over 3, as 2 was a significantly higher quality, heavier muscled carcass that would yield meatier, trimmer retail cuts with more consumer appeal. 2 displayed a much higher degree of more evenly distributed marbling. In addition, 2 was more youthful in its appearance as evidenced by rounder, redder ribs and softer, whiter buttons. 2 also featured a larger, more oval-shaped ribeye, a deeper cushioned round, and a thicker clodded chuck. I concede that 1 was decidedly trimmer over the ribeye, lower rib, rump, and round. 1 also had an advantage in cod, kidney, and heart fat.

THIRD 3/4 I preferred 3 over 4 as 3 would yield a significantly higher percentage of high value, trimmed primals. 3 displayed less external fat over the round, rump, loin edge, and was especially trimmer over the ribeye and lower rib. In addition, 3 was trimmer over the chuck and displayed less cod and kidney fat. I admit 3 had a larger ribeye and a thicker, clodded chuck, partially due to fat.

FOURTH 4 I admit 4 was acceptable in terms of cutability. However, it displayed the narrowest, lightest muscled round in the class, which, coupled with the lowest quality in the class, would make it the least desirable to the profit-minded retailer.

PORK JUDGING

Judging of pork carcasses and cuts relies, to a very high degree, on cutability. There is no standard method for yield and quality grading pork, but the same principles apply. Pork carcasses or cuts with high cutability and quality are preferred over fat, low quality carcasses and cuts.

JUDGING PORK CARCASSES

Hams and loins are the most highly priced pork cuts. Therefore, carcasses with the highest percentage of lean in the ham and loin, with acceptable quality are preferred. Carcasses are split and may be judged ribbed (loin muscle cut and displayed at the 10th rib) or not ribbed. When carcasses are ribbed, fat thickness should be evaluated 3/4 the distance over the eye muscle, similar to beef carcasses. When carcasses are not ribbed, judges must make estimates of the size of the loineye and total carcass muscle based on width and thickness of the loin and ham, as well as the amount of lumbar lean, depth of chine, or overall carcass thickness. See Figure 15-17 for parts of a pork carcass.

Trimness is evaluated primarily by the amount of backfat present, either at the 10th rib, or an average. Other indications of carcass trimness include fat over the ham collar, the ham-loin junction, elbow pocket, jowl, sternum, and belly edge.

Quality can be easily decided in ribbed carcasses by evaluating the color (grayish pink is ideal), firmness, and marbling of the exposed loin muscle. In unribbed carcasses, color and firmness can be evaluated in the lumbar lean, over the ham face, and between the ribs where feathering (fat streaking) is an indication of marbling.

PORK TERMINOLOGY AND REASONS

Again, success in giving pork reasons relies on the proper use of terminology. Study and memorize the parts of a pork carcass, as well as the parts of a ham.

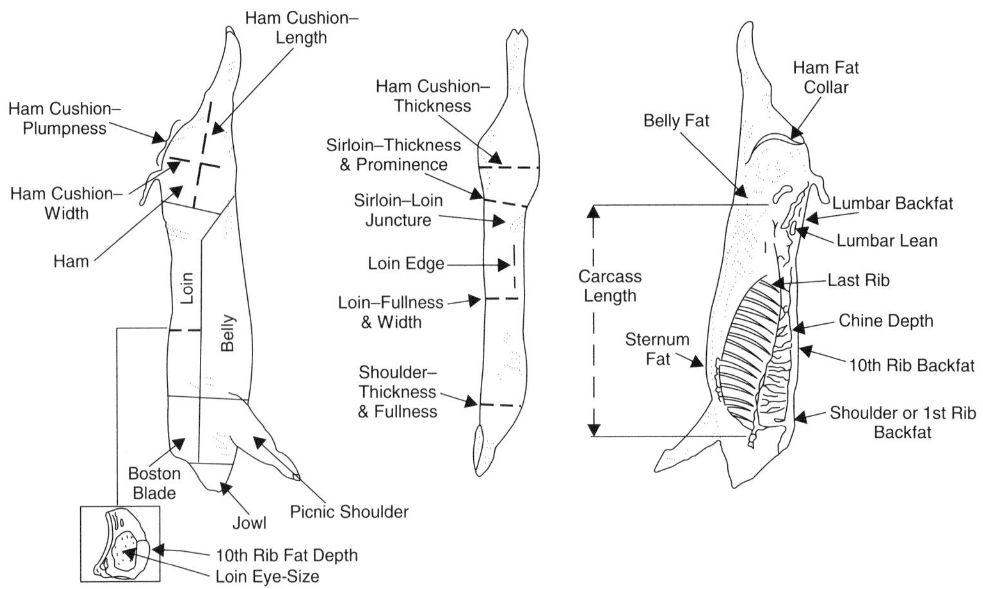

Figure 15-17. Parts of a pork carcass.

PORK CARCASSES

Use Figure 15-17 to assist with the following section. The following are places to look for muscling differences in pork carcasses and some descriptive terms to use in a set of reasons:

ham—fuller, meatier, thicker, more bulging, wider, plumper

sirloin—fuller, plumper, meatier

lumbar lean—greater exposed area, larger area

loin—fuller, wider, deeper chined

loin eye (if ribbed)—larger, shapelier

shoulder—meatier, thicker fleshed, thicker

The following are places to look for trimness differences in pork carcasses and some descriptive terms to use in a set of reasons:

backfat—trimmer at the lumbar, last rib, and first rib (if not ribbed), loineye (if ribbed)

ham collar, flank, belly, sternum, jowl—trimmer

elbow pocket—more definition

ham-loin junction—more definition

The following are places to look for quality differences in pork carcasses and some descriptive terms to use in a set of reasons:

ANIMAL SCIENCE FACTS

A plastic or cellophane grid is one way to measure rib eye or loineye areas. Simply lay the grid on the cut surface of the eye muscle. Count the dots and divide the number counted by 10 (for beef grids) or 20 (for pork and lamb grids) to calculate area in square inches.

A more accurate way to determine the area is with a compensating polar planimeter. This machine is used to outline a tracing of an eye muscle. The area (in square inches) can be calculated by subtracting the initial reading from the final reading.

loineye (if ribbed)—finer textured, more evenly dispersed marbling, higher degree of marbling, firmer textured lean, more ideal grayish-pink color

lumbar lean—higher degree of marbling

belly—fuller, firmer, thicker (these are important quality characteristics for bacon production)

rib feathering—higher degree, greater quantity, more extensive

JUDGING PRIMAL CUTS

Fresh hams are the most commonly judged primal cut of pork. As with carcasses, the ham that best combines muscling and trimness, with acceptable quality wins the class. See Figure 15-18 for the parts of a fresh ham.

HAMS

Use Figure 15-18 to assist with the following section. The following are places to look for muscling differences in fresh hams and some descriptive terms to use in a set of reasons:

shank—shorter (indicates more muscle)

rump or butt face—deeper, wider, greater amount of exposed lean

forecushion or knuckle—fuller, meatier, plumper

cushion or center section—longer, wider, deeper, plumper, more bulging

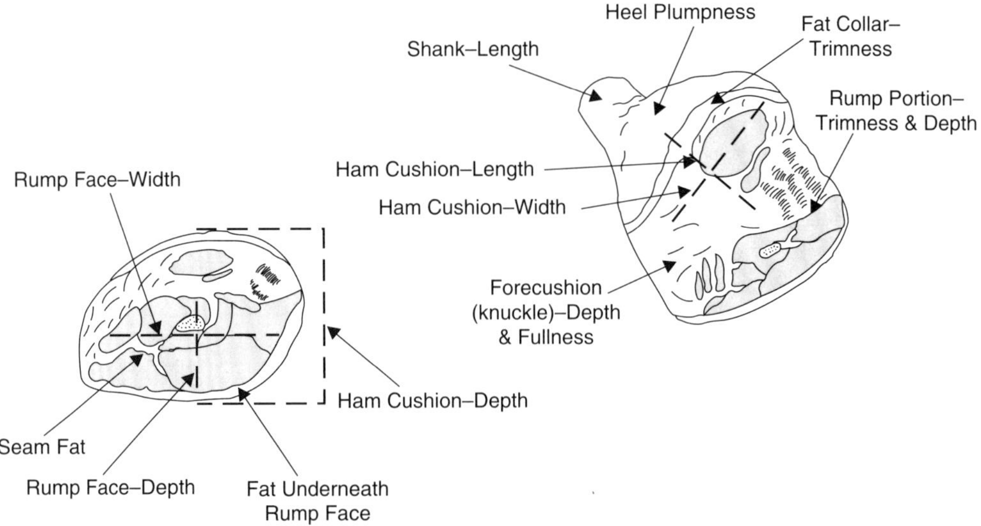

Figure 15-18. Parts of a fresh ham.

The following are places to look for trimness differences in fresh hams and some descriptive terms to use in a set of reasons:

under the butt face, along the rump face, over the collar, over the cushion—trimmer

seam fat—more visible

The following are places to look for quality differences in fresh hams and some descriptive terms to use in a set of reasons:

marbling in the rump face—more, a higher degree, more visible

texture of exposed lean—firmer, finer textured lean

color of exposed lean—more uniform grayish-pink

GENERAL INTRODUCTORY PORK TERMS

Carcasses

A higher yielding carcass

Carcass that would yield a higher percent muscle

Carcass that would yield a higher percent ham and loin

Carcass with a higher lean to fat ratio

Carcass yielding a higher percentage of the trimmed four lean cuts

Hams

Ham that would yield a higher percentage of muscle

Ham that would yield a higher percentage of center cut slices

Ham with a higher lean to fat ratio

SAMPLE PORK REASONS

Pork Carcasses Placing: 1–2–3–4

FIRST 1/2 I placed 1 over 2 as 1 combined muscling and trimness to a higher degree. 1 was thicker through the center portions of the ham and significantly wider through the sirloin. In addition, 1 was leaner at the last rib and displayed less sternum fat. I admit that 2 was trimmer over the ham collar and exhibited more rib feathering.

SECOND 2/3 2 placed over 3 as 2 was simply a higher cutability carcass with a higher lean to fat ratio. 2 possessed less fat at the first rib, last rib and last lumbar and was particularly leaner over the loin edge and at the ham collar. Furthermore, 2 displayed a shorter-shanked ham and more exposed lumbar lean. I concede 3 did exhibit more marbling in the lumbar lean.

THIRD 3/4 3 more completely combined muscling and quality to a higher degree than 4. 3 displayed a decidedly thicker-cushioned ham, a thicker sirloin, and a more prominent shoulder. Furthermore, 3 displayed a more desirable grayish-pink color in the lumbar lean, as well as more rib feathering. I admit 4 was trimmer at the jowl and over the ham collar.

FOURTH 4 I placed 4 last as it was the lightest muscled, poorest quality carcass in the class. 4 was narrowest in the shoulder, loin and ham. Also, 4 showed the poorest colored exposed lean and would be the least appealing to the quality-conscious consumer.

LAMB JUDGING

As with pork, the judging of lamb carcasses and cuts relies to a very high degree on cutability.

JUDGING LAMB CARCASSES

Some lamb carcasses are split between the 12th and 13th rib, as are beef carcasses, exposing the ribeye. Others are not ribbed. When carcasses are ribbed, fat estimates or measurements should be made $^1/_2$ the distance over the eye.

Lambs are yield graded using the 12th rib measurement. Simply plug the 12th rib fat depth into the formula:

$$(\text{Fat depth} \times 10) + .4 = \text{yield grade}$$

Other areas to check for external fat deposit include the shoulder, sirloin, dock, leg, cod or udder, flank or breast. The expensive cuts of lamb are from the hindsaddle, or the portion of the carcass from the 12th rib back. Therefore, differences in muscling of the hindsaddle should be evaluated with care. Muscling differences are most easily seen in the ribeye (ribbed carcasses), dimensions of the leg, and width of the loin, sirloin and rump (unribbed carcasses).

The color of lamb should be bright reddish pink. Other parameters showing quality include flank streaking and rib feathering.

Lambs are also quality graded using a combination of overall conformation and flank streaking. A vast majority of the lambs that are graded reach choice or prime. However, an age parameter is in place so that only young lambs can be graded.

As lambs mature to one year of age, their bones become more calcified. When the front legs of a young, immature lamb are removed, the resulting red, porous bone left showing is called a ***break joint***. As lambs approach 12 to 15 months of age, the break joints ossify and, at some point, will no longer come apart at the break joint. The place further down on the leg at which older sheep's front legs are removed from the carcass are called ***spool joints***. Spool joints are bone white in color. If one break joint and one spool joint are present, the carcass is called a lamb carcass. See Figure 15-19.

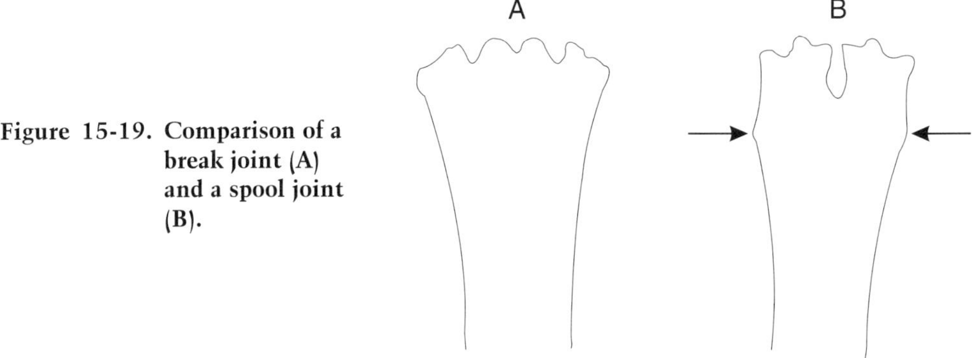

Figure 15-19. Comparison of a break joint (A) and a spool joint (B).

LAMB TERMINOLOGY AND REASONS

Lambs have a few different parts than those studied previously in beef and pork. Study and memorize the parts in Figure 15-20.

LAMB CARCASSES

Use Figure 15-20 to assist with the following section. The following are places to look for muscling differences in lamb carcasses and some descriptive terms to use in a set of reasons:

leg—fuller, meatier, thicker, more bulging, wider, plumper

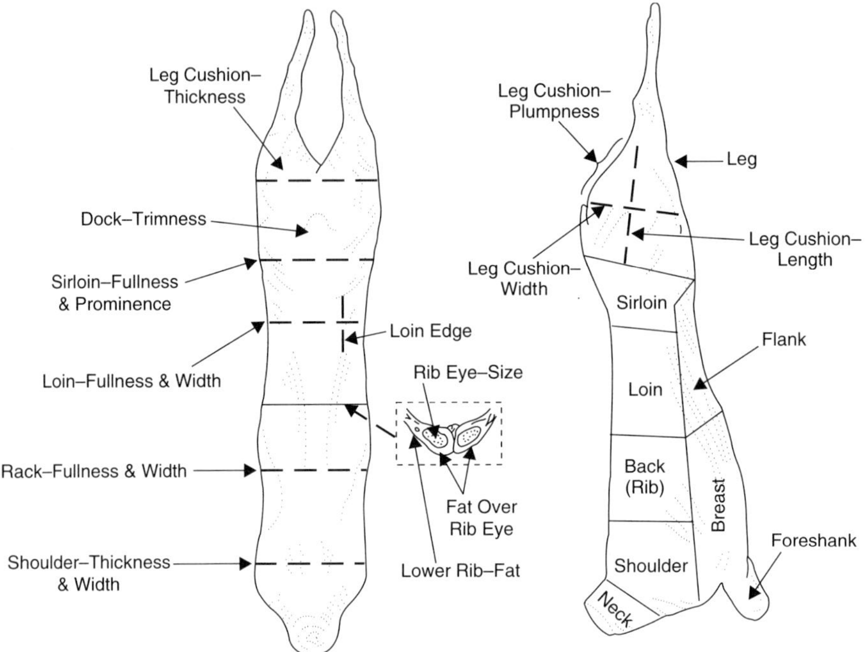

Figure 15-20. Parts of a lamb carcass.

sirloin—fuller, plumper, more prominent

rack or rib—fuller, wider

rib eye (if ribbed)—larger, shapelier

shoulder—fuller, thicker

The following are places to look for trimness differences in lamb carcasses and some descriptive terms to use in a set of reasons:

ribeye, lower rib, leg cushion, sirloin, dock, loin edge, rack, shoulder, internal, cod or udder, breast, flank—trimmer, displayed less fat

The following are places to look for quality differences in lamb carcasses and some descriptive terms to use in a set of reasons:

ribeye (if ribbed)—more youthful, brighter colored lean, firmer textured lean, higher degree of marbling

rib feathering—higher degree, greater quantity, more extensive

flank streaking (lacing)—greater primary and secondary flank streaking, more finely dispersed

JUDGING PRIMAL CUTS

Primal cuts of lamb are judged sparingly. Terms are similar to those used for beef primal cuts (but substituting leg for round, etc.)

GENERAL INTRODUCTORY LAMB TERMS

Carcass that would yield a higher percentage trimmed leg and loin.

Carcass that would yield the highest percentage boneless, trimmed retail cuts.

Carcass that would yield a higher percent hindsaddle.

SAMPLE LAMB REASONS

Lamb Carcasses Placing: 1–2–3–4

FIRST 1/2 1 placed over 2 as 1 was a significantly heavier muscled carcass that would yield a higher percent hindsaddle. 1 displayed more inside and outside flare in a heavier muscled leg, a wider sirloin, and a decidedly larger ribeye. In addition, 1 displayed a more youthful appearing rib eye and possessed more primary and secondary flank streaking. I admit 2 was somewhat trimmer over the leg, dock, cod, and breast.

SECOND 2/3 2 placed over 3 in a near decision as 2 was a trimmer, higher cutability carcass. 2 exhibited less fat at the ribeye, over the dock, leg, and shoulder. Furthermore, 2 had a longer cushioned leg. I concede 3 possessed a larger ribeye, and was a higher quality carcass, as evidenced by a higher degree of marbling at the rib eye and more rib feathering.

THIRD 3/4 3 handily placed over 4 as 3 more ideally combined trimness and muscling into a higher cutability carcass. 3 was trimmer at the rib eye, over the sirloin, and over the leg. Furthermore, 3 had decidedly less pelvic fat. Moreover, 3 was meatier at the rib eye, over the sirloin, and possessed a wider, thicker cushioned leg. I admit 4 showed more primary flank lacing and a higher degree of marbling at the eye.

FOURTH 4 I realize 4 showed the brightest colored lean and the highest degree of marbling; however, 4 simply combined trimness and muscling to the lowest degree. 4 exhibited the most fat over the rib eye, loin, dock, leg, and shoulder. Also, 4 displayed the smallest rib eye in the class. Accordingly, 4 would yield the lowest percentage of closely trimmed retail cuts.

SUMMARY

Meat is judged on a combination of muscle quantity, leanness, and muscle quality. Carcasses are broken down into primal, sub-primal, and then retail cuts. IMPS cuts are standardized primal and subprimal cuts used in the institutional and restaurant trade. Tender cuts of meat should be cooked with high, dry heat for a short duration. Tougher cuts require a longer period of moist heat for optimum flavor and tenderness. Yield grades are used to differentiate between lean and fat carcasses. Quality grades attempt to predict the flavor and tenderness of meat. Age is an important consideration when quality grading beef and lamb. A knowledge of comparative terminology of carcasses and cuts is very useful to meat judges.

CHAPTER SELF-CHECK

Term		Definition
___ lean-to-fat ratio	1.	taken from primal cuts
___ cutability	2.	estimate of cutability
___ primal cuts	3.	measurement of fat cover $^3/_4$ the distance over the ribeye at 12th rib
___ subprimal cuts	4.	measures lean to fat
___ Institutional Meat Purchasing Specifications (IMPS)	5.	large primary carcass cuts like loin, brisket
___ yield grades	6.	amount of muscle compared to fat
___ quality grades	7.	point at which lambs front legs are removed from carcass
___ preliminary yield grade (PYG)	8.	internal fat
___ kidney, pelvic, and heart fat or KPH	9.	cartilage located along split surface of carcass over rib
___ buttons	10.	point at which older sheep's front legs are removed from carcass
___ break joint	11.	estimate of marbling and age
___ spool joint	12.	standardized primal and sub-primal cuts

QUESTIONS AND PROBLEMS FOR DISCUSSION

1. Three primary factors influence the judging of meat. Name them.
2. List the primal cuts of lamb.
3. Primal cuts taken from the ______ are the most valuable.
4. Write the IMPS series number for fresh beef.
5. Differentiate between methods for cooking a steak and a roast.
6. Yield grade indicates _______.
7. Quality grade indicates _______ and _______.
8. ___ percent internal fat is considered average for beef carcasses.
9. Name the nine marbling categories.
10. Maturity scores in beef carcasses range from ___ (youngest) to ___ (oldest).
11. The highest quality grade is _______.
12. Trimness in a pork carcass is primarily evaluated at the ___ rib.
13. Ideal color of fresh pork is _____________.
14. Presence of a break joint in a lamb carcass indicates the age to be __________________.
15. The most expensive cuts of lamb are located __________.

ACTIVITIES

1. Tour a local slaughter house. Have employees discuss their duties.
2. Visit the meat counter of a grocery store. Identify wrapped cuts of meat.Contrast cutability and quality.
3. Cook steaks using both the high heat, short time, dry method and the low heat, long time, moist method. Compare the taste.

LABORATORY ACTIVITY

YIELD GRADING

Purpose

To practice yield grading cattle

Materials

paper and pencils for notes
calculator
untrimmed ribeye steak (cut around the 12th rib)
actual size acetate photocopy provided by instructor

Procedure

1. With the ruler provided on the edge of the acetate, measure the fat thickness of the steak $3/4$ the distance over the ribeye muscle. See Figure 15-5. Record your measurement.
2. Lay the grid on the ribeye muscle of the steak and count the dots lying within the ribeye muscle. Divided the number counted by 20. Repeat three times. Record the average number as "square inches of ribeye."
3. Calculate the yield grade of the steak using the measured "square inches of ribeye" under each of the following three assumptions:
   a. 600 pound carcass, 2.5 percent KPH
   b. 725 pound carcass, 3.5 percent KPH
   c. 550 pound carcass, 2.0 percent KPH

Round your answers to the nearest tenth of an inch.

Analysis

1. Which of the carcasses would require the largest ribeye area?
2. In which of the three carcasses would the measured ribeye most likely reduce the yield grade?
3. Which two carcasses would receive a yield grade deduction for KPH?
4. Rank the three carcasses from lowest to highest yield grade.

Application of Laboratory Activity

USDA beef graders are extremely proficient at estimating yield grades with a single glance at a carcass. Professional graders simply look at carcass weight and make snap decisions about additions or deductions for ribeye and KPH.

Beef is sold by whole yield grade numbers. For example, a yield grade 2 may have an actual yield grade of 2.0 to 2.9. Most slaughter cattle are yield grade 2 and 3.

Chapter 16

LIVESTOCK JUDGING

Phenotypically yours

INTRODUCTION

Livestock judging is the art and science of evaluating the phenotype of live animals. Phenotype gives indications of the utility of livestock on a farm, ranch, or feedlot that performance data cannot. For example, performance data give no clues to the structural correctness of an animal, which may affect the longevity in a breeding herd. Ideally, a combination of performance data and visual appraisal should be used to make placing decisions. Unfortunately, data are not always available. Often, visual evaluation is the only method available for livestock selection.

Figure 16-1. (Courtesy, Jasper S. Lee)

OBJECTIVES

1. Market Animal Evaluation
   a. Describe an ideal market hog, steer, and lamb
   b. Place classes of market hogs, steers, and lambs
2. Breeding Animal Evaluation
   a. Describe ideal male and female breeding stock
   b. Place classes of male and female breeding stock

TERMS

blind nipples
bowlegged (front legs)
bowlegged (rear legs)
buck-kneed
bucks
calf-kneed
cowhocked
finish
flat nipples
knock-kneed
pigeon-toed
pin nipples
posty
sickle-hocked
splay-footed

ANIMAL SCIENCE FACTS

Individual animals in a judging contest may have identification points that should be referenced in a set of reasons. For example, a heifer that has had her ear tag pulled out, leaving a split in the ear could be referred to as "the split-eared heifer." Likewise, males and females should be differentiated in a mixed class of market animals. Color patterns that differ from the norm should also be identified. Littermates can be identified in a swine class by the ear notches. Using identification points helps the listener better follow a set of reasons and makes the contestant seem more observant and knowledgeable about the class.

MARKET ANIMAL EVALUATION

Market animals are often evaluated, bought, and sold solely by phenotype. Livestock buyers make decisions based on the apparent amount of salable muscle in a live animal. Many individuals have perfected their abilities to estimate the relative amounts of muscle and fat. Livestock buyers make a living based on this ability. Judging any market animal relies on three parameters: muscle, fat, and structural correctness. The following is a discussion of places to look on live market animals for indications of these parameters.

BEEF

As with meats judging, proper terminology is essential to describing a market steer or heifer. Memorize the parts presented in Figure 16-2.

Evaluating muscle content in market steers and heifers is not difficult. As with all market animals, the highest priced cuts of meat originate from the rear half of the animal. Therefore, select steers or heifers that are widest over the top, rump, and rear quarter.

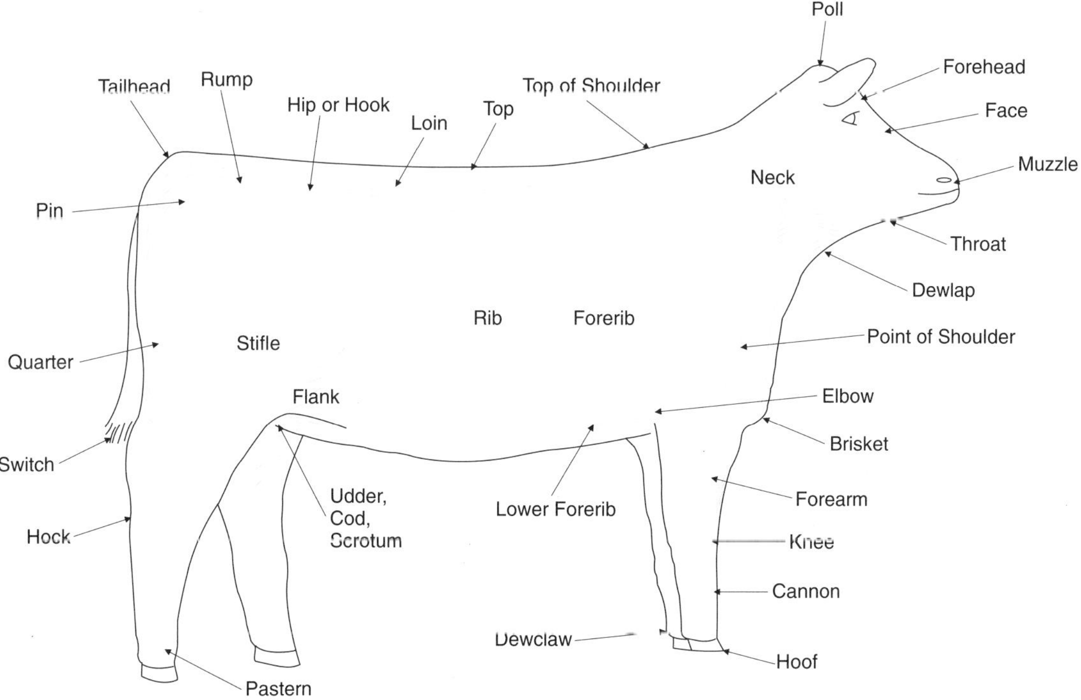

Figure 16-2. Parts of a beef animal.

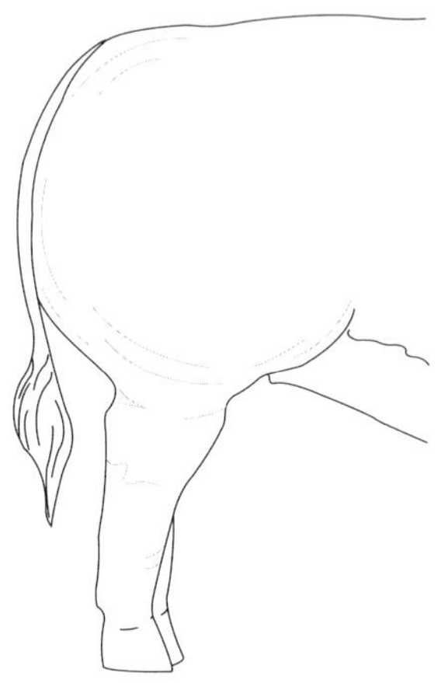

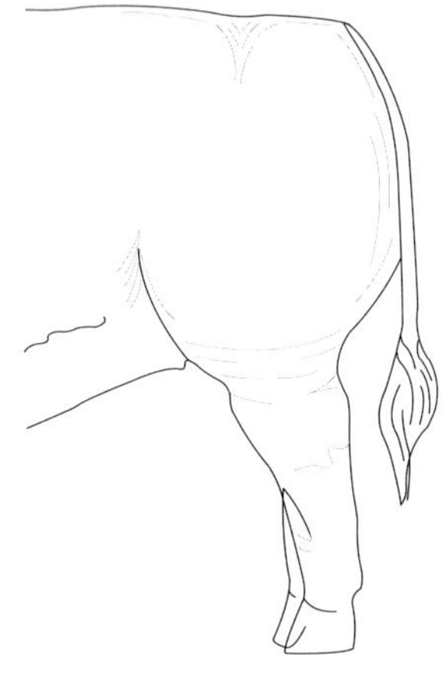

Figure 16-3. Two types of muscle design in beef cattle.

Muscle pattern and design are also important. Muscle should be long, smooth, and thick, not short, bunchy, and round. See Figure 16-3.

Terms used to describe muscle content in market cattle follow.

Desirable	Undesirable
thicker made	narrower made
heavier muscled	lighter muscled
longer, smoother muscle design	shorter, tighter muscle design
thicker quartered	
deeper quartered	shallower quartered
wider topped	narrower topped

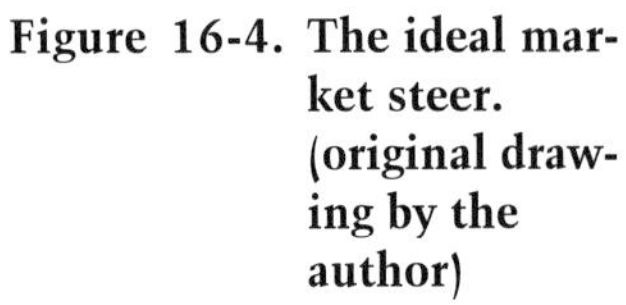
Figure 16-4. The ideal market steer. (original drawing by the author)

Desirable	Undesirable
thicker stifled	
pushed more stifle on the move	
wider rumped	narrower rumped
more total muscle mass	
more muscle dimension	
rail a carcass with a larger ribeye	

Fat thickness on market steers and heifers must be adequate, but not excessive. Remember, a certain amount of exterior fat must be present to assure adequate marbling. Too much exterior fat results in lower yield grading, low cutability carcasses. Fat may be observed on a live steer or heifer most easily by observation of a deep, full brisket, an overall smooth appearance, and patches of fat beside the tailhead. Some judging contests allow market cattle to be handled to determine correct ***finish*** (fat cover). Experienced judges handle cattle by placing four fingers together and running them along the upper and lower rib cage. Fat cover will "squish" out from under the fingers, while unfinished or underfinished cattle will feel hard and bony to the touch. Some cattle will have adequate fat cover over the upper rib, but will be free of finish over the lower rib. These cattle are referred to as unevenly finished.

The following are terms used to describe finish on market cattle.

Desirable	Undesirable
more ideally finished	wastier
smoother in his/her finish	patchy, uneven finish
should rail a carcass with	
–a higher lean to fat ratio	
–a more desirable yield grade	
more apt to grade choice	
trimmer, cleaner patterned	
trimmer middled	wastier middled

Structural correctness incorporates frame size, outline, and balance, as well as feet and legs. The ideal market steer should be moderately framed, well-balanced, and boxy in appearance with a strong, straight topline. Feet and legs should be spaced evenly under each corner of the box. The rump should be level. The neck should be long and clean, and the point of the shoulder should be neatly laid-in. The ideal market steer should be long bodied, deep bodied, and have good spring of rib. Use the steer in Figure 16-4 as your example of an ideal market steer and compare all other market cattle to that mental image.

The description of feet and legs requires a knowledge of correct skeletal design and the ability to recognize when that design is not quite right. From the side, rear legs can be ***posty*** (too straight), or ***sickle-hocked*** (too curved). From the rear, legs can be bowed in (***cowhocked***) or out (***bow-legged***). Front legs can also have similar problems. When viewed from the side, front legs that are too straight are called ***buck-kneed***, Too much curvature is called ***calf-kneed***. When viewed from the front, legs can be ***bowlegged*** (bow out), ***knock-kneed*** (knees too close together), ***splay-footed*** (feet turned out), or ***pigeon-toed*** (feet turned in). See Figure 16-5 for pictures of common rear leg faults and Figure 16-6 for common front leg faults.

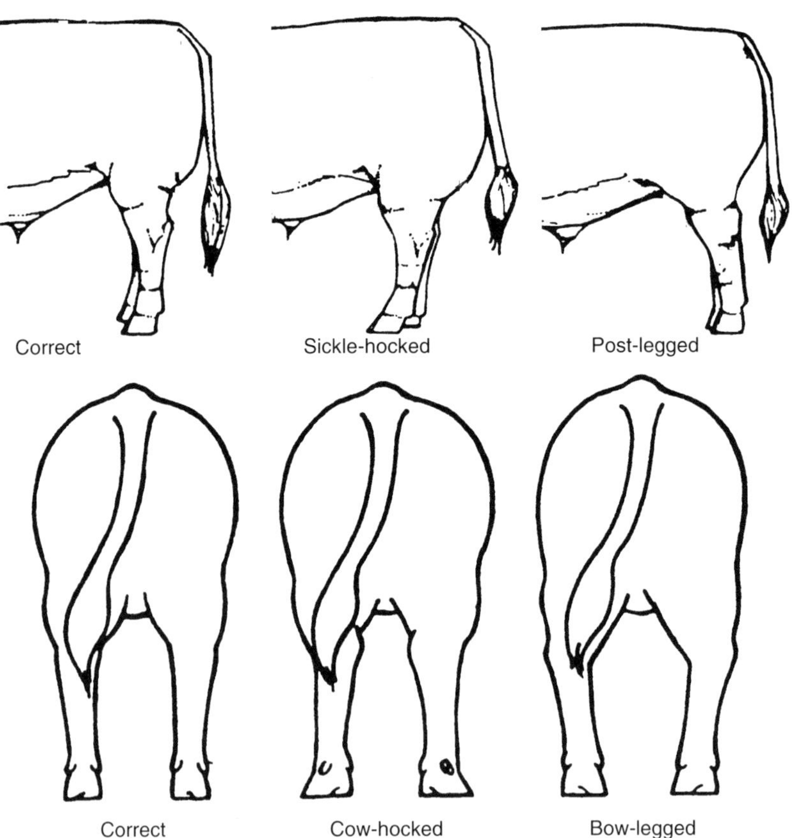

Figure 16-5. Common rear leg faults.

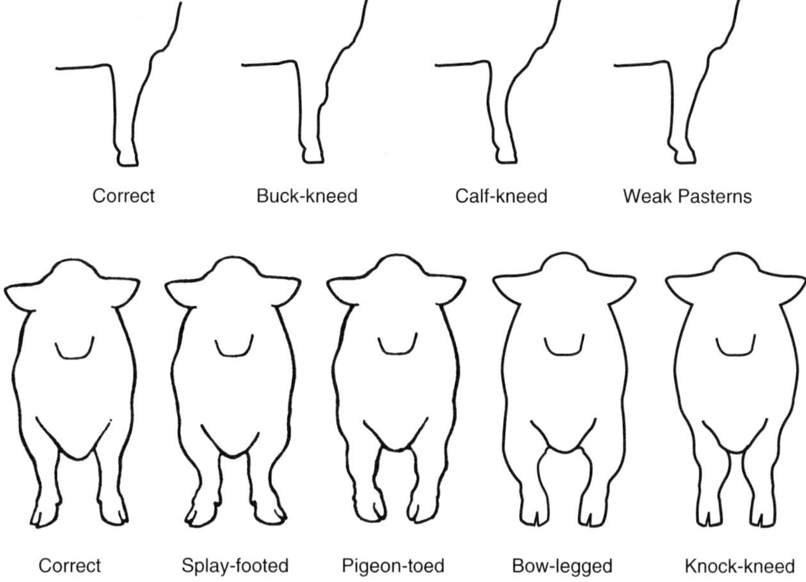

Figure 16-6. Common front leg faults.

The following are terms used to describe structural correctness in market cattle.

Desirable	**Undesirable**
Frame/outline	
better balanced	off balance
more eye appealing	
fresher appearing	staler appearing
larger framed	smaller framed
straighter lined	slack framed
later maturing	earlier maturing
stronger topped	weaker topped
more structurally correct	structurally incorrect
fault free	ill structured
longer bodied	shorter bodied
Capacity	
bigger volumed	had less volume
more arch and spring of rib	flatter ribbed, deeper ribbed
shallower ribbed	
deeper flanked	shallower flanked

Desirable	Undesirable
Front ends	
more extension through his/her front end	shorter fronted
smoother, tighter shouldered	open shouldered
laid-in more neatly about the the shoulder	coarser fronted
longer, leaner neck	
tighter fronted	
Rump	
squarer, leveler rump	narrow, steep rump
wider/longer from hooks to pins	narrower from hooks to pins
wider rump	narrower rump
Feet/Legs/Movement	
heavier boned	finer boned
stood on greater length of cannon	
more desirable set to the hock	posty, sickle-hocked
stronger pasterns	weaker pasterns
stood squarer when viewed from the rear/front	cowhocked, toes out
stood/tracked wider behind	narrower tracking
longer strided	shorter strided
mover freer and easier	shorter strided
moved with more strength of top	roached his top on the move
moved with more levelness of rump	dropped his pins on the move
General	
more progressive	conventional
performance oriented	lacks performance
growthier	
more dimensional	

Sample reasons for market steers

Placing: 1–2–3–4

In my top pair of heavier muscled steers, I preferred 1 to 2 as 1 was a trimmer, cleaner patterned steer that was more ideal in his finish. 1 was a longer necked, tighter shouldered steer that was stronger in his top and more nearly level through his rump. He also handled smoother over his lower rib and should be more apt to rail a carcass reaching the choice grade. I realize that 2 was a heavier boned, deeper bodied, more ruggedly designed steer.

Nonetheless, it was 2 over 3 in my middle pair. 2 simply overpowered 3 in total muscle dimension. 2 was a thicker topped steer that carried his muscle mass into a wider, more expressively muscled rump and a deeper, thicker quarter. In addition, 2 was a wider tracking steer that pushed more stifle on the move. He should rail a carcass with a larger ribeye. I must admit that 3 was a trimmer middled, tighter fronted steer, but he lacked the sheer muscle volume of my top pair.

Moving to my bottom pair of lighter muscled, narrower tracking steers, I preferred 3 to 4 as 3 was a tighter framed, more structurally correct steer. 3 was a longer necked, tighter shouldered, stronger topped steer that had a more desirable set to his hock. Thus, 3 moved with a longer, more ground covering stride.

I fully realize that 4 was more performance oriented, deeper ribbed, and deeper flanked. But, 4 placed last in the class as he was the most structurally incorrect, off-balance steer. He lacked the overall style and symmetry of the steers placed above him. Thank you.

SWINE

Figure 16-7 illustrates the parts of a market hog. Study them with care.

Muscle evaluation of market hogs relies on an estimate of thickness. However, thickness caused by fat looks nearly the same as thickness caused by muscle. It takes a practiced eye to tell the difference. Often, the leanest, heaviest muscled pig will not be the thickest pig in the class.

Places to look for muscle on a market hog naturally include the loin, rump, and ham. Hogs thick in these areas containing high-priced cuts will probably have wide shoulders and chests also.

The following are terms used to describe muscle content in market hogs.

Desirable	**Undesirable**
heavier muscled	lighter muscled
more expressively muscled	flatter muscled

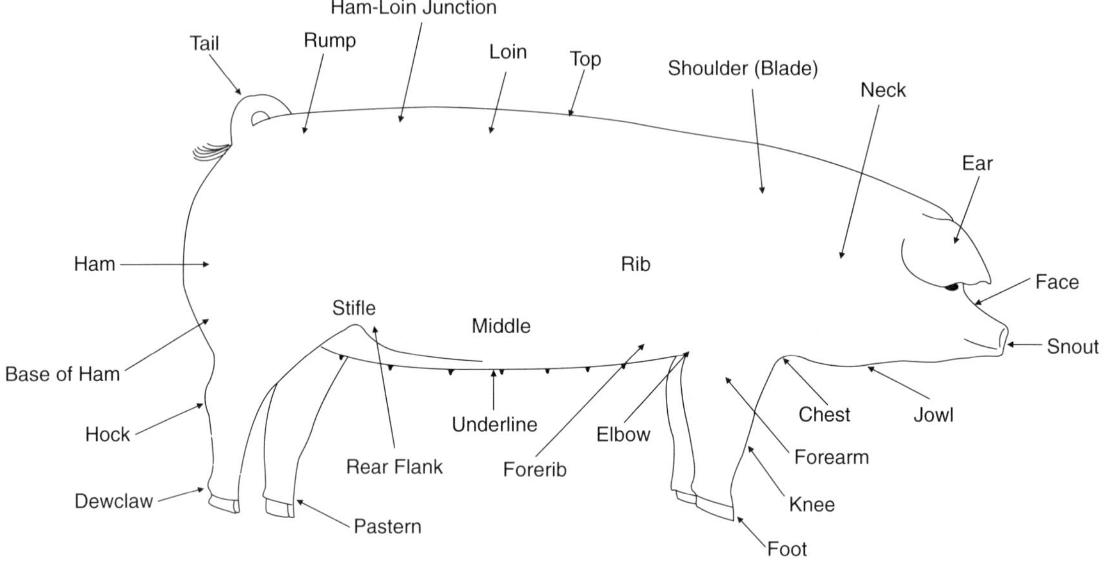

Figure 16-7. Parts of a hog.

Desirable	**Undesirable**
more natural turn to his/her top	
pushes more stifle on the move	narrower stifled
more total muscle mass	less total muscle mass
wider rumped	narrower rumped
more muscle dimension from stifle to stifle	
more flair and dimension to the lower $^1/_3$ of the ham	narrower through the lower $^1/_3$ of the ham
should rail a carcass with a higher lean to fat ratio	
should rail a carcass with a larger loineye	
should rail a carcass with a higher percent lean ham and loin muscle	

The ability to visually evaluate fat on market hogs is probably the most important and difficult skill learned by hog judges. Market hogs are not handled to evaluate fat cover, so visual clues are the only indication of relative fatness.

Fortunately, visual clues to fatness exist. Pigs carrying too much fat can be distinguished by a square shape over their top when viewed from the rear. The shoulder blade will be made almost invisible by a thick fat layer. Additionally, pigs carrying too much fat will appear sloppy in the jowl, at the base of the ham, and in the flank. A roll of fat may appear in the elbow pocket when the pig's front leg is extended to the rear.

An exceptionally lean pig will show a definite ham-loin junction and a more rounded shape to the top when viewed from the rear. The shoulder blade will be obvious when the pig moves. A lean pig will appear trim in the jowl, flank, and at the base of the ham.

The following are terms used to describe relative leanness in market hogs.

Desirable	Undesirable
leaner designed	
raw made	
leaner topped	plainer topped
cleaner over the loin edge	squarer over the loin edge
more natural shape his/her top	
trimmer in his/her lower one third	wastier in his/her lower one third
showed more blade action on the move	
should rail a carcass with a greater lean value	

Structural correctness of market hogs also deserves a great deal of attention. A correctly designed pig should be large framed, long and deep bodied, level topped, and naturally wide from front to rear. See Figure 16-8 for an ideal market hog.

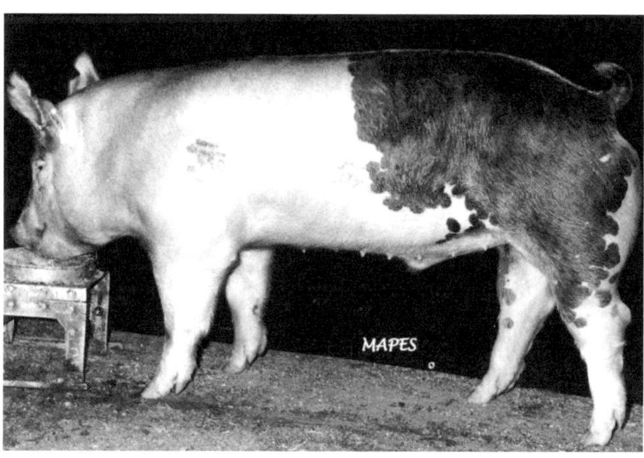

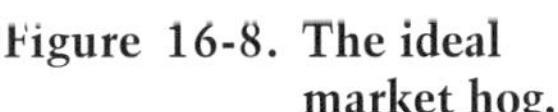
Figure 16-8. The ideal market hog.

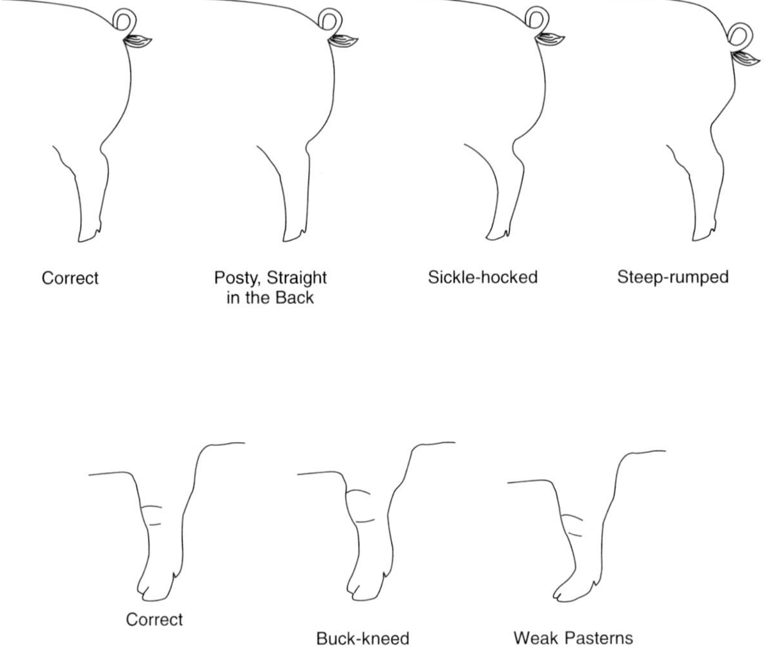

Figure 16-9. Common leg faults of swine.

Feet and legs should show a great deal of flex at all joints when the pig moves. See Figure 16-9 for pictures of common leg structure faults.

The following are terms used to describe structural correctness in market hogs.

Desirable	Undesirable
Frame/outline	
larger framed	smaller framed
bigger outlined	smaller outlined
longer sided	shorter sided
leveler topped	higher topped
better balanced	more off balance
more functional	
looser made	tighter
more rugged	more frail
sounder made	poorer structured
more durable	less durable

Desirable	Undesirable
Capacity	
wider sprung	narrower made
deeper ribbed	shallower ribbed
deeper chested	tighter ribbed
wider through his/her chest floor	
wider based	narrower based
bolder sprung	
deeper flanked	tighter flanked
more expanded through his/her rib	constricted in his/her rib
looser middled	tighter middled
Rump	
longer rumped	shorter rumped
leveler rumped	steep rumped
leveler in his/her rump design	
wider rumped	narrower rumped
squarer rumped	
Feet/Legs/Movement	
more confinement adaptable	less confinement adaptable
more curvature to his/her knee	buck-kneed
more desirable slope to his/her shoulder	straighter shouldered
more flex of hock	straighter hocked
more cushion in his/her pastern	
more animated in his/her movement	
looser strided	tighter wound
longer strided	shorter strided
wider tracking	narrower tracking
more flexible	tighter wound
heavier boned	finer boned
stood on greater substance of bone	
General	
growthier	
higher performing	poorer performing
more functional	
more producer oriented	
more complete	
more fault-free	

Sample market hog reasons

Placing: 2–3–4–1

In my top pair, I preferred 2, the blue butt barrow over 3, the red gilt, as 2 was a more producer oriented, confinement adaptable barrow that should go to the rail and hang a carcass with a higher percent muscle. 2 was a wider sprung, deeper ribbed barrow that was thicker over his top, through his rump, and showed more natural thickness from stifle to stifle. In addition, he stood on more substance of bone and moved with more flex and animation to his hock. I grant that 3 was trimmer through her lower one third.

Nevertheless, it was 3 over 4 in my middle pair. 3 was a leaner made, more expressively muscled gilt that should rail a higher cutability carcass. 3 showed more blade action on the move, displayed a more natural turn to her top and was more expressive through the center and lower portions of her ham. I realize 4 was more nearly level through his rump and moved with more curvature and cushion to his knee and pastern, but he was planer in his top and narrower based.

4 placed over 1 in my bottom pair of black barrows. 4 more closely followed my top barrow in his kind being a looser designed, bigger scaled, wider made barrow from end to end. 4 was a wider chested, deeper ribbed barrow that moved out with a longer, looser stride. Additionally, he was a wider topped, wider rumped barrow that stood on a greater diameter of bone.

Granted, 1 was leaner over his loin edge and would hang a carcass with less fat trim. However, 1 was the poorest performing, narrowest made, lightest muscled, shallowest bodied pig in the class that had the least to offer the performance minded producer. Thank you.

LAMBS

The parts of a market lamb are shown in Figure 16-10.

Muscle thickness in lambs is easiest to detect through handling. Many judging contests allow contestants to handle all lambs. A heavily muscled lamb will visually appear wide from directly behind the shoulders, through the rump, and stand naturally wide when viewed from the rear. When handled, a heavily muscled lamb should feel wide and thick over the top directly behind the shoulders. The width should be maintained through the loin, and get progressively wider through the rump. The rump and leg should feel thick, full, and square. When grasped directly behind the last rib, the loin should be deep, as well as wide. The length of loin can

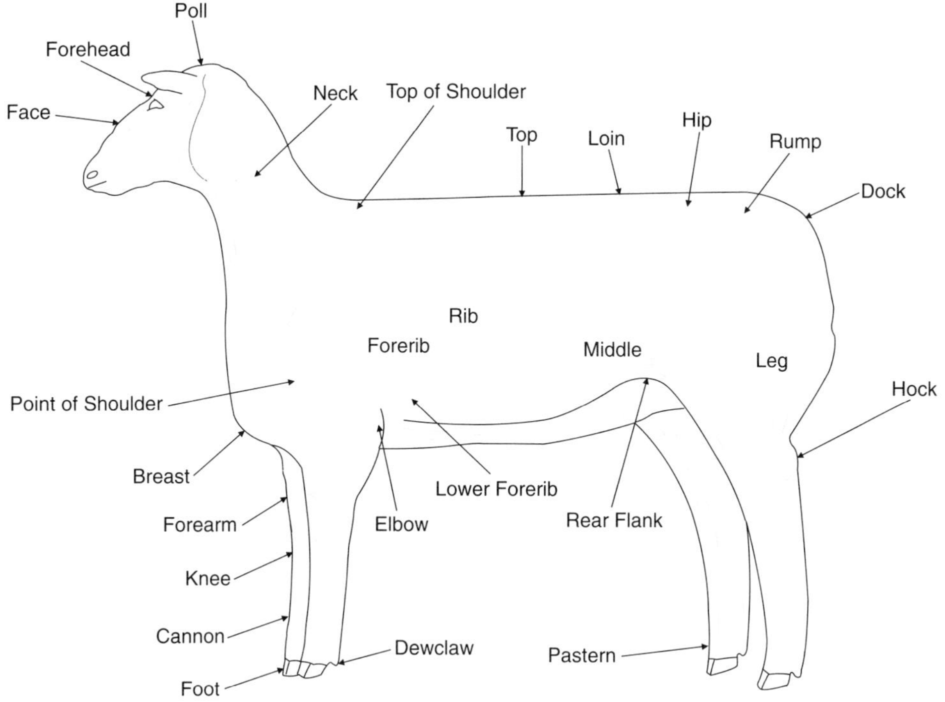

Figure 16-10. Parts of a sheep.

be evaluated by using the hands to measure the distance between the last rib and the hip.

Each lamb should be handled with a predetermined routine so differences among lambs can be remembered. The four fingers should be extended and kept together. The topline should be handled first (with one hand) from behind the shoulder, progressing to the rump. Next, with both hands, determine the depth and length of loin. Last, with both hands, grasp the leg as close to the lamb's flank as possible. Touch the tips of your fingers on the inside of the leg and note the distance between the thumbs on the outside of the leg. This distance can be used to compare leg muscle among lambs.

The following are terms used to describe muscle content in market lambs.

Desirable	**Undesirable**
heavier muscled	lighter muscled
thicker made	
more expressively muscled leg	narrower, lighter muscled leg

Desirable	Undesirable
more dimensional leg	
heavier muscled hindsaddle	lighter muscled hindsaddle
handled with a longer loin/deeper loin/thicker rump/heavier muscled leg	
should hang a carcass with a higher percent hindsaddle	
should rail a carcass with a heavier muscled leg	

Fat determination can also be accomplished by handling a market lamb. Sometime during the handling process, gauge the amount of fat that squishes out from under the fingers when a lamb is handled over the upper and lower rib. Carefully pinching the rear flank of the lamb for fullness can also give a good approximation of the relative fat cover over the rib. A lamb that has a full, fat-filled flank is probably fat over the rib. Remember, ideal fat cover is .10 to .15 inches. Anything over .20 inches is excessive. Accurately determining the amount of fat on market lambs takes considerable practice.

Fat can also be estimated by visual appraisal. Fat lambs will appear smooth, while especially trim lambs will appear lumpier.

The following are terms used to describe fat differences in market lambs.

Desirable	Undesirable
trimmer made	wastier
trimmer, cleaner patterned	overfinished
cleaner conditioned	heavier conditioned
harder handling	softer handling
handled with more trimness over the upper and lower rib/at the 12th rib	
met my hand with ____	
should hang a trimmer carcass	should hang a wastier carcass
should hang a higher cutability carcass	

Structure of market lambs is most similar to that of cattle. Therefore, the terminology used is also similar. An ideal market lamb should be

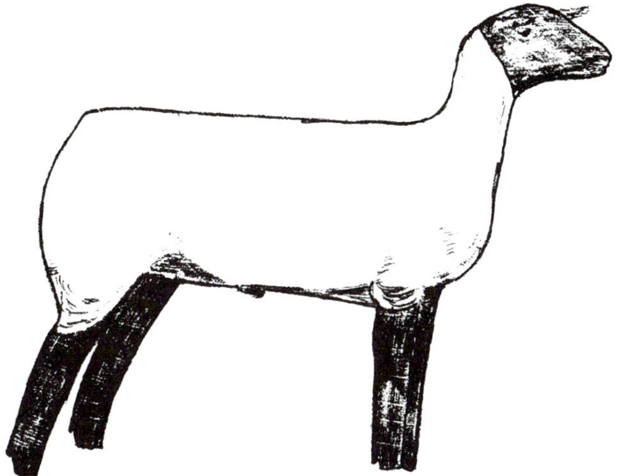

Figure 16-11. The ideal market lamb. (original drawing by the author)

wedge-shaped. The front end should be trim, the neck long and lean, and the shoulders tight and neatly laid-in. The topline and dock should be straight and level when viewed from the side. The body should be deep, but not wasty. The loin and the leg should be much wider than the front end. See Figure 16-11 for an example of an ideal market lamb.

Feet and legs should be squarely placed with the front legs spaced slightly closer together than the rear legs. Bone should be heavy, but not coarse. Feet and leg faults are similar to those described for cattle.

The following are terms used to describe structural correctness in market lambs.

Desirable	Undesirable
Frame/outline	
better balanced	off balance
more eye appealing	
larger framed	smaller framed
more size and scale	
tighter framed	slack framed
later maturing	earlier maturing
stronger topped	weaker topped
more structurally correct	structurally incorrect
fault free	ill structured
longer bodied	shorter bodied

Desirable	Undesirable
Capacity	
wider sprung	narrower made
higher capacity	tighter ribbed
trimmer middled	wastier middled
Front ends	
fresher appearing	staler
more extension through his/her front end	shorter fronted
smoother, tighter shouldered	open shouldered
laid-in more neatly about the shoulder	coarser fronted
longer, leaner neck	pelty about the neck
tighter fronted	
smoother neck-shoulder junction	ewe necked
Dock	
squarer, leveler dock	narrow, steeper dock
wider/longer dock	
wider rump	narrower rump
Feet/Legs/Movement	
heavier boned	finer boned
more desirable set to the hock	posty, sickle-hocked
stronger pasterns	weaker pasterns
stood squarer when viewed from the rear	cowhocked
stood wider behind	stood narrower behind
stood squarer in his/her foot placement	cowhocked, bowlegged
General	
more progressive	conventional
performance oriented	lacks growth and performance
growthier	
more dimensional	
stouter made	
more upstanding	
growthier	

Sample market lamb reasons

Placing: 4–3–2–1

In my top pair of wethers, I place 4 over 3 as 4 was a trimmer handling, cleaner patterned lamb that should go to the rail and hang a higher cutability carcass. 4 was longer in his neck, more neatly laid-in at the point of his shoulder, and handled harder over his top. I admit 3 was a heavier muscled lamb that was deeper and wider in his loin, but he was a softer handling, wastier fronted lamb.

Nonetheless, I preferred 3 to 2 in my middle pair because 3 outdistanced 2 in terms of balance and muscle. 3 was a more upstanding, eye appealing lamb that was longer in his hindsaddle. In addition, he handled with a deeper loin, a wider, squarer dock and a more expressively muscled leg. He should rail a heavier muscled carcass. I grant 2 was more correct in his finish, but he was an off-balance, coarser fronted lamb that lacked the muscle dimension of my top pair.

Moving to my bottom pair of ewe lambs, I placed 2 over 1 because 2 was more correctly finished and should rail a trimmer, higher cutability carcass. 2 handled harder over her top and was trimmer in her flank.

I realize 1 was a taller fronted, more upstanding lamb, but she was the wastiest, heaviest conditioned, lightest muscled lamb that should hang the least profitable carcass in the class. Thank you.

BREEDING ANIMAL EVALUATION

Evaluation of breeding livestock often includes performance data. However, when data are not available, breeding livestock should be selected based on their apparent ability to be the parents of ideal market animals. Therefore, muscle, fat and structural correctness should receive significant attention. In addition, breeding soundness, body capacity, and visual clues to performance must be included in the evaluation.

BEEF

Breeding cattle should have some of the same components as an ideal market steer. The basic structure should be about the same, with small refinements discussed in a set of reasons. For example, instead of keying in on the potential carcass cutout of the animal in question, the animal's femininity, masculinity, body depth, structure, and performance should

ANIMAL SCIENCE FACTS

The discussion of testicular development may seem in poor taste when giving reasons on a class of bulls, rams, or boars. However, testicle size is economically important in terms of a sire's fertility. Larger testicles produce more sperm. Therefore, males with larger testicles should produce a larger amount of semen, providing a better chance of impregnating females. Large volumes of sperm are also desirable from males being collected for artificial insemination. In addition, large testicle size is an indication of earlier sexual maturity for a sire's daughters.

be emphasized. Breeding cattle must be relatively fault-free in their makeup and designed with added body depth and volume to perform in all management conditions—even when feed resources are limited. Most judging classes of breeding cattle will be heifer classes, so most of the following discussion will be focused on females. However, classes of bulls are sometimes evaluated in judging contests.

Heifers

See Figure 16-12 for the ideal beef heifer.

As with market cattle, muscle dimension and pattern are important to breeding heifers. Use the same terms used to describe muscle volume in market cattle (do not use the carcass terms).

Relative body condition is used to describe heifers, but for a different reason than with market cattle. Condition can give an indication of the ability of a heifer to survive and thrive in various management situations. Heifers should be in good condition, but not exceptionally thin or fat.

ANIMAL SCIENCE FACTS

When confronted with a class of livestock, first impressions are usually the best impressions. Judging contestants too often place a class correctly at first glance, but change their minds further into the allotted time period. When making difficult decisions on a class (or even a certain pair) of livestock, it often pays to go with the first impression.

Figure 16-12. The ideal beef heifer. (original drawing by the author)

The following are terms used to describe body condition in beef heifers.

Desirable	Undesirable
easier keeping	harder doing
easier fleshing	harder fleshing
	heavier conditioned

Structure is more critical to breeding heifers than market cattle. Correctly structured individuals (especially on their feet and legs) are less apt to be culled for unsoundness, and thus, should have a longer life in the breeding herd. Structure terms are the same as those used for market cattle.

Femininity is important to heifers. Femininity is generally recognized by a long, clean, smooth front end.

The following are terms used to describe femininity in beef heifers.

Desirable	Undesirable
broodier	
more feminine fronted	coarser fronted
more angular	
more stylish	

Visual indications of performance include overall body length, depth, and capacity, as well as relative size and weight in comparison to other heifers in the class.

The following are terms used to describe visual indications of performance in beef heifers.

Desirable	Undesirable
more 3-dimensional	
bigger volumed	had less volume
broodier	
roomier	
higher weight per day of age	lower weight per day of age
more producer oriented	
more performance oriented	
more arch and spring of rib	constricted in her rib
more capacious	less capacious
more dimension through the center of her rib	
longer ribbed	shorter ribbed
growthier	lacked do-ability

The following are general introductory terms used to describe beef heifers.

Desirable

better combines performance and eye appeal
better combination of structure and performance
more progressive

Bulls

Most terms used to describe bulls can be gleaned from the terminology lists already presented. However, bulls should be described using terms that indicate the bull's muscle, masculinity, structural correctness, and the kind of progeny he will most likely sire.

The following are terms used specifically to describe bulls.

Desirable	Undesirable
stouter	
more powerful	
greater scrotal circumference	less scrotal circumference
more masculine	

Desirable	Undesirable
more athletic	
more agile	restricted in his movement
should sire calves with:	
added growth and performance	
added muscle volume	
added frame	
added length of body	
etc.	

Sample Angus heifer reasons

Placing: 3–2–4–1

In my top pair, I placed 3 over 2 because 3 better combined femininity and performance into a broodier, more producer oriented package. 3 was a cleaner fronted, longer necked heifer that was deeper ribbed, heavier muscled, and stood down on more substance of bone. In addition, 3 appeared to have a higher weight per day of age. I admit that 2 was more structurally correct, more nearly level from her hooks to her pins, and moved out with a longer, more ground covering stride.

In my middle pair, I preferred 2 to 4 as 2 was a better balanced, more upstanding, later maturing heifer. 2 was laid-in more neatly at the point of her shoulder, was stronger topped, leveler rumped and stood with a more correct set to her hock. Also, 2 tracked wider and squarer both coming and going. I grant 4 was deeper through her forerib and wider at her pins, but she simply lacked the balance, femininity, and structural correctness of 4.

Moving to my bottom pair, it was 4 over 1. 4 was an easier keeping, bigger volumed, more performance oriented heifer that was more rugged in her design. She was a more 3-dimensional heifer that showed more arch and spring of rib, more depth of flank, was more ideal in her condition, and heavier boned.

I realize 1 was more refined and feminine fronted, but she lacked the frame, scale, muscle and do-ability to place any higher in the class. Thank you.

SWINE

Judging breeding swine is very similar to judging market swine. The only differences are that structure, frame size, and body capacity should receive more emphasis than in market classes. Muscle and leanness are also very important.

Gilts

See Figure 16-13 for the ideal breeding gilt. Good replacement gilts should be large framed, long bodied, and late maturing with adequate muscle and bone. High volume, deep ribbed gilts that move easily are preferred. Replacement gilts must also have an underline with at least six functional nipples per side. ***Blind*** (nonfunctional), ***flat*** (functional, but not prominent), and ***pin nipples*** (small, undeveloped nipples) are discriminated against. See Figure 16-14. Underline quality should also be examined in boar classes (particularly boar classes of maternal breeds).

Terms used to describe muscle content and fat cover are the same as those used for market swine (do not use carcass terms).

The following are some additional terms used to describe body volume and performance in breeding gilts.

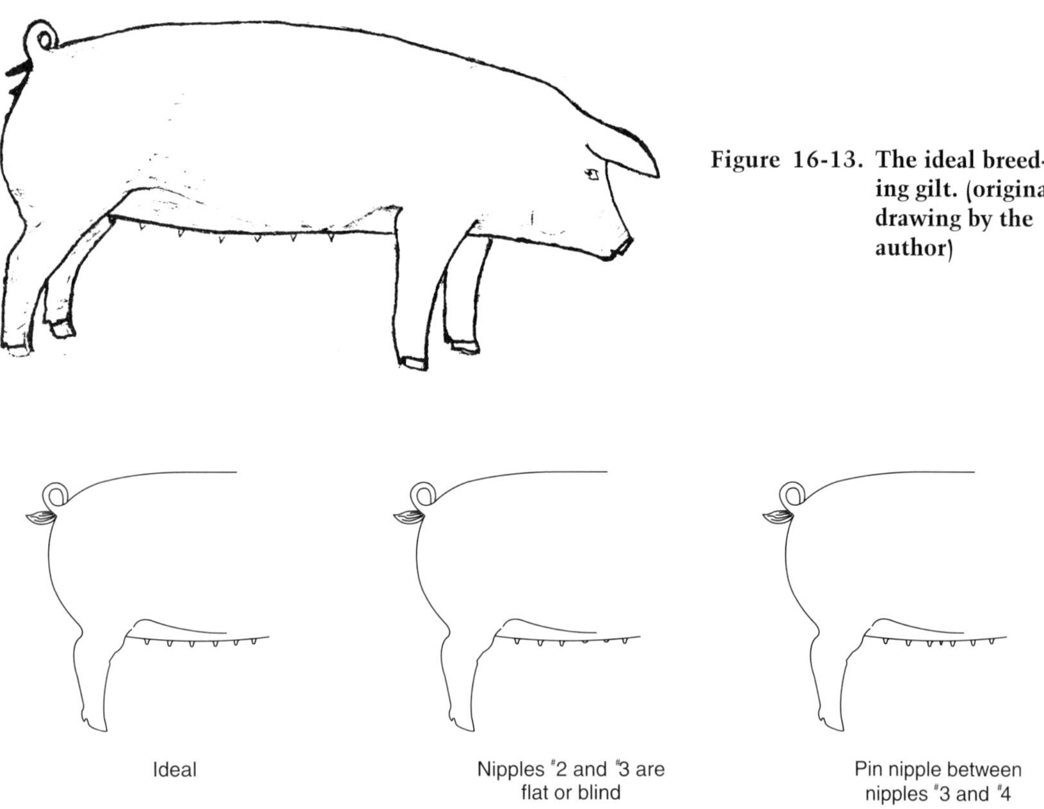

Figure 16-13. The ideal breeding gilt. (original drawing by the author)

Figure 16-14. Common underline faults in swine.

Desirable	Undesirable
more 3-dimensional	
bigger volumed	had less volume
roomier middled	
higher weight per day of age	lower weight per day of age
more performance oriented	
more internal capacity	less capacious
more natural width through the center of her rib	
deeper sided	

Remaining terms concerning structure, as well as feet and legs, can be gleaned from the market terminology lists.

The following are terms used to describe femininity and underlines in gilts.

Desirable	Undesirable
longer fronted	shorter fronted
more feminine fronted	
higher quality underline	poorer quality underline
more desirable teat spacing	unevenly spaced teats
more refined underline	coarser in her teat quality
more numerous underline	less numerous underline
more prominent underline	possessed a pin nipple
	was blunt in her teats

Boars

Except the few following terms, most terminology is the same as that used for market hogs and gilts.

Desirable	Undesirable
more powerful	
heavier structured	
more testicular volume	less testicular volume
potential to sire pigs with:	
added growth and performance	
additional leanness and/or muscle	
more soundness	
more confinement adaptability	

Sample Hampshire gilt reasons

Placing: 4–2–1–3

I started this class with 4 over 2 as 4 was a larger framed, bigger scaled, leaner made, later maturing gilt that appeared to have a higher weight per day of age. 4 was taller fronted, longer necked and especially leaner over her top and through her lower one third. I grant that 2 was a bolder sprung, bigger ribbed gilt, but she was more conventional in her outline and earlier in her maturity pattern.

Nonetheless, I placed 2 over 1 in my middle pair as she was a bolder sprung, roomier middled, heavier muscled, more structurally correct gilt. 2 was wider based, deeper ribbed, deeper flanked, and was heavier muscled from blade to ham. In addition, 2 was looser in her hip and moved with a longer, more ground covering stride. 1 was leaner at her blades, but lacked the muscle, volume, and mobility of the gilt placed above her.

In my bottom pair, I preferred 1 to 3 as she more closely followed the kind of my first place gilt. She was a larger framed, bigger outlined gilt that was particularly leaner from end to end. She showed more blade action on the move, more natural turn to her top and was trimmer in her lower one third. 1 also was more numerous and more prominent in her underline.

I grant 3 stood down on more ruggedness of bone, but she was the smallest framed, earliest maturing, fattest gilt in the class and thus could place no higher. Thank you.

SHEEP

The evaluation of breeding sheep is similar to the evaluation of market lambs. However, structure, volume, performance, and feet and legs should all receive a somewhat higher priority.

Ewes

Ewe classes are judged as lambs (less than one year) or yearlings (greater than one year). Maturity pattern should be emphasized more in ewe lambs than in yearling ewes. The ideal ewe should have the same general characteristics as the ideal beef heifer. She should be trim and feminine fronted but thick, deep, large volumed, and heavily muscled. Feet and legs should be correctly placed and sound.

Muscle and fat evaluation for breeding sheep are normally made without the benefit of handling. Judging contestants must make evaluations based

Figure 16-15. The ideal breeding ewe. (original drawing by the author)

on the appearance of thickness for muscle content and smoothness for fat evaluation.

The following are terms used to describe muscle volume in breeding sheep that have not been handled.

Desirable	Undesirable
longer, smoother muscle design	
thicker made	
more expressively muscled leg	narrower, lighter muscled leg

Body condition is important to breeding ewes for the same reasons as is it for breeding heifers.

The following are terms used to describe fat differences in breeding sheep that have not been handled.

Desirable	Undesirable
trimmer made	wastier
trimmer, cleaner patterned	overfinished
cleaner conditioned	heavier conditioned
more ideal in her condition	
trimmer breasted	wastier in her breast
easier keeping	harder doing
easier fleshing	

Most structure terms are the same as those used for market lambs.

The following are additional structure terms to be used in breeding sheep reasons.

Desirable	Undesirable
Movement	
moved freer and easier	stiffer strided
moved with more strength of top	
moved with more levelness of dock	
longer strided	shorter strided
truer tracking	narrower tracking
sounder footed	

In addition to the capacity terms used for market lambs, the following terms can be used to describe volume and capacity in breeding sheep.

Desirable	Undesirable
more capacious	shallower made
more internal volume and dimension	
more arch and spring of rib	flatter ribbed
bolder spring of rib	constricted in her forerib
more three dimensional	

Ideally, fleeces on breeding sheep should be dense and tight with few black fibers. Fleece quality should be emphasized when judging breeds known for their wool production. Fleece quality differences should be discussed using the following terms.

Desirable	Undesirable
tighter denser fleece	more open fleece
fleece with a finer crimp	cottony fleece
freer from black fiber	fleece with more black fiber

Rams

Besides the sheep terms already given, the following terms can be used to describe classes of rams (also known as ***bucks***).

Desirable	Undesirable
greater scrotal circumference	less scrotal circumference
more powerful	
more masculine	
more ruggedly designed	
stouter made	
more apt to sire lambs with more:	
frame	
growth	
trimness	
performance	
etc.	

Sample Dorset ewe reasons

Placing: 2–1–4–3

In my top pair, I easily placed 2 over 1 as she was a more progressive ewe in terms of frame, muscle, and volume. She was a trimmer, cleaner patterned, taller fronted ewe that was thicker topped, deeper ribbed, and tracked wider behind. I concede 1 was a heavier boned ewe.

Moving to my middle pair of similarly designed, more conventional ewes, I preferred 1 to 4 as she was a cleaner fronted, sounder footed ewe. 1 was longer and leaner necked, laid-in neater at her shoulder, and was smoother at the neck-shoulder junction. Additionally, she was stronger and straighter on her pasterns and tracked with more ease and agility. I grant 4 was more nearly level through her dock and possessed a tighter, denser fleece.

Nevertheless, I confidently placed 4 over 3 in my bottom pair as she was a bolder sprung, more capacious, more performance oriented ewe. 4 showed more arch and spring of rib, was deeper flanked, and heavier muscled as exhibited by a thicker top and wider, squarer dock. 4 also appeared to have a higher weight per day of age.

I admit that 3 was a cleaner fronted ewe, but she simply lacked the performance, volume, and muscle to place any higher. Thank you.

SUMMARY

Market livestock are evaluated on their ability to produce heavily muscled, lean carcasses. Structural correctness, performance, and muscle pattern must also be assessed. Evaluation of breeding livestock relies on some of the same parameters, but structure, volume, performance, and breeding soundness play a much more important role.

Terminology used for reasons is also similar between market and breeding livestock within the same species. However, terminology differs among the livestock species. Livestock judges must memorize the parts of market and breeding livestock. Care must be taken not to mix terminology among species.

Becoming a good livestock judge takes work and practice. Only a limited amount of livestock judging can be learned from a book. The rest must be learned from looking at livestock and evaluating the strengths and weaknesses of individual animals.

CHAPTER SELF-CHECK

___ finish	1. front legs too curved (side view)
___ posty	2. rear legs bowed in (rear view)
___ sickle-hocked	3. front feet turned out
___ cowhocked	4. functional, but not prominent nipples
___ bowlegged (rear legs)	5. rams
___ buck-kneed	6. rear legs bowed out (rear view)
___ calf-kneed	7. front legs too straight (side view)
___ knock-kneed	8. rear legs too straight (side view)
___ splay-footed	9. front legs bowed out (front view)
___ pigeon-toed	10. front feet turned in
___ blind nipples	11. rear legs too curved (side view)
___ flat nipples	12. knees too close together (front view)
___ pin nipples	13. small, undeveloped nipples
___ bucks	14. nonfunctional nipples
___ bowlegged (front legs)	15. fat cover

QUESTIONS AND PROBLEMS FOR DISCUSSION

1. Name the three judging parameters of selecting market livestock.
2. Highest priced cuts of meat are located in the front/rear half of the animal. Select one.
3. Explain where fat cover can be observed in a beef animal.
4. List two terms describing undesirable fat cover in beef.
5. What does posty mean?
6. Name three places to look for muscling on a market hog.
7. When viewed from behind, would a hog with a square or rounded top be fatter?
8. How do judging contestants usually determine muscling in lambs?
9. In which species would you use the terminology heavier-muscled hindsaddle?
10. Give the ideal fat cover for lambs.
11. Describe the front end of a feminine heifer.
12. A gilt should have at least ___ nipples per side.
13. ________ nipples are not functional.
14. Differentiate between ewe lambs and yearlings.
15. Which is more desirable, a dense or open fleece?

ACTIVITIES

1. Have sheep and cattle judges show you how to handle a market lamb and a market steer.
2. Organize and hold a county or regional livestock judging contest.
3. Participate in a livestock judging contest.
4. Visit a livestock producer in your area. Discuss the criteria used to select replacement or slaughter animals.

LABORATORY ACTIVITY

JUDGING PRACTICE

Purpose

To practice judging pictures of market hogs

Materials

pencil
paper
terminology lists from Chapter 16

Procedure

Answer the following questions about the pictured class of market hogs.

1. Rank the four market hogs from leanest to fattest. What indications of fatness did you use?
2. Rank the four market hogs on muscle volume. Remember, thickness can be partially due to fat.
3. Rank the four market hogs on style and balance.
4. Using bone length (either front or rear legs) as a gauge, rank the pigs from largest to smallest in frame size.
5. Which two pigs are the heaviest boned?
6. Between 3 and 4, which is the deepest ribbed?
7. Between 3 and 4, which has the longest rump and the most stifle muscle?
8. Between 1 and 2, which appears wastiest in the lower one third?
9. Write identification points for each pig in the class.

Analysis

1. Using answers to the above questions, rank the pigs as you would a class of market hogs.
2. Write a set of market hog reasons based on your placing. Use terms from the terminology lists.
3. Check the official placing and cuts found in the instructor's manual. Score your own placing.
4. Have your instructor grade your written reasons on a 50-point scale.
5. Practice oral delivery of your set of reasons and give the set to the class.

Application of Laboratory Activity

Livestock judging develops a host of life skills that are useful, even if the student never works with livestock again. Judging teaches students to ask questions important to a particular situation, like, "Which one is the fattest?" Students answer such questions by analyzing the available information to arrive at a logical decision. Oral reasons develop communication skills and encourage a well-presented defense of a deliberated decision.

1

2

3

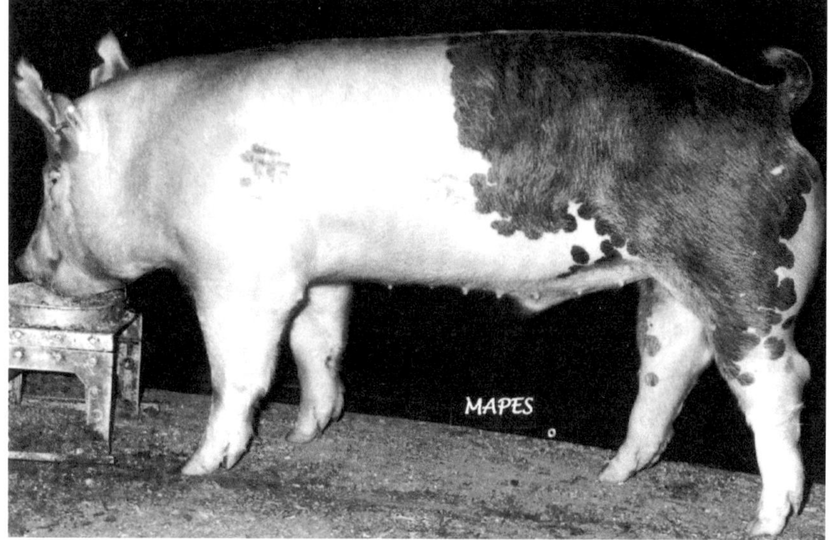

4

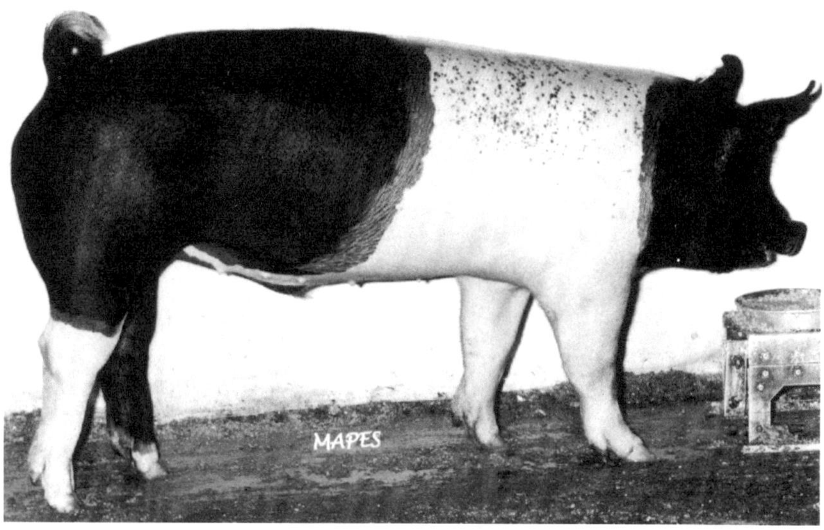

Chapter 17

DAIRY CATTLE JUDGING

Udderly Impressive

INTRODUCTION

Selection of dairy cattle should be based primarily on the cow's ability to produce large quantities of milk for several years. Thus, the milk factory, or udder is extremely important when evaluating dairy cattle type. The rest of the cow can be viewed as an udder support system. Feet and legs must be correct so the udder can move about easily, possibly over rough terrain for the production life of the animal. The body should be wide and deep to hold feed that the udder eventually converts to milk. In plain terms, dairy cattle judging centers around the mammary gland.

Figure 17-1. (Courtesy, National FFA Organization)

OBJECTIVES

1. Describe the desirable traits of an ideal dairy cow
2. Score a cow using the Dairy Cow Unified Score Card
3. Differentiate between type and linear classification
4. Judge and give reasons on a class of dairy cows

TERMS

colored breeds
Dairy Cow Unified Score Card
linear classification
predicted transmitting ability for type (PTAT)
type classification

ANIMAL SCIENCE FACTS

The Dairy Cow Unified Score Card was revised four times to reflect the needs of progressive dairy producers.

TYPE CLASSIFICATION

As with any livestock species, a knowledge of the external parts precedes further critical discussion. See Figure 17-2 for the parts of a dairy cow.

Dairy cows can be evaluated by comparing each cow to an ideal dairy cow, then assigning a score. The Purebred Dairy Cattle Association made this comparison easier by establishing the ***Dairy Cow Unified Score Card*** in 1943. See Figure 17-3. The current score card divides the dairy cow into five major traits, each consisting of several elements. Each element is evaluated to arrive at a final score for the trait. Point values for the five traits total 100 (frame-15, dairy character-20, body capacity-10, feet and legs-15, udder-40). Traits and the corresponding point values changed in 1994 to reflect producers' emphasis on cow longevity. Feet and legs now form a separate trait category worth 15 points. Earlier, feet and legs fit in the category of general appearance. Along with that, the udder category, formerly mammary system, increased in point value to 40.

Breed associations have trained classifiers who visit farms and classify registered cattle. Presently, the Holstein Association provides this service for some of the smaller ***colored breeds***. Using the Dairy Cow Unified Score Card, classifiers rate registered cattle. Classified cows fit into cate-

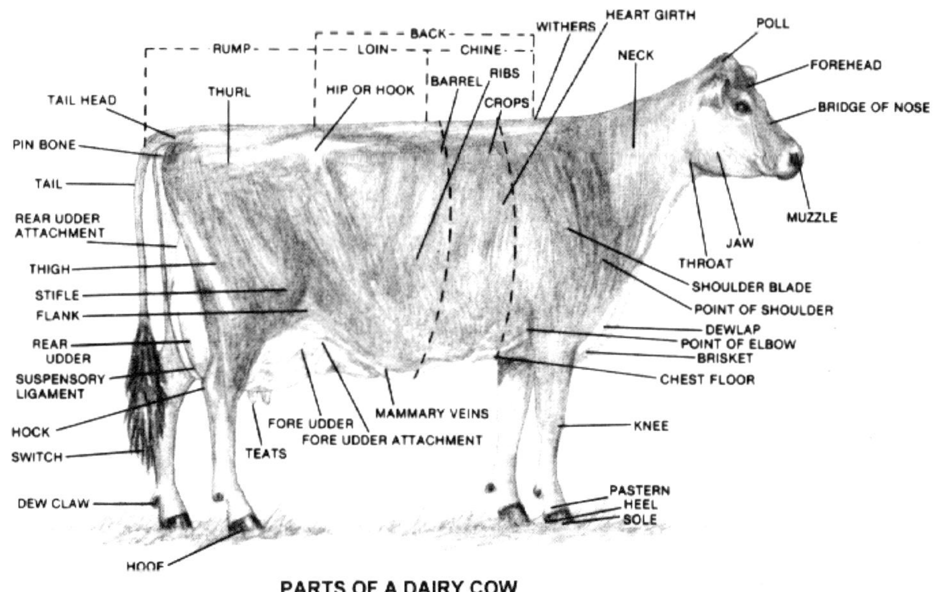

Figure 17-2. Parts of the dairy cow taken from the Dairy Cow Unified Score Card.

DAIRY COW UNIFIED SCORE CARD

Copyrighted by The Purebred Dairy Cattle Association, 1943. Revised, and Copyrighted 1957, 1971, 1982, and 1994.

Breed characteristics should be considered in the application of this score card

MAJOR TRAIT DESCRIPTIONS	Perfect Score
There are five major classification traits on which a classifier bases a cow's score. Each trait is broken down into body parts to be looked at and ranked.	
1) Frame - 15% The skeletal parts of the cow, with the exception of feet and legs, are evaluated. Listed in priority order, the descriptions of the traits to be considered are as follows: **Rump** - long and wide throughout with pin bones slightly lower than hip bones. Thurls need to be wide apart and centrally placed between hip bones and pin bones. The tailhead is set slightly above and neatly between pin bones, and the tail is free from coarseness. The vulva is nearly vertical. **Stature** - height, including length in the leg bones. A long bone pattern throughout the body structure is desirable. Height at the withers and hips should be relatively proportionate. **Front End** - adequate constitution with front legs straight, wide apart and squarely placed. Shoulder blades and elbows need to be firmly set against the chest wall. The crops should have adequate fullness. **Back** - straight and strong; the loin - broad, strong, and nearly level. **Breed Characteristics** - overall style and balance. Head should be feminine, clean-cut, slightly dished with broad muzzle, large open nostrils and a strong jaw is desirable. Rump, Stature, and Front End receive primary consideration when evaluating Frame.	15
2) Dairy Character - 20% The physical evidence of milking ability is evaluated. Major consideration is given to general openness and angularity while maintaining strength, flatness of bone and freedom from coarseness. Consideration is given to stage of lactation. Listed in priority order, the descriptions of the traits to be considered are as follows: **Ribs** - wide apart. Rib bones are wide, flat, deep, and slanted toward the rear. **Thighs** - lean, incurving to flat, and wide apart from the rear. **Withers** - sharp with the chine prominent. **Neck** - long, lean, and blending smoothly into shoulders. A clean-cut throat, dewlap, and brisket are desirable. **Skin** - thin, loose, and pliable.	20
3) Body Capacity - 10% The volumetric measurement of the capacity of the cow (length x depth x width) is evaluated with age taken into consideration. Listed in priority order the descriptions of the traits to be considered are as follows: **Barrel** - long, deep, and wide. Depth and spring of rib increase toward the rear with a deep flank. **Chest** - deep and wide floor with well-sprung fore ribs blending into the shoulders. The Barrel receives primary consideration when evaluating Body Capacity.	10
4) Feet and Legs - 15% Feet and rear legs are evaluated. Evidence of mobility is given major consideration. Listed in priority order, the descriptions of the traits to be considered are as follows: **Feet** - steep angle and deep heel with short, well-rounded closed toes. **Rear Legs**: **Rear View** - straight, wide apart with feet squarely placed. **Side View** - a moderate set (angle) to the hock. **Hocks** - cleanly molded, free from coarseness and puffiness with adequate flexibility. **Pasterns** - short and strong with some flexibility. Slightly more emphasis placed on Feet than on Rear Legs when evaluating this breakdown.	15
5) Udder - 40% The udder traits are the most heavily weighted. Major consideration is given to the traits that contribute to high milk yield and a long productive life. Listed in priority order, the descriptions of the traits to be considered are as follows: **Udder Depth** - moderate depth relative to the hock with adequate capacity and clearance. Consideration is given to lactation number and age. **Teat Placement** - squarely placed under each quarter, plumb and properly spaced from side and rear views. **Rear Udder** - wide and high, firmly attached with uniform width from top to bottom and slightly rounded to udder floor. **Udder Cleft** - evidence of a strong suspensory ligament indicated by adequately defined halving. **Fore Udder** - firmly attached with moderate length and ample capacity. **Teats** - cylindrical shape and uniform size with medium length and diameter. **Udder Balance and Texture** - should exhibit an udder floor that is level as viewed from the side. Quarters should be evenly balanced; soft, pliable and well collapsed after milking.	40
TOTAL	100

Figure 17-3. Trait descriptions and scoring of the Dairy Cow Unified Score Card. (Courtesy, The Purebred Dairy Cattle Association)

gories according to their scores: 90 to 100 excellent, 85 to 89 very good, 80 to 84 good plus, 75 to 70 good, 70 to 74 fair, less than 70 poor. This process is known as ***type classification***. Herds receive a summary score indicating the overall rating of the herd. Classification assists dairy producers with merchandising and selection of breeding replacements and sires. An objective opinion of cattle allows better decision making on the part of producers.

The first trait in the Dairy Cow Unified Score Card, "frame," can receive up to 15 points. The elements of rump, stature, front end, back, and breed characteristics comprise this category.

The second trait, "dairy character," assesses the physical characteristics that should allow the cow to milk well. Ribs, thighs, withers, neck, and skin texture are all elements used to score dairy character. A perfect dairy character score would earn 20 points.

The third, and least heavily weighted trait, "body capacity" can count for up to 10 points. Body capacity measures the volume of the cow's body. The evaluation of two elements, the barrel and the chest, provide indicators for body capacity.

"Feet and legs," the fourth trait counts for 15 points. Foot angle and rear legs (from both a side and rear view) determine the points scored in this category.

The most important of the five traits, the "udder" is analyzed from all angles. Udder depth, teat placement, rear udder, udder cleft, fore udder, teats, as well as udder balance and texture tally up to 40 points in this category.

Side 1 of the Dairy Cow Unified Score Card gives excellent descriptions of the ideal elements for each trait. Study and memorize the elements.

Side 2 of the Dairy Cow Unified Score Card pictures cows from all six of the major dairy breeds. See Figure 17-4. Use these pictures as the ideal dairy cow of each breed. Ideal breed characteristics are listed beneath the pictures. At the bottom of the card, a guide of defects and conditions, along with degrees of discrimination, is given. Use the guide to assess the importance of any physical flaws you may encounter. See Figure 17-4.

LINEAR CLASSIFICATION

Linear classification carries dairy type traits into the realm of genetics. Linear classification gives dairy producers a method to utilize the heritability of individual type traits. Linear traits are scored on a basis of –3 to +3. Teat placement, teat width, rear legs from side, rear legs from rear, and foot angle are some traits considered in linear evaluation. Producers can select a potential sire by evaluating his score for an individual trait and then matching the sire's stronger type traits with a dam's weaknesses. These linear trait scores are combined to calculate ***predicted transmitting ability for type (PTAT)***, an overall indicator. Most PTAT scores usually range from .5 to 2.5. A 2.18 PTAT bull would be a better type sire selection than a bull with a PTAT of .5. Resulting positive changes in

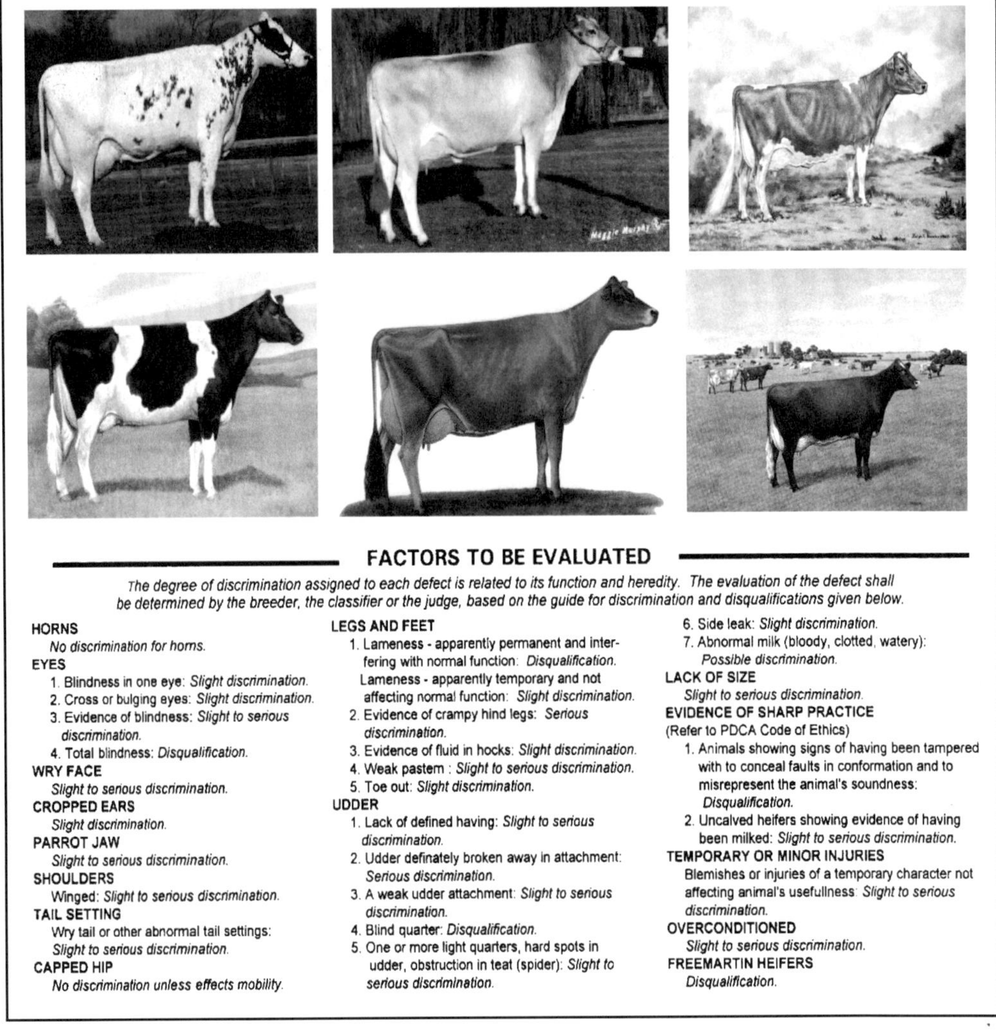

FACTORS TO BE EVALUATED

The degree of discrimination assigned to each defect is related to its function and heredity. The evaluation of the defect shall be determined by the breeder, the classifier or the judge, based on the guide for discrimination and disqualifications given below.

HORNS
No discrimination for horns.

EYES
1. Blindness in one eye: *Slight discrimination.*
2. Cross or bulging eyes: *Slight discrimination.*
3. Evidence of blindness: *Slight to serious discrimination.*
4. Total blindness: *Disqualification.*

WRY FACE
Slight to serious discrimination.

CROPPED EARS
Slight discrimination.

PARROT JAW
Slight to serious discrimination.

SHOULDERS
Winged: *Slight to serious discrimination.*

TAIL SETTING
Wry tail or other abnormal tail settings: *Slight to serious discrimination.*

CAPPED HIP
No discrimination unless effects mobility.

LEGS AND FEET
1. Lameness - apparently permanent and interfering with normal function: *Disqualification.* Lameness - apparently temporary and not affecting normal function: *Slight discrimination.*
2. Evidence of crampy hind legs: *Serious discrimination.*
3. Evidence of fluid in hocks: *Slight discrimination.*
4. Weak pastern : *Slight to serious discrimination.*
5. Toe out: *Slight discrimination.*

UDDER
1. Lack of defined having: *Slight to serious discrimination.*
2. Udder definately broken away in attachment: *Serious discrimination.*
3. A weak udder attachment: *Slight to serious discrimination.*
4. Blind quarter: *Disqualification.*
5. One or more light quarters, hard spots in udder, obstruction in teat (spider): *Slight to serious discrimination.*
6. Side leak: *Slight discrimination.*
7. Abnormal milk (bloody, clotted, watery): *Possible discrimination.*

LACK OF SIZE
Slight to serious discrimination.

EVIDENCE OF SHARP PRACTICE
(Refer to PDCA Code of Ethics)
1. Animals showing signs of having been tampered with to conceal faults in conformation and to misrepresent the animal's soundness: *Disqualification.*
2. Uncalved heifers showing evidence of having been milked: *Slight to serious discrimination.*

TEMPORARY OR MINOR INJURIES
Blemishes or injuries of a temporary character not affecting animal's usefullness: *Slight to serious discrimination.*

OVERCONDITIONED
Slight to serious discrimination.

FREEMARTIN HEIFERS
Disqualification.

Figure 17-4. Ideal Dairy Cows, as pictured, and Factor to be Evaluated, in the Dairy Cow Unified Score Card.

structural or udder types from selection based on linear evaluation can add years to the life of the average cow in a given herd. See Figure 17-5.

JUDGING DAIRY COWS

Classes of dairy cows are evaluated much the same as individual dairy cows. Completing a Dairy Cow Unified Score Card for each cow and then comparing the final scores is the simplest way to place classes of dairy

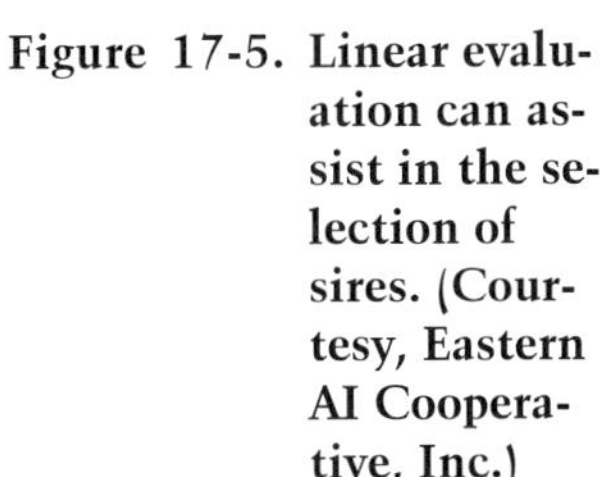
Figure 17-5. Linear evaluation can assist in the selection of sires. (Courtesy, Eastern AI Cooperative, Inc.)

cows. Reasons should evolve from the differences found in the elements of the scorecard.

Terminology used in reasons for dairy cattle originate from the Dairy Cow Unified Score Card, as well as the parts of the dairy cow. Positive comparison terms should be used while negative comparisons should be avoided when possible. An example of a positive term would be "longer bodied." The comparative negative term would be "shorter bodied." A breakdown of the scorecard traits, with terms used to describe the elements, follows.

FRAME

The ideal dairy cow is large framed, long bodied, and correctly structured. The shoulder, neck, and head should be clean, feminine, and angular. The topline should be straight and strong. Hooks and pins should be wide, and the rump should be level.

General terms

exhibits more style, balance, and symmetry
smoother blending throughout
taller, longer, more upstanding
more harmonious blending of body parts
more strength and power
more angular
milkier

Size and scale

taller
shows more size and scale
more upstanding
longer bodied
framier

Head

more alert
broader through the muzzle
more feminine about the head
longer, more balanced head
cleaner, more refined head

Front end (shoulders)

smoother through the front end
tighter and smoother at the point of the shoulder
more neatly laid in at the shoulders
blends more smoothly from neck to shoulders
tighter and fuller at the point of the elbow
shoulders blend more smoothly into the body wall
sharper over the withers
fuller at the crops

Topline and rump

straighter over the topline
cleaner over the topline
harder down the topline
stronger in the chine
stronger and harder in the loin
more nearly level from hooks to pins
smoother tailhead
neater tail setting
wider through the (hooks, pins, thurls, rump)
more level through the chine and loin
more refined tailhead

higher chine
more prominent about the hooks and pins
longer from hooks to pins
cleaner over the rump

DAIRY CHARACTER

Dairy character describes femininity and overall dairyness. Cows with good dairy character look sharp and angular through the front end and free from excessive flesh over the top and in the hindquarters. Cows carrying an abundance of flesh (muscle and/or fat) are not considered to have good dairy character because they are storing nutrients "on their back" instead of converting them to milk "in the bucket."

General terms

sharper, cleaner-cut, more angular throughout
excels in dairy character
cleaner and more feminine throughout

Neck

longer, leaner neck
more clean-cut head and neck
trimmer neck, throat, dewlap, and brisket
more refined through the head and neck

Thighs

more incurving in the thighs
cleaner, flatter and thinner in the thighs

Body

sharper over the withers
cleaner over the topline
displays more prominent vertebrae
more prominent hooks and pins
cleaner rump
flatter and cleaner bone
more open rib

FEET AND LEGS

Common flaws in feet and legs are much the same as those in beef cattle. A correctly set hock (side view) is vital to a cow's longevity. Correct pastern angle combined with a deep heel serve to wear the hoof evenly and naturally, decreasing the need for costly hoof trimming.

Rear leg stance

straighter legs from side view (rear view)

stands more squarely on her rear legs

shows more moderate set to her hock

has stronger pasterns

cleaner in the hocks

Feet

has a deeper heel

has a more correctly shaped hoof

Front legs

stands straighter on her front legs

Movement

tracks straighter on her rear legs

moves more gracefully

walks out more correctly on her front feet

BODY CAPACITY

The amount of nutrients that go into a cow ultimately affect the amount of milk she can produce. A cow must have space to hold the nutrients (feed) she eats. Body capacity is evaluated by the width and depth of rib, and the size of the barrel.

General capacity terms

wider chest
deeper chest
deeper barrel
displays more spring of rib
deeper in the flank
more overall depth and capacity
more depth of heart
deeper in the fore rib
deeper in the rear rib
more depth of barrel
more depth and width of body

UDDER

As mentioned previously, the udder is the most important single trait upon which dairy cows are judged. In a dairy cattle judging contest, the cow with the best udder wins the class most of the time. Therefore, the ability of dairy judges to analyze udders and identify differences among udders becomes critical to success.

Ideally, an udder should be deep, to provide capacity for milk production. However, the udder should not drop below the hock. Udders that drop too far are more susceptible to bacterial invasions and are more prone to physical damage, such as teat tramping. Young cows will have higher, tighter udders than older cows.

Teats should be placed squarely beneath each quarter and should hang perpendicular to the ground. The distance between the front teats and the distance between the rear teats should be the same. Teats should be of uniform size, intermediate length, and moderate diameter.

The rear of the udder should be attached high when viewing the cow from the rear. A high rear udder attachment signifies added udder volume. The width of the udder should be uniform from top to bottom. The median suspensory ligament should be plainly visible, dividing the udder in half from side to side. The lateral suspensory ligament separates the rear and fore udder. Cows without obvious ligaments tend to have udders that drop too low, leaving the udder subject to injury. Alternately, a weak or broken suspensory ligament can cause the teats to jut or strut outward.

The fore udder should blend smoothly into the body wall and show a strong attachment to the body wall. In addition, the fore udder should be deep and moderately long, but not disproportionate to the rest of the udder. All quarters should be evenly balanced, and the udder floor should be level when viewed from the side. Moreover, udder veining is an indication of the amount of blood passing through the udder, so veining should be prominent.

General terms

more shapely, symmetrical udder
more balanced udder
more level udder floor
udder carried higher above the hock
more capacious udder
more youthful mammary system
more apparent quality and texture of udder

Fore Udder

longer fore udder attachment
smoother blending fore udder
blends more smoothly into the body wall

Rear Udder

higher, wider rear udder
more uniform width of rear udder
more defined halving
stronger median suspensory ligament
fuller rear udder

Teats

closer front teat placement
front teats more centrally placed
teats hanging more nearly plumb
teats more evenly spaced beneath quarters

more desirable teat size

more desirable teat shape

shorter teats

teats more neatly set on the udder floor

Sample Reasons for Holstein 4-year-old cows

Placing: 3–2–1–4

3 and 2 show a definite advantage in dairy character over 1 and 4. I placed 3 first and over 2 because of her advantage in mammary system. 3 displayed a higher, wider rear udder and a more defined suspensory ligament, as viewed from the rear. Furthermore, 3's udder is more nearly level when viewed from the side and has shorter teats. I grant that 2 is sharper in her chine and more incurving in her thighs.

In my middle pair, 2 easily placed over 1 on her overall general appearance, strength, and power. 2 is taller in her front end and harder down her topline. 2 is also tighter in the shoulder, longer and cleaner in her neck, and more refined about the head. She is deeper in the forerib and barrel, and more open in her rib with cleaner, flatter bone. In mammary, 2 is higher and wider in her rear udder. I concede that 1 is deeper in the heel.

In my final placing, 1 edges the white cow as she displays a more youthful mammary system by carrying her udder higher above the hocks, having a longer, smoother fore udder, and a stronger suspensory ligament. 1 also is more refined and angular about her front end.

Although 4 is a longer bodied cow, I placed 4 last in this class for being too deep in the udder and coarse throughout.

For these reasons, I placed this class of Holstein 4 year old cows 3–2–1–4.

JUDGING DAIRY HEIFERS

Heifers are evaluated on the same general traits as cows in production with the notable exception of the udder. Stature, style, dairy character, and structural correctness are the factors upon which placings should be made. Terms are the same as those used for cows in production. See Figure 17-6 for a picture of a dairy heifer.

Figure 17-6. Evaluation of heifers includes consideration of stature, style, dairy character, and structural correctness. (Courtesy, Mississippi State University)

Sample reasons for Guernsey Intermediate Heifers

Placing: 1–3–4–2

In my top placing, 1 places over 3 on her obvious advantage in body capacity. 1 exhibits more spring of rib and is deeper in both heart and barrel. Also, 1 is wider in the chest floor, wider at the pins, and neater at her tailhead. I concede 3 is longer, cleaner necked, and shows more breed character about the head.

In my middle pair, 3 follows 1's pattern of style and straightness of lines to place over 4. 3 is taller fronted, longer bodied, harder and higher in her chine, and lower at her pins. Furthermore, 3 is cleaner and more angular about her front end, being especially cleaner about the dewlap and smoother at the point of her shoulder. I recognize 4 is deeper in her fore rib and tracks straighter on her rear legs.

ANIMAL SCIENCE FACTS

FFA members can compete in two career skills events involving dairy: dairy food judging and dairy cow judging contests. Other FFA members act as handlers during the National Dairy Judging Contest held annually in Kansas City, Missouri.

In my final placing, 4 logically places over 2 with her advantage in dairy character and feet and legs. 4 is a cleaner, flatter boned heifer with more open ribbing and more incurving thighs. 4 stands straighter on her rear legs, when viewed from the side, has stronger pasterns, and is deeper in her heel. I grant that 2 is wider at her hooks.

Although 2 is a long bodied, neat fronted heifer, she lacks the strength, power, body depth, and correctness of hock found in the other three heifers.

For these reasons, I placed this class of Guernsey Intermediate Heifers 1–3–4–2.

SUMMARY

Dairy cattle are judged with an emphasis on the traits needed to produce milk and to remain productive over several years. The Dairy Cow Unified Score Card lists and assigns relative weighting of these traits. The udder category receives the highest point value. Linear classification is used to predict the ability to genetically alter type traits. Dairy terms are distinctly different from those used for livestock judging classes. A working vocabulary of the parts of the dairy cow is essential to good reasons. Heifers are judged on the same traits as cows in production, except for the udder.

CHAPTER SELF-CHECK

___ Dairy Cow Unified Score Card	1. scoring traits from –3 to +3
___ colored breeds	2. an excellent cow earns 90 to 100 points using this method
___ type classification	3. used to type classify
___ linear classification	4. summary score of linear evaluation
___ predicted transmitting ability for type (PTAT)	5. Guernseys, Jerseys, Brown Swiss, Aryshires, Milking Shorthorns

QUESTIONS AND PROBLEMS FOR DISCUSSION

1. When judging dairy cows using the Dairy Cow Unified Scorecard, the _____ receives the most consideration.
2. List the five traits found on the Unified Dairy Cow Score Card.
3. Write perfect scores for the Unified Dairy Cow Score Card traits.
4. Type classification conducted by a trained classifier rates registered dairy cows on a scale worth up to ____ points.
5. List two defects that can lead to disqualification during classifying.
6. Two breeds list mature cows as weighing 1,500 pounds. Name those two breeds.
7. Differentiate between a cow's heart girth and barrel.
8. Which is located closer to a cow's head, the hooks or the pins?
9. Linear classification rates individual traits on a score of ___ to ___.
10. The summary of linear traits compiles to the ___________.
11. Why should dairy cows not carry an abundance of condition or flesh?
12. Jutting or strutting teats indicate a weak ____________.
13. Why does the udder not receive primary attention when judging heifers?
14. List two career skills contests in which FFA members interested in dairy can participate.
15. The Dairy Cow Unified Score Card has been revised most recently in ____.

ACTIVITIES

1. Using the Dairy Cow Unified Score Card, classify dairy animals.
2. Participate in a dairy judging contest.
3. Ask your extension dairy specialist to speak to your class about dairy judging.
4. Contact the Holstein Association (Address in Chapter 10). Inquire as to when a classifier will be in your area. Ask to observe.

LABORATORY ACTIVITY

JUDGING PRACTICE

Purpose

To practice judging pictures of Guernsey cows

Materials

pencil
paper
terminology lists from Chapter 17

Procedure

Answer the following questions about the pictured class of Guernsey cows.

1. Between 1 and 2, which cow has the smoothest fore udder attachment?
2. Between 1 and 2, which cow has the deeper fore rib?
3. Between 1 and 2, which cow has the steeper foot angle?
4. Between 1 and 2, which cow has a higher rear udder attachment?
5. Between 2 and 4, which is cleaner fronted?
6. Between 2 and 4, which cow carries her udder higher off the ground?
7. Between 3 and 4, which cow is deeper ribbed?
8. Between 3 and 4, which cow is wider at the pins?

Analysis

1. Using answers to the above questions, rank the cows.
2. Write a set of reasons based on your placing. Use terms from the terminology lists.
3. Check the official placing and cuts found in the instructor's manual. Score your own placing.
4. Have your instructor grade your written reasons on a 50-point scale.
5. Practice oral delivery of your set of reasons and give the set to the class.

Application of Laboratory Activity

Visual evaluation and scoring of dairy cows are very important to purebred breeders of dairy cattle. Type traits are nearly as important as production in many instances, and perhaps rightly so. When a producer incurs the expense of raising a dairy heifer from calfhood until she enters production, the producer wants that cow to remain a member of the milking herd for many years. Many traits that are visually evaluated are directly linked to longevity. For example, the visual evaluation of udder height can be directly associated with longevity. Many cows are culled because the udder falls too low and is damaged, or impedes the cow's movement.

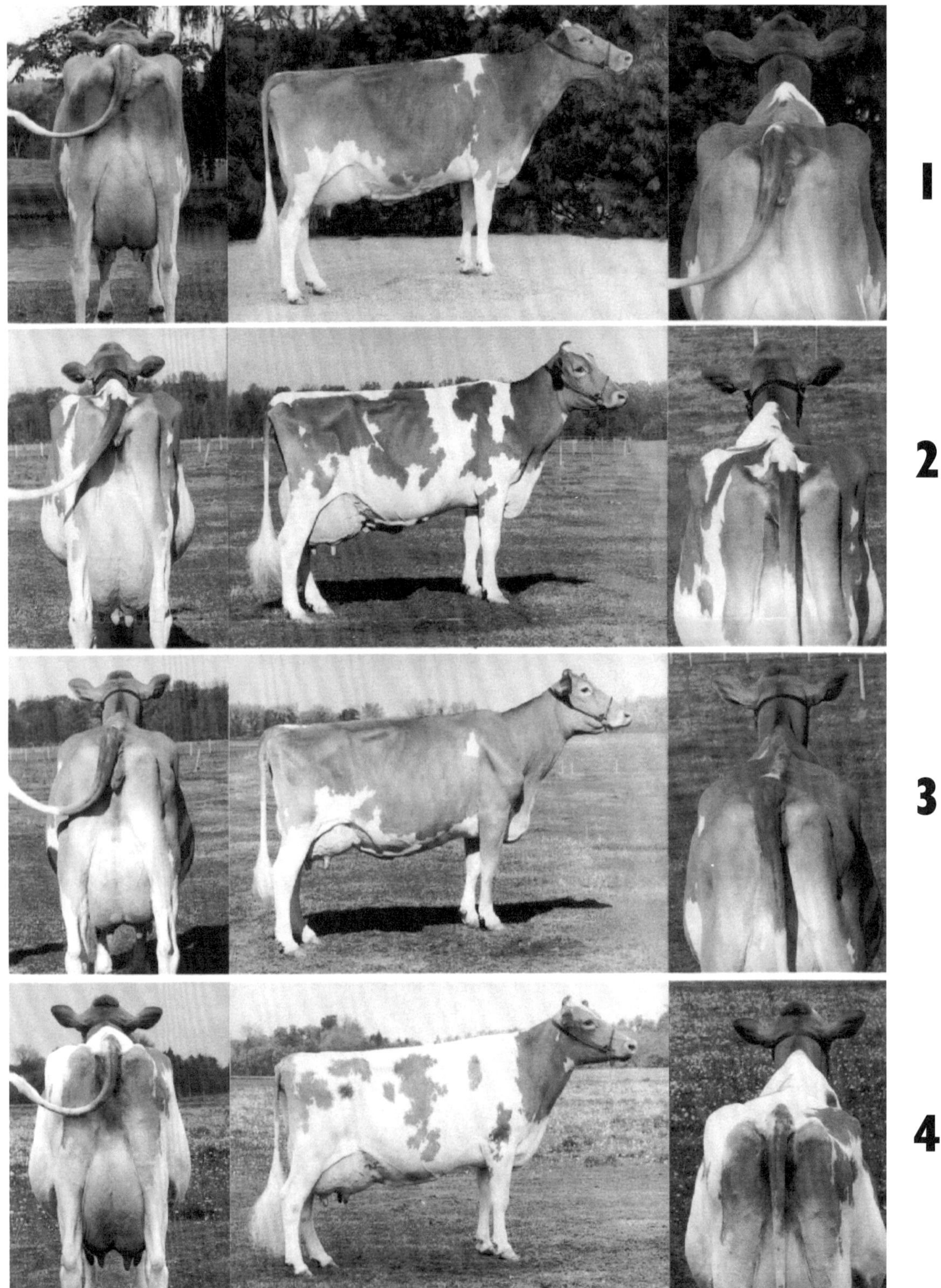

(Reprinted by permission from Hoard's Dairyman magazine. Copyright 1995, W. D. Hoard and Sons Company.)

UNIT IV

Industry

Chapter 18

SWINE INDUSTRY

Hog Wild!

INTRODUCTION

Most hog production occurs in the Midwestern Grain States. Iowa leads all states in hog production. In recent years, the swine industry has changed dramatically. More producers are entering into contractual agreements with large companies, while fewer are maintaining their independent status. Hog production ranks fifth in total commodity income for the United States. Swine operations can be classified by type or size of operation. However, most larger producers run farrow to finish barns. The swine industry offers a variety of job opportunities ranging from herdmanagers to food scientists to advertising specialists.

Figure 18-1. (Courtesy, Mississippi State University)

OBJECTIVES

1. Give an overview of the swine industry in the United States.
2. Differentiate between farrow and finish operations.
3. Identify geographic areas of sow and finishing operation concentration.
4. Discuss current trends in the swine industry.
5. Discuss one issue facing the swine industry.
6. List three careers in the swine industry.
7. Name two organizations that play supporting roles in the swine industry.

TERMS

check-off
farrow to finish
independents
integrated
seedstock

ANIMAL SCIENCE FACTS

Since 1992, North Carolina has moved from sixth place among all states in hog production to second place. Iowa remains in the number one position.

ANIMAL SCIENCE FACTS

Hogs generate over 10 billion dollars in gross receipts each year in the United States.

INDUSTRY OVERVIEW

The trend toward fewer but bigger farming operations is perhaps most evident in the swine industry. Thirty years ago, over one million American farms raised an average of 50 head of hogs per year. In 1994, less than 210,000 farms averaged 285 hogs per year. Moreover, 37 percent of the nation's hogs are currently housed on farms with inventories reaching over 2,000. Another 18 percent are on farms with numbers in the 1,000 to 2,000 head range. The remaining 45 percent of hogs are on farms with less than 1,000 head. These small farms account for 94 percent of all swine operations. In other words, just 6 percent of all swine operations contain 55 percent of the United States' swine inventory. See Figure 18-2.

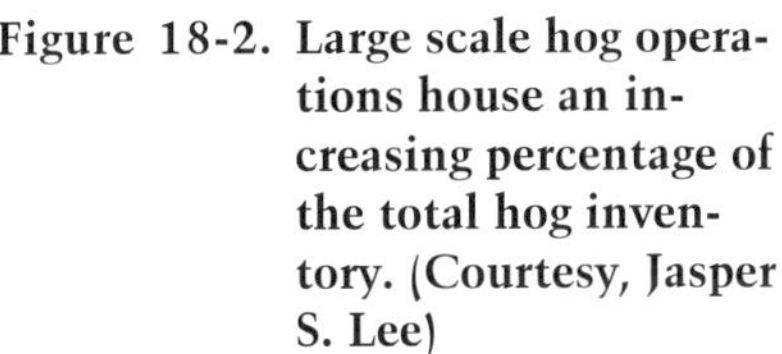

Figure 18-2. Large scale hog operations house an increasing percentage of the total hog inventory. (Courtesy, Jasper S. Lee)

Traditionally, the swine industry has been centered in the Corn Belt where feed is readily available. Since feed accounts for 65 percent of total swine production costs, the Corn Belt remains the heart of hog finishing operations. However, recent years have seen sow operations moving from the Corn Belt to states where little or no anti-corporate farm sentiment exists. North Carolina led the way in rapid swine production growth. In 1994, 86 percent of all North Carolina swine were located on facilities holding 2,000 or more hogs. Most of these large operations are highly ***integrated*** with owners controlling many aspects of production. Large North Carolina operations average over 7,000 head, outdistancing the national average for large farms by more than 3,000. Other states following the North Carolina model are: Arizona, Colorado, Missouri, Mississippi, Pennsylvania, Texas, and Utah.

As swine operations increase in size, so do the slaughter houses that process pork. The ten largest packing firms can each process over 15,000 hogs per day. This number is sure to increase as expansion in the industry continues.

INDUSTRY ORGANIZATION

The swine industry can be classified by operation type or size. Four basic types of operations exist. ***Seedstock*** producers or companies provide registered and nonregistered or planned crossbred breeding stock. See Figure 18-3. Producers purchase seedstock hogs to better the genetics, and therefore, the profitability of their herd. Farrowing operations maintain sows and sell young pigs to finishers. In addition, finishing operations feed hogs until slaughter—five to six months of age and 240 to 270 pounds. Many producers have ***farrow to finish*** operations where hogs remain from birth to slaughter.

Figure 18-3. Seedstock companies provide breeding stock to improve herd genetics. (Courtesy, Jasper S. Lee)

As discussed in the swine industry overview, swine operations can also be segmented by size. The usual size breakdown includes operations of less than 1,000 head; 1,000 to 2,000 head; and over 2,000 head.

INDUSTRY ISSUES

Profitability remains the key issue in the swine industry. When inflation adjustments are made, the real price of hogs for 1993 and 1994 averaged

40 percent less than 1970. Some increases in productivity, such as number weaned and annual amount of pork produced per head of breeding stock, have offset the deflated price. Nevertheless, the depressed market of 1994 and 1995 hit the industry hard. ***Independents*** (not aligned with a company) are joining ranks to take advantage of the economies of scale. Alignment activities include marketing, using similar genetics, finishing hogs together from joint off-site nurseries, as well as other ventures.

Environmental concerns surrounding manure handling face the swine industry. Most swine manure is stored in pits and lagoons where little breakdown can occur. See Figure 18-4. When the manure is spread, odor and runoff become issues. Ground water pollution can result from manure runoff and excessive leaching through the soil.

CAREER PROFILE

Federal Meat Inspector

Meat inspectors determine the wholesomeness and safety of carcasses and processed meats. Inspectors reserve the right to reject a carcass or cut that shows evidence of possible contamination. Inspectors can also close packing plants that fail to meet government standards. (Courtesy, Mississippi State University)

Figure 18-4. Proper manure handling concerns swine producers. (Courtesy, Jasper S. Lee)

Because of the increase in the number of large swine corporations, an anti-corporate farm movement has affected the industry. While existing across the nation, anti-corporate farm sentiment centers in the Corn Belt States. This movement has encouraged large company-owned farms to locate in peripheral states, where the benefits of jobs brought from large swine operations outweigh any potential disadvantages.

CAREER OPPORTUNITIES

CAREER PROFILE

Swine Producer

Swine producers perform daily tasks associated with pork production. These jobs include mixing feed, feeding pigs, breeding, farrowing sows, vaccinating, marketing, and record keeping. Exceptional swine producers know and implement the sciences of reproduction, genetics, nutrition, and behavior. In addition, they make sound business decisions based on the current economics of the swine industry. (Courtesy, National FFA Organization)

Career opportunities exist for personnel interested in the swine industry. While some industry jobs require a college education, many can be obtained with a high school diploma. Owners often hire workers and unit managers to conduct the daily operation of barns. Tasks for these employees can include checking feeders and waterers, treating and culling sick pigs, record-keeping, breeding, farrowing, and, even, barn repair. Many swine companies also employ supervisors to provide coordination and leadership to specific divisions like finishing or genetics. Most large companies employ a herd veterinarian as well.

Sales personnel market swine feeds manufactured by a multitude of feed companies. Also, interested individuals may choose to work for one of the many swine housing or equipment companies.

Due to the integration of the industry, transportation to slaughter may be provided by the producer or an independent trucking company with dispatchers and drivers. Packing firms, too, can be owned by the producer. However, most firms operate independently. Thousands of employees are needed to process the 90 million head of hogs slaughtered each year!

The swine industry relies heavily on researchers to provide insight into more efficient swine production. Extension agents, government workers, and teachers disseminate this new information to other swine industry personnel. Furthermore, communication specialists help producers market their product to consumers with advertising slogans, such as, "Pork . . .The Other White Meat."

CAREER PROFILE

Swine Geneticist

Swine geneticists design and implement planned crossbreeding programs using available breeds. Outstanding animals are identified, tested, and selected through the use of computerized performance data in conjunction with visual evaluation. Only the best animals are retained as parents of the next generation. (Courtesy, Jasper S. Lee)

SUPPORTING ORGANIZATIONS

From a slate of producer nominees, fifteen individuals are appointed by the U. S. Secretary of Agriculture to sit on the National Pork Board. This group administers funds gleaned from the pork ***check-off*** program, which began in 1986. The intent of the check-off program is to finance education, research, and product promotion. Pork check-off monies are collected at the rate of 45 percent of gross price. Part of the collected check-off monies is returned to the states for promotion, using a variable percentage scale based on number of hogs sold. See Figure 18-5.

Figure 18-5. Pork check-off dollars assist the swine industry in the development of promotional campaigns designed to entice consumers to choose pork. (Courtesy, Mississippi State University)

Housed in Des Moines, Iowa, with the National Pork Board, the National Pork Producers Council provides a voice for producers across the country. Along with representatives elected from state affiliate groups, the National Pork Producers Council maintains a hired staff. This organization sponsors the World Pork Expo, which is held annually in June. Participants from around the world attend this industry-oriented event. The National Pork Producers Council also hosts the Pork Forum held in the spring. Producers voice current concerns and set industry policy at this conference. The National Pork Council and the state affiliates are supported by pork check-off dollars.

Lastly, the Pork Industry Group (PIG) Directors, elected at the Pork Forum, provide input to the Pork Industry Group Staff of the National Livestock and Meat Board located in Chicago.

SUMMARY

The swine industry has undergone tremendous change. Large swine farms account for only 6 percent of the nation's operations while housing over 55 percent of all hog inventories. While many nontraditional states are gaining in sow numbers, the Corn Belt States remain the heart of swine finishing operations. The swine industry can be organized by either type or size of operation. Major issues face the swine industry. Profitability tops the issues list with manure management and corporate farms also being concerns. Several careers in the swine industry exist for potential employees. While some jobs in the swine industry require a college education, many can be obtained with a high school diploma. The National Pork Board and The National Pork Producers Council support the industry in many ways. One such activity is the organization of the World Pork Expo held annually in June.

CHAPTER SELF-CHECK

___ integrated	1. operations with sows and finishing barns
___ seedstock	2. dollars collected for research, education, and promotion
___ farrow to finish	3. not aligned with a company
___ independents	4. own several aspects of industry
___ check-off	5. breeding stock

QUESTIONS AND PROBLEMS FOR DISCUSSION

1. Which state ranks number one in hog production?
2. Thirty years ago over ______________ American farms produced an average of 50 hogs per year.
3. Smaller farms of less than 1,000 head house ____ percent of United States' swine inventory.
4. Which state has moved from sixth to second place in hog production?
5. Name the type of hog operation that sells breeding stock?
6. Hogs are finished for slaughter at approximately ______ months of age.
7. Acceptable finishing weight for hogs ranges from ____ to ____ pounds.
8. List two environmental concerns surrounding manure handling.
9. ____________ hog producers operate without company affiliation.
10. How many hogs are slaughtered each year in the United States?
11. The pork check-off program began in _____.
12. Pork check-off dollars are collected at a rate of _______ of gross price.
13. In what city are the National Pork Board and The National Pork Producers Council located?
14. What large-scale swine industry event is held annually in June?
15. United States' hog production generates ____ billion dollars in gross receipts each year.

ACTIVITIES

1. Research the recent price trends in the hog market. Report your findings to the class.
2. Conduct a taste test of fresh and cured ham. Make a score sheet to use for rating texture, flavor, moistness, etc.
3. Contact a company that offers contracts for swine. Ask them for information about their contractual arrangements with producers.

LABORATORY ACTIVITY

PROMOTING PORK

Purpose

To develop promotional materials to encourage consumers to eat pork (presentations can be adaptable to National FFA Sales Contest)

Materials

Note: materials will vary from group to group

poster board

glue

magazines

markers

stencils, etc.

Procedure

1. Have each student develop a seven-minute sales presentation featuring pork or pork products.
2. Make posters promoting pork using materials provided by instructor.

Analysis

1. Have students evaluate each other. Score presentations on the opening, presentation that emphasizes pork features and benefits, and closing.
2. Conduct a contest with posters. Display winners in your community.

Application of Laboratory Activity

Good sales are essential for survival in the agricultural marketplace. However, producers are sometimes shy about promoting their commodities. Knowing one's customer and having a sales objective can help pork promoters prepare effective sales materials. Sales presentations and display posters are two means used to educate the public about pork.

After completing this lab activity, students can tailor their presentations for competition in the National FFA Sales Contest. The seven-minute sales presentation, along with an objective test and practical skills assessments, help students develop the public relations tools necessary for success in agricultural sales and marketing.

Chapter 19

BEEF INDUSTRY

All Hat, No Cattle?

INTRODUCTION

Beef animals are raised throughout the United States because of their ability to convert forages to meat and the low overhead costs needed for production. Cattle production ranks first in the United States, accounting for 22 percent of all agricultural commodity income. Although larger beef cattle operations can be segmented into four basic production types, many farmers raise beef from calving until slaughter for personal consumption and extra income. However, the beef industry does not stop at the slaughter house. Thousands of employees are needed in the packing, inspection, distribution, promotion, and retailing of beef.

Figure 19-1. (Courtesy, National FFA Organization))

OBJECTIVES

1. Give an overview of the beef industry in the United States.
2. List four production segments of the beef industry.
3. Identify geographic areas of beef cows and cattle feeding concentration.
4. Explain the current beef inventory trend.
5. Discuss one issue facing the beef industry.
6. List three careers in the beef industry.
7. Name two organizations that play supporting roles in the beef industry.

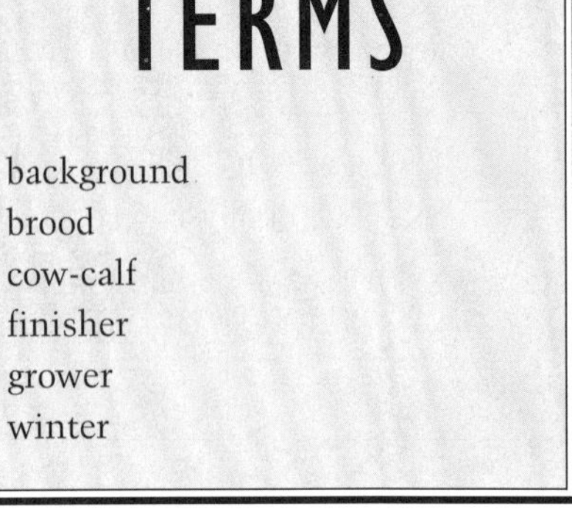

TERMS

background
brood
cow-calf
finisher
grower
winter

ANIMAL SCIENCE FACTS

Cattle and calves generate over 37 billion dollars in gross receipts each year in the United States.

INDUSTRY OVERVIEW

Cattle generate more dollars than any other agricultural enterprise. Twenty-two percent (22%) of all commodity receipts comes from the cattle industry. Texas ranks first in gross receipts for cattle and calves with Nebraska, Kansas, Colorado, and Iowa following in places second through fifth, respectively. These five states account for 50 percent of total cattle production in the United States. See Figure 19-2.

Figure 19-2. Cattle generate 22 percent of all commodity receipts. (Courtesy, Jasper S. Lee)

Over 600,000 beef producers in the United States average approximately 1,000 acres for owned and rented land. Total United States land utilized for beef production remains over 600 million acres. Most beef cows and calves are located in the southeastern and western states, while the Plains States remain the center of feedlot operations. More small, part-time beef producers operate in the East, while western producers tend to run large full-time operations.

The beef industry is in the midst of an expansion period. Cattle inventories are expected to increase until the late 1990s. Typically, beef cattle numbers cycle every ten years.

INDUSTRY ORGANIZATION

Although some producers raise beef cattle from birth to slaughter, most producers can be loosely organized into four types of operations: seedstock, cow-calf, grower, finisher. Seedstock operations provide breeding animals

CAREER PROFILE

Feedlot Manager

Feedlot managers understand the business of finishing beef cattle. They keep careful track of the costs (feeder calf prices, feed ingredient costs, labor, etc.) associated with the finishing enterprise as well as available marketing channels. Proficient feedlot managers are exerts at disease prevention and treatment. (Courtesy, Texas A&M University)

for cow-calf producers. Seedstock producers develop purebreds (registered or nonregistered animals of a specific breed) or planned crosses to improve such genetic traits as carcass quality or feed efficiency.

Cow-calf operations maintain ***brood*** (mother) cows for an annual calf crop. Weaning of these calves occurs at 400 to 600 pounds. At weaning, animals are normally processed (vaccinated, castrated, implanted, dehorned, if necessary, etc.). Calves then move from the cow-calf producer to either a grower operation, where they mature until approximately 900 pounds, or directly to a feedlot. The main objective of ***grower*** operations, sometimes called stocker operations, is to grow the animals until they are more mature for finishing. See Figure 19-3. Growers either ***winter*** the animals for placement on spring pasture, or ***background*** them on a diet intended for more rapid gain and, subsequent, feedlot placement

Figure 19-3. Grower/stocker operations prepare cattle for feedlot finishing. (Courtesy, Jasper S. Lee)

without additional pasturing. Grower/stocker cattle go from the grower to the ***finisher*** (feedlot) for slaughter on a high grain diet. Beef cattle are generally marketed at 14 to 24 months of age weighing 1,100 to 1,300 pounds. The size of feedlots can vary from less than 100 to over 100,000 head.

INDUSTRY ISSUES

Since the death of several fast food customers, due to an *E. coli* poisoning incident from undercooked hamburger, food safety has rushed to the forefront of beef industry issues. A national Blue Ribbon Task Force for food safety is investigating safe meat handling and cooking procedures. In addition, industry-wide educational efforts have encouraged consumers to cook hamburger thoroughly (until gray in the center and juices run clear), and properly clean utensils and surfaces that have contact with raw or undercooked meat.

A second issue facing the beef industry is customer demand for product quality and consistency. Some beef industry personnel believe the recent increases in the consumption of poultry is due in part to the quality and consistency of poultry products. The beef industry has responded with Quality Assurance Programs coordinated nationally and statewide by various cattlemen's associations and beef councils. The program's goal is to assist the beef industry in delivering consistent, high-quality beef to consumers. See Figure 19-4.

CAREER PROFILE

Animal Scientist

Animal scientists can earn their living by conducting either basic or applied research. Basic research investigates animal physiology. New drugs and treatments of human diseases are often first discovered by basic researchers working with animals. Applied research involves practical study in areas such as nutrition that directly benefit producers (Courtesy, National FFA Organization)

Figure 19-4. Consumers demand flavorful and consistent cuts of beef. (Courtesy, Mississippi State University)

Increased fees for use of public land have become a geographical issue within the beef industry. Western range herds depend heavily on public lands for grazing. If costs of public land usage become prohibitive, profitability of range herds will be adversely affected.

CAREER OPPORTUNITIES

Career opportunities in the beef industry begin with the producer. Large beef farms and ranches employ a variety of individuals. Managers and/or supervisors oversee day-to-day operations and direct the activities of farm/ranch laborers. Other farm/ranch employees may maintain records or work specifically with herd genetics. Personnel in feed and equipment businesses play supporting roles within the beef industry. In addition, educational consultants and veterinarians assist farmers/ranchers in efficiently producing beef. Once the animal is ready for slaughter, other industry workers move the beef through processing to the consumer's table. Livestock buyers and sellers arrange for transfer of animals among the types of operations and ultimately to the packer for slaughter with firms, such as Iowa Beef Processors, Inc. Many positions are available to individuals interested in the packing industry. These jobs include federal inspection and beef grading positions. After the slaughter process has been completed, packaged beef is transported to retail markets, such as grocery

CAREER PROFILE

Feed Dealer

This feed dealer is helping a customer choose the best type of feed for his operation. Cattle feed dealers need to be knowledgeable about the nutritional demands of cattle at different stages of growth and development. They play an important supporting role in the production of beef cattle in the United States. (Courtesy, James Leising)

stores, hotels, and restaurants. Salespeople working for packers initiate this transaction.

Besides producers, packers, marketers, and retailers, a variety of supporting individuals and companies facilitate the efficient production and sale of beef. Other supporting personnel include animal scientists, educators, and veterinarians.

SUPPORTING ORGANIZATIONS

The Beef Board located in Denver, Colorado, oversees the dispersal of the beef check-off monies targeted to the research, education, and promotion of beef. The beef check-off program began in 1986. To financially support the check-off program, one dollar per head is collected each time an animal is sold. Members of the Beef Board are appointed by the United States Secretary of Agriculture. Total number of board members fluctuate with national beef cattle inventory. In 1995, 107 individuals sat on the Beef Board.

Figure 19-5. The National Live Stock and Meat Board was formed in 1922.

The United States Beef Breeds Council serves as an umbrella organization to unite the many beef breed associations. Chairship of this group rotates among the associations. The U.S. Meat Export Organization also works to further beef and other meat exports. This organization is industry driven, but is partially supported by government funding.

The planned merger of two industry giants, the 100-year-old National Cattlemen's Association (lobbying arm of the beef industry) and the beef function of the National Livestock and Meat Board (promotion arm of the industry formed in 1922) will have begun by January 1996. This union will coordinate the efforts of the industry organizations in hopes of a more effective and efficient voice for beef producers. See Figure 19-5.

On a state level, 44 of 50 states have beef councils supported in part by check-off dollars. These councils also serve a promotional function. In addition, many states also have cattlemen's associations, which act in advocacy capacities. In some states, the beef council and the state cattlemen's association are housed together.

SUMMARY

Cattle and calves ring up cash receipts accounting for 22 percent of total agricultural commodity income. The beef industry can be segmented into four types of operations: seedstock, cow-calf, grower, and finisher. Beef animals are raised across the country with heavy concentrations of cow-calf operations in the Southeast and West, and feedlots in the Plains States. Issues facing the beef industry include concerns for food safety, delivery of consistent quality and products, and increased fees for use of public lands. Jobs within the industry span from ranch managers to food scientists. Several groups, organized nationally and statewide, provide support for the beef industry.

CHAPTER SELF-CHECK

___ cow-calf	1. feedlot operator
___ grower	2. mother cows with babies
___ finisher	3. readied for the feedlot
___ brood	4. producer between cow-calf and finisher
___ winter	5. placed on spring pasture
___ background	6. mother

QUESTIONS AND PROBLEMS FOR DISCUSSION

1. Which state ranks first in gross cattle receipts generated?
2. How many farms/ranches produce beef?
3. Most feedlots are centered in this region. Name the group of states.
4. List four segments of the beef industry.
5. Seedstock operations provide breeding stock to ____________ producers.
6. What occurs during calf processing?
7. Finished beef cattle weigh in the ____________ range.
8. True or False? *E. coli* poisoning can cause death.
9. Fees for use of public lands for grazing have increased/decreased. Select one.
10. Name the beef packing firm mentioned in this chapter.
11. Who oversees the handling of beef check-off dollars?
12. List three uses of beef check-off dollars.
13. Who appoints the Beef Board?
14. What organization unites beef breed associations?
15. How many states have state-organized beef councils?

ACTIVITIES

1. Contact the beef council in your state. Ask for information about the beef industry and the possibility of a student internship.
2. Track steer prices over the last 50 years. USDA, state departments of agriculture, and extension offices can assist in finding prices. Does a price cycle exist? If so, why do you think the cycle occurred?
3. Research the by-products of beef. Make a display using a collection of these products.

LABORATORY ACTIVITY

COMPARING REGULAR, LEAN, and EXTRA-LEAN GROUND BEEF

Purpose

To compare the amount of lean meat found in one pound of regular, lean, and extra-lean ground beef.

Materials

one pound each of regular, lean, and extra-lean ground beef
nine paper plates
unit price of each type of ground beef
hot plate
frying pan
glass measuring cup
wooden spoon
paper towels
scale for weighing ground beef

Safety Precaution

Participants will be browning ground beef during this lab. Exercise caution when using the hot plate. Keep flammable materials away from hot plate. Use caution when pouring grease from browned beef. Hot grease can cause severe burns! Do not pour grease in a drain. Doing so can clog the drains. Wash hands and all utensils with antibacterial soap and hot water after lab. Also, do not eat any ground beef unless it is thoroughly cooked and all juices run clear. Cooked ground beef should be gray throughout.

Procedure

1. Weigh exactly one-pound samples of regular, lean, and extra-lean ground beef.
2. Place samples on paper plates.
3. Note unit price of each sample.
4. Using frying pan, wooden spoon, and hot plate, thoroughly brown regular ground beef until no pink areas exist. Stir beef constantly with wooden spoon.
5. When browning is completed, turn off hot plate and carefully pour all liquefied grease into glass measuring cup.

6. Record amount of grease.
7. Remove beef from pan and place on a fresh paper plate.
8. Blot remaining fat from meat.
9. Reweigh sample. Record your results.
10. Complete steps 4-9 with lean and extra-lean beef.

Analysis

Reminder: Uncooked weight should be exactly one pound!

Recording of Data

1. Regular 2. Lean 3. Extra-Lean

Amount of grease poured from cooked meat.

1. 2. 3.

Weight of cooked beef sample

1. 2. 3.

Questions

1. Rank the samples in order of highest to lowest fat content.
2. Figure the unit price of each cooked sample.
3. Which product will you buy? Why? Would you consider buying different ground beef products for varying recipes? Why?

Application of Laboratory Activity

In recent years, health conscious consumers are demanding leaner meats. Industry has responded by producing leaner beef. Furthermore, supporting organizations have assisted cattle producers in meeting market needs by developing low fat meat recipes and providing nutritional information to consumers. Monies to finance such promotional projects come from the beef check-off program, which requires a fee from the sale of each beef animal. Several slogans, like "Real Food for Real People," developed from check-off money, have coaxed consumers to eat beef.

Also, understanding the economics of unit prices can help consumers make wise choices. Selecting meat cuts based on package price may not always yield the best buy. Unit prices allow consumers to see the price per pound of each cut. Fat and bone are also included in unit pricing.

Chapter 20

DAIRY INDUSTRY

Get in the Mooed!

INTRODUCTION

The dairy industry is perhaps the most labor intensive of all livestock operations. Cattle require milking two to three times per day, seven days a week. Income generated by dairy commodities is second only to that generated by beef cattle. California ranks first in dairy production. Like many other agricultural industries, the dairy industry is changing. Fewer dairy producers own more cows, which produce more milk per farm. As health-conscious consumers demand lower fat diets, dairy food scientists have responded with a variety of specialty products. Fat-free yogurt, cream cheese, and frozen dairy desserts now line our grocery shelves.

Figure 20-1.

OBJECTIVES

1. Give an overview of the dairy industry in the United States.
2. Explain the changing demographics of the dairy industry.
3. Discuss how dairies may be classified by size.
4. Describe profitability concerns in the dairy industry.
5. Name two organizations that play supporting roles in the dairy industry.
6. List three careers in the dairy industry.

TERMS

cooperatives
cwt.
intensive pasture management
price floor

ANIMAL SCIENCE FACTS

Dairy products generate twenty billion dollars of gross receipts each year in the United States.

INDUSTRY OVERVIEW

California has surpassed Wisconsin to become the number one state in dairy production. Large farms in the Southwest have changed the demographics of the dairy industry. See Figure 20-2. Although traditional dairy states like Wisconsin, New York, and Pennsylvania remain major players, competition from large Southwestern dairies challenges the status quo. Forty years ago, two million dairy farms operated in the United States. Today, far fewer remain in operation. Currently, 130,000 dairy farms averaging 330 acres still exist.

Figure 20-2. Large dairies take advantage of economies of scale. This is the Maddox Dairy in Fresno, CA. (Courtesy, Jasper S. Lee)

INDUSTRY ORGANIZATION

Dairy farms can be organized by the number of cows they maintain: less than 80, 80 to 500, over 500. The economies of scale enjoyed by larger operations increasingly endanger smaller operations. For example, inflated machinery prices force smaller dairies to continue with older, less efficient equipment. Many small dairies of less than 80 milking cows survive only with second incomes generated from off-farm sources.

Dairies of 80 to 500 milking cows can also find themselves in financially difficult situations. These producers need the same equipment as larger operations, but have fewer cows to offset the additional equipment costs. Some mid-size herds respond by implementing ***intensive pasture management*** (closely monitored rotational grazing), which can reduce feed costs by one-half. Intensive pasture management leads to lower veterinary

Figure 20-3. Intensive pasture management can cut feed costs in half. (Courtesy, Jasper S. Lee)

costs due to improved foot health and lower mastitis occurrences. See Figure 20-3.

Dairies with 500 or more milking cows have the financial advantages of operating on a large scale. While smaller dairies are often family-run businesses, many large-scale dairies are corporate owned. These businesses often employ managers to oversee specific divisions, such as crops or cows. Furthermore, some large California farms are taking advantages of such economical and nontraditional feedstuffs, as citrus pulp and almond hulls.

CAREER PROFILE

Breed Association Executive Secretary

Most breed associations employ executive secretaries who work for the membership performing tasks sùch as promotion and marketing. Executive secretaries often organize and run livestock shows and sales, which showcase their breeds. They also attend to the daily business associated with running an effective organization.

INDUSTRY ISSUES

Profitability tops the list of dairy industry concerns. The country's dairy farmers continue to produce a milk surplus of 1 to 2 percent per year. Although government programs guarantee milk ***price floors*** (lowest price), in recent years, surpluses have been minimal and the government has not purchased milk. Government price floors are established every five years in the U.S. Farm Bill. The price floor of $12.50 ***cwt.*** (per hundred pounds), established in the 1985 Farm Bill, fell to $9.85 cwt. in the 1990 Farm Bill.

Figure 20-4. Retail costs of milk have risen despite a static wholesale market. (Courtesy, Edward W. Osborne)

Wholesale milk prices averaged $12.50 cwt. in both 1975 and 1995. During the same period, the inflation rate of dairy inputs increased an average of 5 percent per year. The retail price of milk was $18.50 cwt. in 1975, and $30.00 cwt. in 1995. With producers receiving only the wholesale price of $12.50 cwt. in 1995, dairy processors and retailers realized profits of $17.50 cwt. See Figure 20-4.

CAREER OPPORTUNITIES

Most jobs in the dairy industry are located in sales. Feed, machinery, equipment, and technical sales and service personnel are needed to provide

CAREER PROFILE

Milker

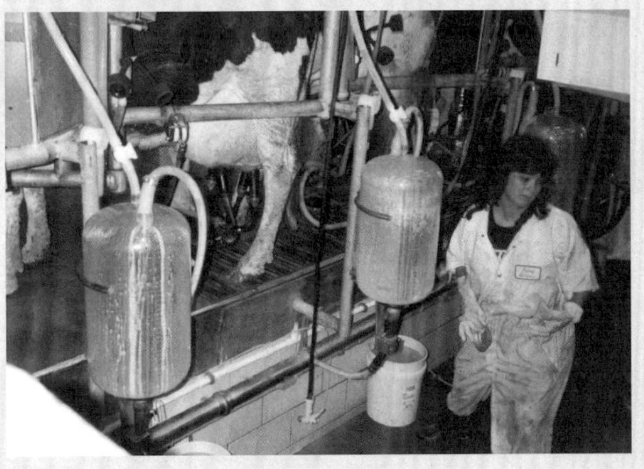

Larger dairies hire milkers on a full-time basis. These employees usually milk cows in shifts, performing their duties in milking parlors. Good milkers apply accepted techniques by carefully washing udders, attaching and removing milking equipment in a timely manner, and treating mastitis cases. Milkers usually wash and disinfect milking equipment as well. (Courtesy, Mississippi State University)

CAREER PROFILE

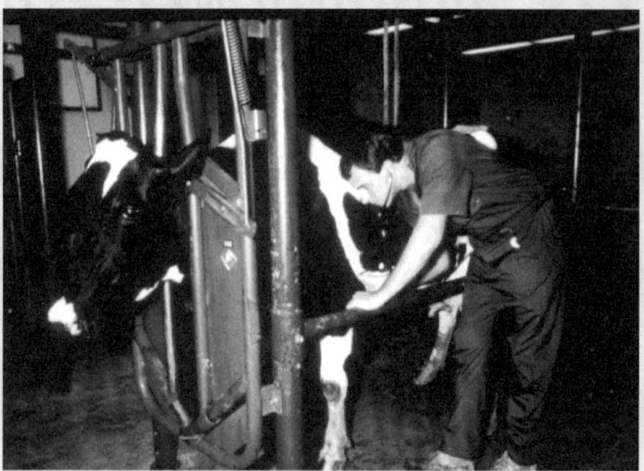

Veterinarian

Veterinarians tend to the health of both large and small animals. Veterinarians work with producers and pet owners to prevent, diagnose, and treat diseases and injuries. Large animal veterinarians often work long and unusual hours. Some veterinarians own their own practice, while others work for a larger, multi-vet practice. (Courtesy, USDA)

Figure 20-5. Dairy promotions urge consumers to use dairy products, such as yogurt. (Courtesy, Edward W. Osborne)

dairy farmers with supplies and services for efficient operation. For example, dairy semen can be purchased from breeding ***cooperatives*** (members share in ownership and operation) who can send artificial insemination technicians to service cattle in heat. Additional job opportunities exist in cooperatives that purchase milk from farmers, process dairy products, and market them to retailers. Milk cooperative personnel include truckers, field service individuals, managers, inspectors, as well as processing plant operators, supervisors, and marketing specialists.

Potential dairy employees may also want to consider becoming DHIA workers who provide extensive record-keeping for producers. Those interested in communications may wish to seek careers in breed associations or with dairy-oriented media and promotion organizations. See Figure 20-5. Dairy food scientists constantly research new products that will appeal to health-conscious consumers. In addition, veterinarians provide valuable herd health services for dairy farmers, such as ration development, herd health maintenance, and disease prevention.

SUPPORTING ORGANIZATIONS

Several organizations support the dairy industry at the national level. The National Dairy Board and United Dairy Industry Association (U.D.I.A.) united in 1994 to form Dairy Management Incorporated, which is located in Illinois. This group represents almost the entire country's dairy industry. Dairy Management Incorporated promotes research, education, and the use of milk products. Before it merged with the U.D.I.A., the National Dairy Board consisted of dairy producers appointed by the U.S. Secretary of Agriculture. This group supervised the expenditure of dairy check-off dollars. Like other commodity groups, the dairy check-off program began in 1984 as an effort to finance milk promotion. Today, check-off monies are collected at a rate of 15 cents cwt. of milk with five cents going to national efforts and ten cents remaining with local agencies deemed qualified by the USDA.

ANIMAL SCIENCE FACTS

United States' consumption of fluid milk has decreased while intake of cheese has increased.

Seventeen states maintain milk marketing boards funded through licensing fees. These boards regulate the prices of milk moving from producers to consumers. Ensuring an adequate supply of milk is the primary goal of these boards. Board members are usually appointed by the state Governor with state senate approval. State milk marketing boards can choose to be associated with the International Organization of Milk Control Agencies. This group meets on an annual basis to discuss issues surrounding milk supply and pricing.

The National Milk Producers Federation, made up of dairy cooperative and handlers representatives, serves to unite processors within the dairy industry.

The Purebred Dairy Cattle Association (located in Brattleboro, Vermont, with the Holstein association) functions as an umbrella group for the six major dairy breeds.

SUMMARY

Like most agricultural commodities, the dairy industry is changing. Large southwestern dairies have dramatically affected the makeup of the industry. Fewer dairies, each with more cows, now operate. Dairy farms can be organized into three groups: 80 or fewer cows, 80 to 500, or more than 500. Poor profitability plagues the dairy industry. Milk prices paid to dairy farmers have remained the same for many years. Meanwhile, the prices consumers pay for milk have increased substantially. While many job opportunities exist in the dairy field, sales careers are most prevalent. Several associations play supporting roles in the dairy industry.

CHAPTER SELF-CHECK

___ cwt.	1. members share in ownership
___ intensive pasture management	2. set lowest price received
___ price floor	3. rotational grazing
___ cooperatives	4. per hundred pounds

QUESTIONS AND PROBLEMS FOR DISCUSSION

1. Which state ranks number one in dairy production?
2. Forty years ago, over ________ million dairy farms operated in the United States.
3. Current dairy farm size averages _____ acres.
4. Explain how dairies can be segmented by number of milking cows maintained.
5. How can intensively managed pastures cut veterinary bills?
6. Name one nontraditional feedstuff used by some large California dairies.
7. What does cwt. mean?
8. The government milk price floor was set at _____ in 1985 and _____ in 1990.
9. The retail price of milk was _____ in 1975 and _____ in 1995.

10. The biggest area of employment in the dairy industry is _____.
11. Consumption of fluid milk has increased/decreased. Select one.
12. The dairy check-off program began in 19___.
13. The _______________________________ serves as an umbrella group for breed associations.
14. Name the state where the national Holstein association is located.

ACTIVITIES

1. Contact both large commercial and a smaller family-run dairies. Report on the differences between the two businesses.
2. Trace the production of milk from a cow to an ice cream cone. List all the jobs involved. Prepare a presentation to a group of younger students.
3. Visit a grocery store. List the dairy products offered for sale. How many products have been recently developed?

LABORATORY ACTIVITY

FINDING THE FAT IN MILK

Purpose

To differentiate fluid milk samples according to fat content

Materials

small cups (three per student)
marker
skim, 2 percent, and whole milk samples (a small amount of each sample per student)
copies of the labels from the milk samples for each student

Safety Precautions

Lactose intolerant students should not taste the milk samples. Also, keep the samples at a cool temperature.

Procedure

1. Code the bottom of the cups so like samples are marked similarly. (Example—all skim milk marked with a star)
2. Distribute one of each sample per student
3. Taste, touch, and observe samples.
4. Rank samples in order from highest to lowest fat content.
5. Write a paragraph justifying the rank.

Analysis

Using the label from each milk product, conduct the following calculations.

1. List the number of calories in an 8-ounce serving of each product.
2. Figure the number of calories coming from fat in an 8-ounce serving of each product. One gram of fat equals 9 calories.
3. Calculate the number of daily calories you need to maintain your weight. Multiply your weight by 15 (average figure for both genders and a range of ages).
4. Current nutritional advice dictates that less than 25 percent of daily calories should come from fat. Further consider that most students need four servings of milk to meet minimum daily nutrition requirements. Explain what products and amounts you would choose to include milk in a diet.

Application of Laboratory Activity

The dairy check-off program takes monies from producers' milk checks to be used toward advertisement. Many recent advertising campaigns have targeted the nonfat/lowfat advantages of skim and 2 percent milk. Producers can benefit from consumers desire to have healthful milk products readily available.

Take a walk through the dairy department at a grocery store. Note the many new dairy products on display. Who would have thought that fat-free cheese products could be developed? Dairy food scientists research and develop new foodstuffs in response to consumer demand.

Chapter 21

SHEEP INDUSTRY

I Am the Black Sheep of the Family!

INTRODUCTION

Compared to other livestock species, initial setup in the sheep industry is relatively inexpensive. Sheep utilize forages and move from conception to a salable market lamb in eight to nine months. Furthermore, sheep produce both wool and meat for human consumption. Still, sheep inventories have dropped by over two-thirds during the last thirty years. In January of 1995, the United States Department of Agriculture stated a national inventory of 8,895,000 sheep. Moreover, average consumption of lamb and mutton continues at less than 2 pounds annually, significantly lower than the 7-pound annual average in the early 1940s. Sheep populations are concentrated in the western states where larger flocks graze range land. Smaller eastern farm flocks supply lambs to the East Coast metropolitan market. General decline tops the list of sheep industry issues. The American Sheep Industry Association speaks on behalf of the sheep producers and industry personnel.

Figure 21-1. (Courtesy, National Lamb and Wool Grower)

OBJECTIVES

1. Give an overview of the sheep industry in the United States.
2. Differentiate between farm and range flocks.
3. Explain current issues and trends in the sheep industry.
4. Name two organizations that play supporting roles in the sheep industry.
5. List three careers in the sheep industry.

jockey
niche market
predators
procurers
tariffs

ANIMAL SCIENCE FACTS

United States' wool use fell off sharply in the 1960s, but rebounded slightly in recent years.

INDUSTRY OVERVIEW

The western range states account for much of the nation's sheep population. Even so, range sheep often take a backseat to other agricultural commodities. For example, Colorado ranks first in the nation for sheep cash receipts (94 million dollars), but sheep stand eleventh among all Colorado agricultural commodities. Other top five sheep states include: Texas (63 million dollars), California (54 million dollars), South Dakota (31 million dollars), and Wyoming (28 million dollars). Wool production is not included in sheep cash receipts. In 1994, over 68 million pounds of wool was shorn from 8,882,400 sheep leaving average fleece weight just less than 8 pounds. See Figure 21-2.

Figure 21-2. In 1994, over 68 million pounds of wool was produced in the United States. (Courtesy, Jasper S. Lee)

INDUSTRY ORGANIZATION

Western range flocks account for three quarters of the United States' sheep inventory. Generally, range flocks are characterized by at least 1,000 head. In range flocks, sheep are most often the primary enterprise.

Skilled flock management is essential for an efficient and profitable operation. Outside range country, smaller farm flocks usually maintain fewer than 100 head. Sheep tend to be secondary to other farm enterprises in these operations. Therefore, management is not as critical to overall farm success. As discussed in chapter 11, profitable hothouse lambs, fed indoors over the winter and marketed at Easter, have become a ***niche market*** (specialized) in the sheep industry. See Figure 21-3.

Figure 21-3. Hothouse lambs are a profitable niche market product. (Courtesy, Jasper S. Lee)

INDUSTRY ISSUES

General decline threatens the sheep industry. Four main factors have contributed to this fall. The aforementioned declines in inventory and the consumption of lamb and mutton have caused great industry concern. In addition, the wide availability of synthetics has provided wool formidable competition in fabric selection. As a result, American wool use fell from an average of more than 5 pounds annually during the World War II era, to a current rate of 1 pound or less. Also, importation of wool and lamb has increased greatly since the 1950s.

Figure 21-4. Predators accounted for over $17 million in sheep losses in 1994.

Two other problems plague the sheep industry. Like cattle ranchers, sheep producers face increased fees for leasing government range land. This input cost increase greatly affects flock profitability. Flock profits are also reduced by ***predators*** (animals killing the sheep). In 1994, animals lost to predators amounted to 17.7 million dollars. See Figure 21-4.

CAREER OPPORTUNITIES

Like the other meat industries, career opportunities are available in sheep production. However, these jobs are available on a far more limited basis, as lamb and mutton account for less than 1 percent of all meat consumed in the United States. Owners often employ shepherds who live with the flocks in range grazing areas for months at a time. Other people shear sheep for a living. Skilled shearers can remove an entire fleece in a few minutes. Careers as ***procurers*** who ***jockey*** (buy and sell) sheep can also be found. These experts must be skilled at selection and economics if they are to turn a profit for themselves and their clientele.

Additional positions are sometimes offered by coopera-

CAREER PROFILE

Wool Grader

Wool graders work for the government or wool buying firms. The graders evaluate the quality of grease wool before it is processed. Wool graders identify high quality by examining the fineness, crimp, and cleanliness before assigning a grade. Wool is then priced to the producer based on the wool grader's determination. (Courtesy, National Lamb and Wool Grower)

CAREER PROFILE

Extension Agent

Extension agents provide producers with information and training in emerging technologies. Such new technologies can have a dramatic impact on the profitability of agricultural operations. To stay abreast of these new technologies, extension agents must regularly attend workshops and training sessions. (Courtesy, Illinois Farm Bureau)

Figure 21-5. Seventy-five to eighty percent of the wool used in the United States is imported. (Courtesy, Jasper S. Lee)

tives, which collectively market industry products for better producer profits. Other opportunities are available in lamb and wool grading and processing. These individuals help assure quality products for consumers. As with other livestock commodities, support personnel, such as extension agents, educate producers in newly developed industry technology.

SUPPORTING ORGANIZATIONS

The American Sheep Industry Association is located in Englewood, Colorado. This federated association, with 50 state affiliates, provides a

CAREER PROFILE

Livestock Hauler

Livestock haulers work independently or are employed by producers or packers to transport livestock safely. Good livestock haulers possess an excellent driving record, as well as a knowledge of animal behavior as it pertains to loading and unloading livestock from trucks. Many haulers maintain records on driving time, expenses, and payloads using a laptop computer.

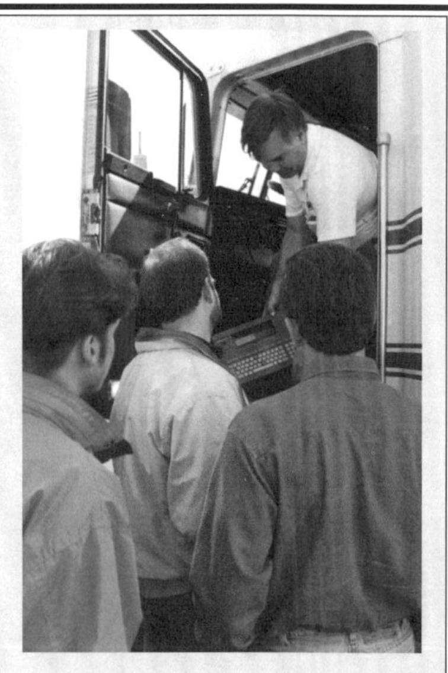

unified voice for the total sheep industry. Moreover, the association serves a host of purposes including research, education, promotion, and lobbying. A producer board, consisting of representatives from state affiliates, advises and oversees program direction. The names of state affiliate associations vary widely.

The American Sheep Industry Association is one benefactor of an incentive program resulting from the 1954 Wool Act. Portions of import ***tariffs*** (taxes on imported wool) are returned to producers as an incentive to compete with foreign wool. Today, roughly 75 to 80 percent of all wool is imported. See Figure 21-5. Many states maintain additional promotion programs.

Two other national organizations support the sheep industry. The National Lamb Feeders Association, headquartered in Bristol, Illinois, sponsors educational programs for producers. In addition, the American Sheep Industry Women cooperate with the other groups in advocacy work.

SUMMARY

The sheep industry can be organized into large range flocks, usually numbering over 1,000 head, and much smaller farm flocks ranging from just a few to less than 1,000 head. Sheep production centers in the western range states. General industry decline is due to a drop in wool use and consumption of lamb and mutton, development of synthetic fabrics, and importation of wool and lamb. Although careers in the sheep industry exist, they are not as plentiful as the other livestock species. The American Sheep Industry Association serves as the voice for the sheep industry.

CHAPTER SELF-CHECK

___ tariffs	1. act as buyers and sellers
___ niche market	2. act of buying and selling
___ procurers	3. kill sheep as a natural behavior
___ jockey	4. specialty mart
___ predators	5. taxes on wool imported into the United States

QUESTIONS AND PROBLEMS FOR DISCUSSION

1. In 1995, sheep inventory in the United States numbered ______.
2. Consumption of lamb and mutton in the United States averages less than _____ pounds per year.
3. ________ ranks first among states in receipts generated by sheep and lambs.
4. In 1995 __ million pounds of wool were produced in the United States.
5. Range flocks usually number more than _____ head.
6. List factors that have contributed to the sheep industry decline.
7. Predators cost the sheep industry ____ dollars in 1994.
8. Lamb and mutton account for less than ___ percent of the meat consumed in the United States.
9. What does a procurer do?
10. Where is The American Sheep Industry Association located?
11. True or False? The American Sheep Industry Association is affiliated with 45 state-organized groups.
12. The Wool Act of 19___ provides an incentive for domestic wool programs.
13. What percentage of wool used in the United States is imported?
14. Sheep producers can benefit from educationally oriented programs sponsored by this Illinois based organization. Name the organization.
15. What is the primary function of the American Sheep Industry Women?

ACTIVITIES

1. Co-sponsor a wool spinning demonstration with a home economics class.
2. If lamb is not a common dish in your area, prepare a leg of lamb to sample.
3. Ask a wool grader to come to class to demonstrate grading techniques.

4. Research the different types of wool fabric. Make a display of the fabrics and accompanying qualities.

LABORATORY ACTIVITY

QUALITIES OF WOOL

Purpose

To compare the flammability and wrinkle resistance of wool with other materials

Materials

two 6-inch square fabric samples of the following materials:
- wool
- cotton
- polyester
- rayon
- any wool blend

flame-proof metal pan (cakc pan will work wcll)
long wooden matches
clothes iron
ironing board
bucket of water
fire extinguisher

Safety Precaution

Use extreme caution when attempting to ignite fabrics. Keep a bucket of water and a fire extinguisher near the experiment. Also, make sure a functional fire blanket and/or safety shower is present. Conduct experiment in an area with proper ventilation.

Procedure

1. Select one of the following hypotheses.
   a. Wool fabric ignites quicker than the other samples.
   b. Wool fabric does not ignite as quickly as the other samples.

   c. There is no difference in speed of igniting between wool and the other samples.
2. Carefully, ignite each fabric sample in the flame proof pan.
3. Observe the reaction paying particular attention to how quickly each sample ignites.
4. Record your observations.
5. Select one of the following hypothesis.
   a. Wool fabric is more wrinkle resistant than the other samples.
   b. Wool fabric is less wrinkle resistant than the other samples.
   c. There is no difference in wrinkle resistance between wool and the other fabrics.
6. Attempt to iron a crease in each sample.
7. Rank the samples according to ease of wrinkling.
8. Record your observations.

Analysis of Laboratory Activity

1. Compare your selected hypotheses with the results of the tests.
2. What can you conclude about the flammability and wrinkle resistance of wool.
3. In light of the results of this experience and personal preference, which fabric will you choose for your next garment?

Application of Laboratory Activity

A wool grower's incentive program assists promotional efforts on wool's behalf. Natural fabrics, like wool, have regained popularity. Consumers often think of wool fabrics only in terms of heavy winter clothing. Once enlightened, they appreciate wool's ability to keep the wearer warm (even if wet) in winter and cool in the summer.

When selecting clothing, always examine the tag for fabric content and care instructions. Fabric selections may be easier based on knowledge gained from this laboratory activity.

Chapter 22

HORSE INDUSTRY

Wanna Make a Bet?

INTRODUCTION

During the twentieth century, the function of horses in the United States has changed drastically. After World War I, the gasoline engine began to replace the horse for both work and transportation. Today most horses in the United States are used for sport or pleasure, although some work horses remain active on ranches and in Plain Societies, such as the Amish. The horse industry does not organize as easily as swine, cattle, or sheep. Moreover, interests and career opportunities with horses vary greatly. Recent horse abuse and horsenapping concern the horse industry.

Figure 22-1.

OBJECTIVES

1. Give an overview of the horse industry in the United States.
2. Explain current issues in the horse industry.
3. List three careers in the horse industry.
4. Name two organizations that play supporting roles in the horse industry.

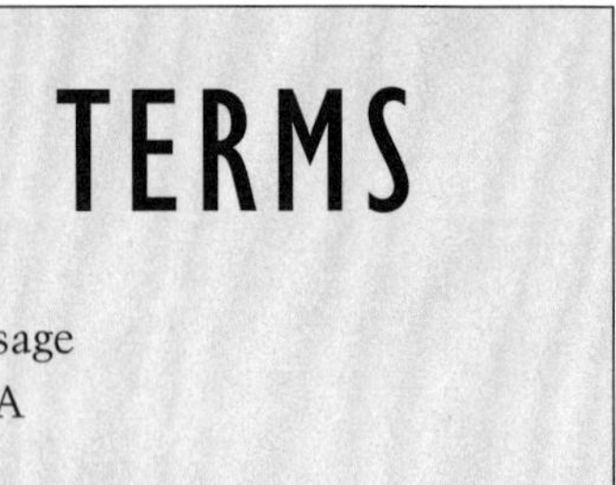

TERMS

dressage
SPCA

ANIMAL SCIENCE FACTS

The Pony Express route, from Missouri to California, began in 1861. It was replaced by the telegraph in less than two years.

Figure 22-2. Although most horses in the United States are used for pleasure, some Plain Societies continue to use them for work and transportation.

INDUSTRY OVERVIEW AND ORGANIZATION

Unlike the meat animal and dairy industries, the horse industry does not neatly organize into type and size of operation. However, as discussed in Chapter 12, horses can be roughly broken in light, draft, and pony categories.

Estimates place over three-quarters of all United States' horses in pleasure situations. The other one-fourth provide income or labor in such arrangements as performance, breeding, or farm work. See Figure 22-3. A precise census of United States' horses is unavailable, but inventories estimated by government and industry place total numbers in the six to ten million range. Over one-third are Quarter Horses. Moreover, the horse industry generates 15.2 billion dollars annually. Horses are raised throughout the country with heavy populations of breeding stock located in the Kentucky Bluegrass region.

Figure 22-3. About 25 percent of the horses owned in the United States are used for tasks other than pleasure.

ANIMAL SCIENCE FACTS

Horses are measured in "hands" equaling four inches. "Hands" of four inches are roughly the size of a fist measured side to side.

INDUSTRY ISSUES

Recent increases in horse abuse and horsenapping top the list of industry concerns. The Horse Protection Act was enacted in 1972. The enforcement arm of this act is the United States Department of Agriculture (USDA). However, delays and difficulties in reporting and enforcement exist, prompting the Society for the Prevention of Cruelty to Animals (***SPCA***) to file complaints with the USDA. Many cases of horse abuse and deliberate maiming have been reported, especially in the Northeastern part of the country. Although some incidents appear to be intended to eliminate outstanding performers from competition, other cases are random vicious acts.

The popularity of horse meat in Europe and Japan has risen substantially. Therefore, the slaughter price for horses has also risen causing a marked increase of horsenapping. In 1995, slaughter prices for select horses in good flesh have risen to upwards of 700 to 900 dollars, an increase of several hundred dollars over the last few years. Approximately nine slaughter houses in the United States process horse meat for human consumption. Abuse of horses on the way to slaughter has also caused industry wide concern.

CAREER PROFILE

Stable Attendant

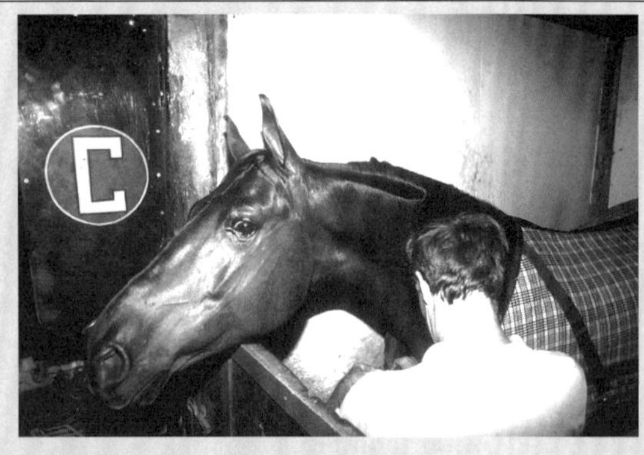

Stable attendants groom, feed, and in other ways care for the well-being of horses. Since they are responsible for rations, they mix feeds. They also inspect horses for sickness or injuries and treat them based on the instructions of a veterinarian. Other tasks include cleaning and maintaining horse facilities and equipment.

In pricier horse circles, guns for hire have killed horses for collection of insurance monies. All the aforementioned incidents show a dark side of the horse industry unknown to many pleasure riders.

CAREER OPPORTUNITIES

Diverse opportunities are available in the horse industry. Horse stables range in size from a few horses boarded for extra income to large-scale operations with a menu of services available to clients. Stables can provide jobs for attendants, barn managers, grooms, trainers, and riding lesson instructors. Farriers and veterinarians are usually contracted to care for horses.

At tracks, owners, exercisers, trainers, jockeys, and other track personnel collaborate in the multi-billion dollar industry of horse racing. Horse racing ranks third in spectator sports behind NASCAR racing and baseball. Both harness and ridden race horses attract almost a billion fans each year.

Along with racing, other equestrians may earn incomes by competing with horses. Rodeo and other performance events provide competitors with an opportunity to win prize money. Only the very accomplished can support themselves with such winnings

CAREER PROFILE

Farrier

A farrier's work involves caring for the hooves of horses. Common tasks include removing old shoes, shaping hoofs, refitting new shoes, and attaching them. Farriers usually work from mobile units on well-outfitted trucks. Many have attended shorts courses at community colleges or universities, or completed studies at one of the country's 50 farrier schools.

SUPPORTING ORGANIZATIONS

Organizations for the horse industry vary as much as the owners themselves. Unlike the meat commodities, no national organizations exist that speak a common voice for and is supported by the whole horse industry. Although, efforts to unify the industry have been made by the American Horse Council.

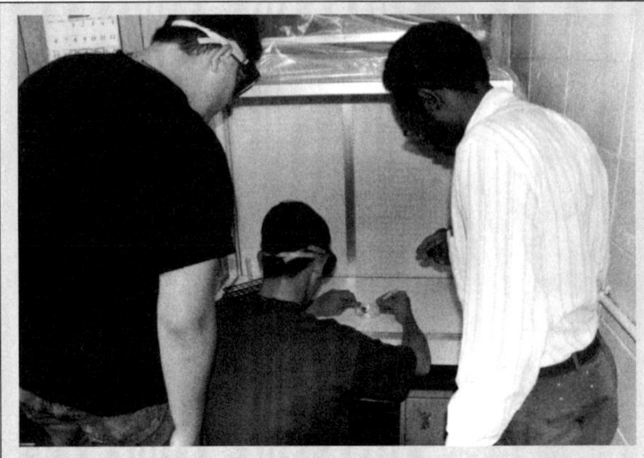

CAREER PROFILE

Agriscience Teacher

Agriscience teachers may be employed on the middle school or high school level. They are responsible for providing classroom and laboratory instruction in the principles of the agricultural sciences. Agriscience teachers should be good communicators willing to work long hours. To stay abreast of new developments, they must be willing to continue their training through workshops and seminars. (Courtesy, Jasper S. Lee)

Most national horse organizations center around like interests. The Professional Rodeo Cowboys' Association and the Professional Rodeo Cowgirls' Association are just two of several groups serving the rodeo industry. In addition, separate event organizations exist for such rodeo sports as barrel racing and team roping. The American Horse Show Association brings together those interested in the exhibition of horses. Many breed organizations sponsor their own show committees. Other equestrians interested in a certain show aspect like jumping, pleasure, youth (4-H or FFA), ***dressage*** (guiding of a horse through unique maneuvers using slight movements), or games form groups to organize show events. See Figure 22-4. Other horse related activities, such as fox hunting and trail riding, have localized groups supporting these interests.

Figure 22-4. Youth organizations, such as 4-H, sponsor jumping events.

Horse racing remains one of the most popular sports in the United States. Furthermore, the horse racing industry generates billions of dollars each year. The Association of Racing Commissioners International, Inc., located in Lexington, Kentucky, brings together the thoroughbred and harness racing industries. This self-contained group attempts to unify racing rules from state to state. They also hold an annual convention and ask for and provide constant feedback to state racing authorities.

SUMMARY

Between six and ten million horses can be found across the United States. These horses are used primarily for pleasure although one-fourth "work" for a living. Moreover, the horse industry brings in over 15 billion dollars per year. Even though, a dark side of the industry exists. Concerns revolving around abuse and horsenapping remain issues in the industry. Several organizations support equestrians. These groups tend to be interest and event oriented. A wide variety of horse related careers are available for employment including trainers, grooms, stable managers, and race course personnel.

CHAPTER SELF-CHECK

___ SPCA	1. guiding of a horse through unique maneuvers using slight movements
___ dressage	2. Society for the Prevention of Cruelty to Animals

QUESTIONS AND PROBLEMS FOR DISCUSSION

1. Horses can be typed into three categories. Name them.
2. Three-quarters of United States' horses are used for _____.

3. The horse industry generates ___ billion dollars each year.
4. Concentrations of brood mares are located in this state. Name the state.
5. Quarter horses account for _______ of the total United States horse population.
6. The Horse Protection Act was passed in 19___.
7. What do the letters SPCA mean?
8. From 1990 to 1995, slaughter prices for horses have increased/decreased. Select one.
9. How many United States slaughter houses kill horses for human consumption?
10. Name one related rodeo association.
11. Describe dressage.
12. The Association of Racing Commissioners International, Inc., is located in this Kentucky city. Name the city.
13. List two youth organizations that organize horse show events.
14. How many inches make a hand?
15. When was the Pony Express initiated?

ACTIVITIES

1. Practice measuring horses figuring "hands." Measure the horse from the bottom of the front hooves to withers (top of shoulder).
2. Prepare reports on famous horse races, such as the Kentucky Derby. Have a "Derby" party complete with regional cuisine. Conduct a simulated race for fun.
3. Invite a farrier to speak to the class. Ask what schooling is needed, as well as the pros and cons of the career.

LABORATORY ACTIVITY

TYING A QUICK RELEASE KNOT*

Purpose

To tie a quick release knot

Materials

rope attached to horse halter
fence rail

Procedure

With one hand, pass the lead rope under, up, and over the fence rail, looping it once over the rail. See diagram A.

2. Pass the loose end of the lead rope to the right, under the lead rope and in an upward direction and over the lead rope (See diagram B) allowing it to fall to the ground (See diagram C).
3. Reach through the newly formed eyelet and grasp the loose end of the lead rope. (See diagram D)

Pull the loose end of the lead rope through the eyelet to form a loop (See diagram E) while simultaneously pulling the knot tight against the loop.

Pull the loose end of the lead rope through the loop and drop it to form a "lock" that prevents the knot from untying. (See diagram F)

To release the knot quickly, grasp the end of the lead rope, and pull firmly.

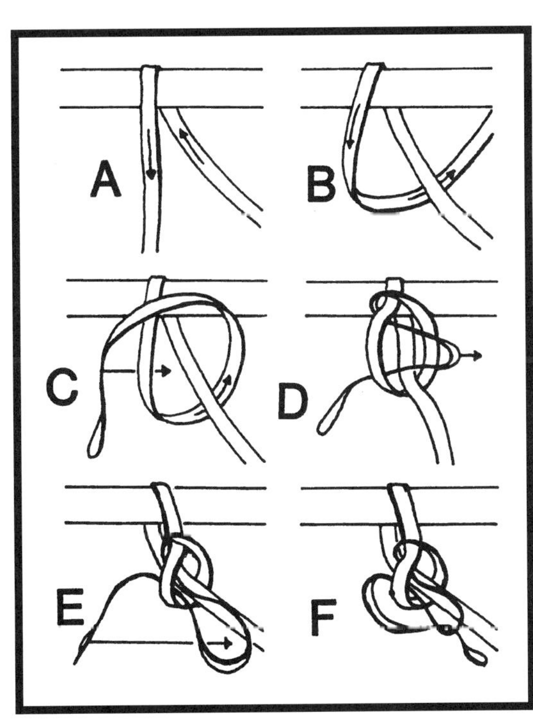

Analysis

Each student should practice tying the knot. Assign a grade based on proficiency and ability to teach others (possibly younger equestrians).

Application of Laboratory Activity

Being able to tie a quick release knot allows livestock handlers to restrain animals while being able to release them quickly in an emergency. This safe knot tying method should be mastered by all stock handlers. Sometimes, animals attempt to break way or get caught in an awkward position. Being able to quickly release them can prevent injuries to both handlers and the animal.

*This lab was adapted from Horse Management Task Sheets written by:

Dr. Marianne Houser, Assistant Professor, The Pennsylvania State University

Glossary

A

Abomasum—the fourth compartment or true stomach in ruminants

Actin—one of the four muscle proteins that comprise a sarcomere

Adipocyte—individual fat cell

Adipose tissue—fat

Aerobic exercise—long-term, less intense exercise (jogging)

Alleles—pairs of genes

All-in all-out production—pigs kept in separate rooms by age

Amino acid—building blocks containing nitrogen

Anaerobic exercise—intense, short in duration exercise (weight lifting)

Androgen—male sex hormone secreted by the testes

Animal rightists—believe that animals should not be used for human benefit

Animal welfarists—believe animals should be well tended

Anthelmintic—worming medicine

Atrophic rhinitis—disease that causes twisted snouts in swine

Average daily gain (ADG)—measures rate of gain

B

Backcrossing (crisscrossing)—uses only two breeds in crossbreeding system

Backfat (BF)—subcutaneous fat

Background—readied for the feedlot

Backgrounding—feeding beef calves an economical diet for the first winter

Baldies—white-faced cattle

Barrow—castrated male swine

Beta agonist—synthetic substance similar to adrenaline

Binomial nomenclature—a scientific system proposed by Carolus Linnaeus for classifying organisms using the genus and species names in Latin

Biosecurity—isolation of a herd

Birth weight EPD (BW EPD)—expected progeny birth weight

Blind nipples—nonfunctional nipples

Boar—intact male swine

Body condition—muscle and fat cover

Bowlegged (front legs)—front legs bowed out (front view)

Bowlegged (rear legs)—rear legs bowed out (rear view)

Break joint—point at which lambs front legs are removed from carcass

Brood—mother

Buck—ram

Buck-kneed—front legs too straight (side view)

Bulls—intact male cattle

Buttons—cartilage located along split surface of the carcass over the rib

C

Calcification—the process of collagen replacement by calcium in the interior of bone

Calf-kneed—front legs too curved (side view)

Capacitation—washing of sperm by vaginal fluid

Cash price—actual price on a given day

Castrated—testicles removed

Cecum—located between the small and large intestine

Cell—the basic unit of life

Cell membrane—the outside of a cell that separates the cell contents from the external environment

Centromere—a central point where two identical sets of chromatids are attached

Cervix—first in path leading from uterus to the outside of the animal, seals off the uterus after pregnancy has been established to guard the fetus against bacteria

Check-off—dollars collected for research, education, and promotion

Chromosome—a small strand of genetic material that resides in the nucleus

Chyme—partially digested material

Clip—fleeces from the flock

Codominance—both genes in pair are expressed

Collagen—fibrous tissue strands that precede bone growth

Colored breeds—Guernseys, Jerseys, Brown Swiss, Aryshires, Milking Shorthorns

Colostrum—first milk from female to newborn

Commercial cattle—crossbred

Compensatory gain—profitable weight gain by thin cattle

Complete dominance—one gene completely masks the presence of the other

Complex carbohydrate—cellulose and hemicellulose

Conception—time of fertilization

Conditioning—associate a behavior with a stimulus

Contemporary group—similarly fed and managed groups of animals

Continental breed—breed from Europe

Cooperatives—members share in ownership and operation

Corpus hemorrhagicum (CH)—blood clot remaining after the egg is released

Corpus luteum (CL)—yellow body at the site of an egg release

Cow-calf—mother cow and calf

Cowhocked—rear legs bowed in (rear view)

Creep feeding—segregated feeding of young animals from adults.

Crop—located at the base of the neck in poultry and serves as a storage area for recently ingested feed

Cross-fostered—piglets moved to other sows with smaller litters

Crossbreeding—mating of unrelated animals

Crutched—wool removed from udder and around vulva

Cull—eliminate or sell

Cutability—measures lean to fat

Cuts—indicates the difficulty of a placing in a judging contest

Cwt.—per hundred pounds

Cytoplasm—the jelly-like substance between the cell membrane and the nucleus

D

Dairy Cow Unified Score Card—used to type classify, divides the dairy cow into five major traits

Days to 230 pounds (DAYS)—measures growth in swine

DHIA—Dairy Herd Improvement Association, a record keeping organization

Displaced abomasum—twisted stomach

Distemper—infection caused by *Streptococcus* bacteria

DNA (Deoxyribonucleic Acid)—a genetic compound that controls inheritance

Docked—tail removed

Draft—work animal

Drench—placing liquid medicine in the throat

Dressage—guiding a horse through unique maneuvers using slight movements

Dressing percentage—carcass percent of live weight

Dry cow—nonlactating (not milking)

Dystocia—calving problems

E

Eared cattle—Brahman or Brahman crossbred cattle

Efferent ducts—site of transported sperm from rete testis

Eliminative—defecating, dunging

Encephalomyelitis—sleeping sickness carried by mosquitos

Endomysium—thin film of connective tissue that wraps each myofiber

Endoplasmic reticulum—a network of membranes that connects the cell membrane to the nucleus and processes all incoming raw materials by breaking them down into simpler components

Epididymis—final collection site for sperm

Epimysium—connective tissue sheath that wraps whole muscles

Epiphyseal growth plate—a thin region at the end of long bones where collagen is formed

Equestrian—horse and rider

Estimated breeding value (EBV)—allows comparison within a group, average equals 100

Estrogen—female sex hormone secreted by the ovaries

Estrus cycle—corresponds with ovarian structures

Ethology—the science of animal behavior related to the environment

Ewe—female sheep

Expected progeny difference (EPD)—compares animals outside the herd

F

Farrier—hoof caretaker for horses

Farrow to finish—operations where hogs remain from birth to slaughter

Feeder calves—young cattle just weaned

FEPDs or flock EPDs—sheep EPD

Feral—animals that were once domestic and have returned to the wild

Fine wool breed—sheep with superior wool quality

Finish—fat cover

Finisher—feedlot operator

Finishing cattle—feedlot animals fed a high grain diet

Flat nipples—functional, but not prominent nipples

Flehman response—curling upper lip in courtship

Flushing—increasing feed prior to breeding

Follicle—developing egg that has not been released

Founder—laminitis or inflamed hoof

Free martin—sterile heifer twin born with a bull

Free-stall housing—allows animals to freely enter and leave

Futures price—expected price at a point in the future

G

Gamete—an animal sex cell

Gene—DNA material that contains hereditary codes

General purpose index (GPI)—evenly weights maternal, growth, and carcass traits

Generation interval—time needed to replace each generation

Genetics—study of heredity

Genotype—an animal's genetic code

Gilt—young female swine

Gizzard—the structure in poultry used to grind coarse feed particles

Glucose—blood sugar

Glycogen—stored glucose, found within the muscle

Golgi bodies—flat, membrane-encased fibers that assist in the processing of raw materials entering a cell

Grade—unregistered animals

Grants—superior characteristics that the second place animal in a pair may possess over first place

Grower—stocker operator—grow animals until the animal is more mature for finishing

H

Hand mating—natural service mating with human assisting

Haploid—the number of paired chromosomes present in every cell

Heifers—young female cattle

Heritability—percentage of trait that results from genetics

Heterosis—increase in performance that exceeds parents' average

Heterozygous—allele that has one copy each of the two possible genes

Homozygous—allele with two copies of identical genes

Hothouse lambs—lambs born in the fall or winter and raised indoors

Hydrochloric acid—strong digestive agent in the stomach

I

Inbreeding—mating of closely related animals

Incomplete dominance—a blending of genes in pair is expressed

Independent producer—not aligned with a company

Individual data—the performance of an individual animal for a specific trait

Infundibulum—funnel to the oviduct

Ingestive—eating

Institutional Meat Purchasing Specifications (IMPS)—standardized primal and sub-primal cuts

Intact—uncastrated

Integrated—an agricultural operation where the owner controls many aspects of production

Intensive pasture management—closely monitored rotational grazing

Intensive rotational grazing—moving from pasture to pasture over short periods

Intermuscular fat—fat stored in the seams between muscles

Intramuscular (IM) fat—fat found inside muscle bundles, more commonly called marbling

J

Jockey—the act of buying and selling

K

Keep-cull—simulates decisions in selecting replacement breeding stock

Ketosis—a metabolic disorder

Kidney, pelvic, and heart fat (KPH)—internal fat

Knock-kneed—knees too close together (front view)

L

Lactating—giving milk

Lactic acid—a secretion (waste product) produced when muscle cells contract

Lean-to-fat ratio—amount of muscle compared to fat

Learned behaviors—positive and negative reinforcement, conditioning

Leptospirosis—disease in swine that causes severe reproductive failure

Leydig cells—produce the hormone testosterone

Linear classification—a method dairy producers use to score the heritability of traits, traits are scored from −3 to +3

Linebreeding—mating of distant relatives

Linnaeus—a Swedish botanist who developed a system for classifying all organisms according to their similarities

Lip prehenders—grasp forage with their lips—sheep

Litter weight at 21 days (LW21)—litter weight of pigs

Longeing—horse exercises by circling handler

Lordosis—sign of heat in swine ("ear bobbing")

Lysosome—a round organelle that digests proteins in a cell through the release of enzymes

M

Macromineral—major mineral needed in the diet in grams per day

Maintenance behaviors—essential behavior for sustenance of life

Marbled—intramuscular fat

Mastitis—an infection and inflammation of the udder

Maternal breeds—excel in pigs born and milking ability

Maternal line index (MLI)—emphasizes maternal records

Maternal milk (MM)—difference in pounds of calf weaned due to milk production

Medium wool breeds—sheep with average fleece quality

Meiosis—the reduction division of sex cells

Metritis—uterine infection caused by retained placenta

Micromineral—trace mineral required in the diet in small amounts

Milk fever—an imbalance of calcium

Mitochondria—small, egg-shaped organelles that manufacture an energy source for the cell

Mitosis—the division of nonsex cells

Monogastric—one- or simple-stomached

Multiple trait indexes—selection for several traits simultaneously

Mycoplasmal pneumonia—chronic disease in swine, coughing in the morning is the most common sign

Myoblast—presumptive myoblast that lose the ability to divide

Myofiber—mature muscle cell

Myosin—one of the four muscle proteins that comprise a sarcomere

Myotube—many myoblast fused together

N

Negative reinforcement punishment for bad behavior

Niche market—specialty market

Nucleus—controls all cell activity

Number born alive (NBA)—pigs born per litter

O

Omasum—the third compartment in the ruminant stomach can also be referred to as manyplies

Outbreeding—crossbreeding

Ovary—primary reproductive organ in females which produces eggs

Oviduct—tube leading to uterus

P

Palominos—yellow or golden colored horses with light colored manes

Parvovirus—swine disease resulting in mummified piglets farrowed

Paternal breeds—excel in growth rate, muscling, and leanness

Penis—male mating organ

Peptidase—enzyme produced by the pancreas

Percent difficult births in heifers (%DBH)—calving ease in dairy animals

Performance testing—details growth rate and other economically important data

Perimysium—connective tissue sheath that wraps bundles of myofibers together

Peristalsis—involuntary muscle contractions

Phenotype—expression of genetic code

Pigeon-toed—front feet turn in

Pin nipples—small, undeveloped nipples

Placenta—membrane in the uterus that protects and nourishes the unborn animal

Polled—naturally hornless

Porcine stress syndrome—swine genetic abnormality causing pale, soft, exudative muscle

Positive reinforcement—reward for good behavior

Post mortem—after death

Posty—rear legs too straight (side view)

Predators—animals that kill as a natural behavior

Predicted transmitting ability (PTA)—dairy equivalent of EPD

Predicted transmitting ability for type (PTAT)—summary score of linear evaluation

Preliminary yield grade (PYG)—measurement of fat cover $^3/_4$ the distance over the ribeye at the 12th rib

Presumptive myoblast—premuscle cell in the embryo

Price floor—lowest price

Primal cuts—large primary cuts, like the loin, brisket

Procurers—act as buyers and sellers

Production type index (PTI)—general purpose index for dairy breeds other than Holsteins

Progeny—offspring

Proventriculus—true stomach in poultry

PRRS (porcine reproduction and respiratory syndrome)—swine disease with reproductive and respiratory components

Pseudorabies—easily spread virus prevalent where large numbers of hogs are raised

PTA$ for cheese yield (CY$)—dollar value of progeny cheese (milk protein content)

PTA$ for milk and fat (MF$)—dollar value of progeny milk production and butterfat content

PTA for milk, fat, and protein (MFP$)—dollar value of milk and contents

Q

Qualitative trait—trait controlled by only a single pair of genes

Quality grades—estimate of marbling and age

Quantitative trait—trait controlled by several pairs of genes

R

Ram—male sheep

Ratio—compares all animals within a herd or flock

Reasons—explanation of the logic for the placing order in a judging contest

Replacement heifers—retained for herd

Rete testis—first collection site of immature sperm

Reticulorumen—the first two stomach parts in a ruminant

Ribosome—manufactures new proteins in the cell

Rigor mortis—muscle stiffness that occurs from lactic acid build-up in the muscles after death

Rotaterminal cross—requires three maternal breeds in crossbreeding system

Rumen—unique stomach chamber in cattle and sheep

Ruminant—animal (cattle and sheep) with a four-part stomach

S

Sarcomere—the segment of the myofiber that shortens when muscles contract and elongates when muscles relax

Sarcoplasmic reticulum (SR)—reservoir of calcium that surrounds each myofiber

Scenario—production situation in which animals will be placed

Scur—small unattached horn

Seasonal breeders—conceives in the fall of the year

Seedstock—registered and nonregistered breeding stock

Selection intensity—top percentage of animals used for production of next generation

Seminiferous tubules—meiosis takes place in males here

Settle—become pregnant

Shipping fever—bovine respiratory disease complex

Sickle-hocked—rear legs too curved (side view)

Simple carbohydrate—starch and sugar used as quick energy

Social behaviors—interaction of two or more animals

Somatic cell count—a test that indicates mastitis and sanitation conditions

Somatotropin (ST)—naturally occurring hormone in all animals that regulates the growth of long bones, muscle, and fat

Sow productivity index (SPI)—combines NBA and LW21 into a gauge of maternal performance

SPCA—Society for the Prevention of Cruelty to Animals

Splay-footed—front feet turn out

Split-sex feeding—barrows and gilts fed separately

Spool joint—point at which older sheep's front legs are removed from carcass

Stanchion—tie-stall barn

Steers—castrated male cattle

Stillborn—dead at birth

Subcutaneous—under the skin

Subprimal cuts—taken from primal cuts

Sulky—cart

Swine erysipelas—disease that causes diamond-shaped skin patches

Symbiotic—mutually beneficial relationship between two species

Synthetic breed—planned crosses of established breeds of sheep

T

Tack—harnessing, riding, and grooming equipment

Tariffs—taxes on products imported into a country

Taxonomy—a system of scientific classification

Terminal crossbreeding—crossbred females bred with heavily muscled males

Terminal sire—father in a crossbreeding system where all progeny are marketed

Terminal sire indexes (TSI)—measures lifetime growth rate in pigs

Testis—primary male reproductive organ

Tetany—muscle exhaustion

Three breed rotational cross—uses one maternal, one paternal, and one combination breed in cross breeding system

Titan—one of the four muscle proteins that comprise a sarcomere, prevents overstretching of a muscle

Tongue prehenders—use tongue to gather forage—cattle

Tropomyosin—one of the four muscle proteins that comprise a sarcomere, helps regulate muscle contraction

Type classification—an excellent cow earns 90 to 100 points using this method

Type-production index (TPI)—general purpose index for Holsteins

U

Urethra—transports sperm to the exterior

Uterus—womb—where egg develops until birth

V

Vagina—canal between the vulva and cervix

Vas deferens—tube from epididymis to urethra

Villi—finger-like projections

Vitelline Block—bars second sperm from egg

Vulva—visible part of the female reproductive tract

W

Weaning weight EPD (WW EPD)—expected progeny weaning weight

Weaning—removal from the dam

Wethers—castrated sheep

Winter—placed on spring pasture

Y

Yearling weight EPD (YW EPD)—expected progeny yearling weight

Yield grades—estimate of cutability

Z

Zona pellucida—barrier surrounding egg

Zygote—fertilized egg cell

Bibliography

Benjamin, G. L. April 1995. "The Changing Hog Sector," *AgLetter, the Agricultural Newsletter of the National Reserve Bank of Chicago.* No. 1863.

Bentley, S. E. 1990. *Farm Operating and Financial Characteristics.* U. S. Department of Agriculture. Statistical Bulletin No. 860.

Bixby, D. E., C. J. Christman, C. J. Ehrman, & D. P. Sponenberg. 1994. *Taking Stock: The North American Livestock Census.* The American Livestock Breeds Conservency. The McDonald & Woodward Publishing Company, Blacksburg, VA.

Boggs, D. L., and R. A. Merkel. 1984. *Live Animal Carcass Evaluation and Selection Manual,* 2nd ed. Kendall/Hunt Publishing Co., Dubuque, IA.

Bryan, K. A. 1994. *Pennsylvania State 4-H Livestock Judging Manual.* College of Agicultural Sciences, Cooperative Extension, The Pennsylvania State University.

Bush, L. J. 1979. *Livestock Feeding.* Oklahoma State University, Stillwater.

Campbell, J. R., and J. F. Lasley. 1985. *The Science of Animals That Serve Humanity.* McGraw-Hill Book Co., New York.

Campbell, J. R., and R. T. Marshall. 1975. *The Science of Providing Milk for Man.* McGraw-Hill Book Co., New York.

Cheeke, P. R. 1993. *Impacts of Livestock Production on Society, Diet/Health and the Environment.* Interstate Publishers, Inc., Danville, IL.

Cole, H. H., and M. Ronning, eds. 1974. *Animal Agriculture: The Biology of Domestic Animals and Their Use by Man.* W. H. Freeman and Co., San Francisco.

Cullison, A. 1979. *Feeds and Feeding,* 2nd ed. Reston Publishing Co., Inc., Reston, VA.

Cunha, T. J. 1977. *Swine Feeding and Nutrition.* Academic Press, New York.

Crampton, E. W., and L. E. Lewis. 1969. *Applied Animal Nutrition: The Use of Feedstuffs in the Formulation of Livestock Rations.* W. H. Freeman and Co., San Francisco.

Crampton, E. W., and L. E. Lloyd. 1959. *Fundamentals of Nutrition.* W. H. Freeman and Co., San Francisco.

Ensminger, M. E. 1991. *Animal Science,* 9th ed. Interstate Publishers, Inc., Danville, IL.

Ensminger, M. E. 1992. *Animal Science Digest.* Interstate Publishers, Inc., Dnaville, IL.

Ensminger. M. E. 1987. *Beef Cattle Science,* 6th ed. Interstate Publishers, Inc., Danville, IL.

Ensminger, M. E. 1990. *Horses and Horsemanship,* 6th ed. Interstate Publishers, Inc., Danville, IL.

Ensminger, M. E. 1992. *Stockman's Handbook Digest.* Interstate Publishers, Inc., Danville, IL.

Ensminger, M. E., and R. O. Parker. 1986. *Sheep & Goat Science,* 5th ed. Interstate Publishers, Inc., Danville, IL.

Ensminger, M. E., and R. O. Parker. 1984. *Swine Science,* 5th ed. Interstate Publishers, Inc., Danville, IL.

Etgen, W. M., and P. M. Reaves. 1978. *Dairy Cattle Feeding and Management,* 6th ed. John Wiley & Sons, New York.

Etherton, T. 1990. Animal Growth and Development. Animal Science 514. The Pennsylvania State University. Class notes.

Fenner, C., Editor, *The Paper Horse.* 1995 Personal communication.

Gillespie, J. R. 1989. *Modern Livestock and Poultry Production,* 3rd ed. Delmar Publishers, Inc., Albany, NY.

Greer, J. G., and J. K. Baker. 1992. *Animal Health: A Layperson's Guide to Disease Control,* 2nd ed. Interstate Publishers, Inc., Danville, IL.

Herren, R. V. 1994. *The Science of Animal Agriculture.* Delmar Publishers, Inc., Albany, NY.

Holstein-Fresian Association of America. 1984. "When You Talk About Holsteins . . . Use the Right Words." Pamphlet. Battleboro, VT.

Ivan, D., Executive Director of Pennsylvania Bee Council. 1995. Personal communication.

Krider, J. L., J. H. Conrad, and W. E. Carroll. 1982. *Swine Production,* 5th ed. McGraw-Hill Book Co., New York.

Lasley, J. F. 1963. *Genetics of Livestock Improvement.* Prentice-Hall Inc., Englewood Cliffs, NJ.

Lee, J. S., C. Embry, J. Hutter, J. Pollock, R. Rudd, L. Westrom, and A. Bull. 1996. *Introduction to Livestock and Poultry Production.* Interstate Publishers, Inc., Danville, IL.

Lister, D., D. N. Rhodes, V. R. Fowler, and M. F. Fuller, eds. 1974. *Meat Animals: Growth and Productivity.* Plenum Press, New York.

McLaren, G. A. 1976. *Biochemistry of Animal Nutrition.* Division of Animal and Veterinary Sciences. West Virginia University, Morgantown.

Mercia, L. S. 1990. *Raising Poultry the Modern Way.* Garden Way Publishing. Storey Communications, Inc., Pownal, VT.

Moody, E. G. 1991. *Raising Small Animals.* Farming Press Books, Ipswitch, U.K.

Nalbandov, A. V. 1964. *Reproductive Physiology: Comparative Reproductive Physiology of Domestic Animals, Laboratory Animals, and Man.* W. H. Freeman and Co., San Francisco.

Neuman, A. L. 1977. *Beef Cattle,* 7th ed. John Wiley & Sons. New York.

Olson, John, Executive Director, The American Sheep Industry Association. 1995. Personal communication.

Pond, W. G., and J. H. Maner. 1984. *Swine Production and Nutrition.* AVI Publishing Company, Inc., Westport, CT.

Purebred Dairy Cattle Association. 1994. Dairy cow unified scorecard.

Romans, J. R., W. J. Costello, C. W. Carlson, M. L. Greaser, and K. W. Jones. 1994. *The Meat We Eat,* 13th ed. Interstate Publishers, Inc., Danville, IL.

Schneider, B. H., and W. P. Flatt. 1975. *The Evaluation of Feeds Through Digestibility Experiments.* The University of Georgia Press, Athens.

Simpson, L., Agricultural Statistician, U.S. Department of Agriculture. 1995. Personal communication.

Strickland, R. P., C. J. Steele, and R. P. Williams. 1992. *Ranking of States and Commodities by Cash Receipts.* U.S. Department of Agriculture. Statistical Bulletin No. 871.

Swartz, D., Regional Dairy Extension Specialist, The Pennsylvania State University Cooperative Extension Service. 1995. Personal communication.

Taylor, R. E. 1984. *Beef Production and the Beef Industry: A Beef Producer's Prospective.* Macmillan Publishing Co., New York.

Taylor, R. E., and R. Bogart. 1988. *Scientific Farm Animal Production: An Introduction to Animal Science.* Macmillan Publishing Co., New York.

Williams, W. F., and T. T. Stout. 1964. *Economics of the Livestock-Meat Industry.* The Macmillan Co., New York.

Index